# *Pests of the West*

# *Pests of the West*

## *Prevention and Control for Today's Garden and Small Farm*

Whitney Cranshaw

FULCRUM PUBLISHING
GOLDEN, COLORADO

Copyright © 1992 Whitney Cranshaw

Book Designed by Ann E. Green, Green Design
Original Cover Illustration by Ben Brown, Copyright © 1992

All Rights Reserved

Library of Congress Cataloging-in-Publication Data

Cranshaw, Whitney.
Pests of the West : prevention and control for today's garden and small farm / Whitney Cranshaw.
p. cm.
Includes index.
ISBN 1-55591-097-1
1. Plant parasites—West (U.S.) 2. Plant diseases—West (U.S.) 3. Garden pests—West (U.S.) 4. Plant parasites—Control—West (U.S.) I. Title.
SB605.U5C73 1992
635'.049'0978—dc20 91-58489
CIP

Printed in the United States of America

0 9 8 7 6 5 4 3 2 1

Fulcrum Publishing
350 Indiana Street, Suite 350
Golden, Colorado 80401

*I would like to dedicate this book to Sue Ballou, my best friend and wife.*

*I would like to also acknowledge the entire class of animals known as Insects, which, although they sometimes wreak havoc in the garden, can be endlessly fascinating to observe.*

*Sue Ballou, set for an invading cabbage butterfly*

# TABLE OF CONTENTS

# ACKNOWLEDGMENTS

Many have helped create this book.

Perhaps most important is the system of cooperatively shared research information that has been developed through the state Agricultural Experiment Stations and Cooperative Extension via the Land Grant educational system. I have drawn heavily from the previously published efforts of my colleagues at Colorado State University and across the country, tempering these with my own observations and experiences.

Several of these publications are worthy of particular note, including *Prevention and Control of Wildlife Damage*, produced cooperatively by Extension wildlife specialists in coordination with the University of Nebraska; and *Weeds of Colorado* and *Insects and Related Pests of Vegetables*, produced by the University of North Carolina. These have been supplemented by scores of USDA technical bulletins and information fact sheets produced by states within the scope of this book. In addition, I found such old standards as *Destructive and Useful Insects* and *Insects of Western North America* to be outstanding references for information on insects. *Vegetable Diseases and Their Control* served as a very useful starting point for information on plant disease. And I must further acknowledge the writings of Howard Ensign Evans, including *Life on a Little Known Planet*, as a source of tremendous information for any bug-watcher.

Most of the artwork used in this book has also been developed for use in USDA or state Cooperative Extension publications. In addition, many illustrations have been borrowed from the California Department of Food and Agriculture Publication 518, *Stone Fruit Orchard Pests: Identification and Control.*

Several individuals also helped me by reviewing sections as they were prepared. Barbara Hyde, Curt Swift, Ken Barbarick and George Beck have all helped straighten me out when I ventured into areas on the fringe of my expertise. Boris Kondratieff, Bill Brown, Howard Schwartz, Jim Feucht and Bob Hammon have all provided invaluable technical assistance in developing sections of this book. Dave Leatherman, Joseph Strauch and Wendy Meyer provided much of the material for Appendix IV regarding insectivorous birds.

Finally, I must thank Bob Baron and the crew at Fulcrum, who have carried me through this project since its inception.

# *How to Use This Book*

Western North America is a unique area to garden. Most areas are blessed by abundant sunshine, but rainfall is typically scarce and all too often may be accompanied by hail. Soils are often less than ideal, a bit on the heavy side, high in clay and tend to the alkaline range. Salts can also be a severe problem.

One difficulty in gardening anywhere is dealing with the occasional pest problem. In the West, we have our share of pests, although fortunately we are often spared many of the problems common to the warmer and wetter areas of the country.

Pests are organisms that cross human interests—or just rub us the wrong way. This is not a natural distinction; "good" bug vs. "bad" bug means nothing in nature. What we choose to call a pest also varies with the situation. For example, honeybees are more than welcome while pollinating fruit trees, but should they choose to nest in the wall of your home, they become pests. On the other hand, most people have trouble even thinking about earwigs, but when they are not resting in the sweet corn, they are probably eating aphids, caterpillars and other pests.

In the West, gardeners experience many pest problems unique to this area. Coyotes stealing melons or an elk napping in the cabbage patch are situations that could not occur in most of North America, but occasionally do in the West. California has its fruit flies, harlequin bugs and root knot nematodes roam the South, and Japanese beetles and gypsy moths are often on the mind of an eastern gardener. None of these are major problems in the West, where soil diseases, grasshoppers and potato psyllids may be the biggest garden headaches.

*Pests of the West* emphasizes the common garden problems of the high plains, Rocky Mountains and intermontane region of the United States and southern Canada. It covers the major insect pests, plant diseases and weeds found in this area of tremendous diversity. The focus of the book is limited to the commonly grown vegetables, fruits, flowers and herbs. Lawns and landscape plants are not included.

This book is organized in four major sections. Chapter 1 focuses on soils, fundamental to any garden. Common soil problems and their correction are discussed.

Chapters 2, 3 and 4 discuss techniques for controlling garden pests. Natural and biological controls, which provide a tremendous amount of often unrecognized "background" plant protection, are brought out first, where they belong. Cultural and mechanical controls follow. The section concludes with the chemical controls—the pesticides (a subject that is further covered in Appendix II).

The first step in managing pests is to properly identify the problem. To this aim Chapter 5 provides an index of common problems associated with vegetables, fruit crops, herbs and flowers grown in the West. Since cultural and environmental conditions are often at the root of garden problems, each crop is preceded by a brief discussion of the plant's growth requirements.

Finally, it's important to learn something about the habits, damage and control of pests.

With this information in mind, you can more effectively manage the problem. Chapter 6 describes the common "bugs" (insects, mites, slugs) found in the area. Chapter 7 deals with diseases caused by microorganisms (fungi, bacteria, viruses) and environmental stresses. Chapter 8 concerns weed control around the vegetable and flower garden, and Chapter 9 involves common problems with mammals and birds.

*Warning:* Reading this book may be hazardous to your gardening enjoyment. *Pests of the West* will help you better understand common garden problems so that you can avoid them. However, because the entire text is dedicated to this rather sordid subject, you may feel overwhelmed at times. You will likely start to notice a few more pests around the garden after reading the book.

Relax. Although you may become more sensitive to some common garden problems, there will be no more (or fewer) just because you read this book. However, you will be in a better position to understand, correct and head off pest problems so that they do not detract from your gardening enjoyment.

*Bill Cranshaw checks out a parsleyworm.*

CHAPTER 1

# *Soils, the Foundation of a Healthy Garden*

When planning a garden to avoid damage by pests, attention to the soil is fundamental. The condition of our garden soils is one of the important determinants of the health of the plants we grow. Good garden soils allow plants to grow vigorously, with minimal stress, as well as avoid or outgrow many disease and insect pest problems.

The soils of the West have formed as the result of constant but slowly evolving changes over the course of millions of years. Unfortunately, one of the forces in making your garden soil could be a recent one: the contractor who built your home. During house construction, topsoils are often stripped and poor-quality subsoils are brought to the surface. Machinery compacts soils and construction debris may be left behind. A gardener may start work with a soil plot that is far from ideal.

Several factors are involved in developing good garden soils. Soil texture can affect how water and roots move through the soil. Soil chemistry determines how nutrients needed for growth are made available to the plant. A discussion of soil problems common to the West is therefore the place to start toward building and maintaining a healthy garden.

## *Soil Texture and Structure*

Soils are a blend consisting primarily of minerals from weathered rock and *organic matter* from plant materials. The minerals form small particles that are classified by size. Extremely fine soil particles are known as *clay*, coarse particles as *sand*, and those that are intermediate are called *silt*. The soil mixture of sand, silt or clay determines the *soil texture*.

Soil texture is one of the most important characteristics of any soil, and problems with poor soil texture are common in the West. Soils high in clay are heavy soils with small soil particles that make it difficult for roots to push through the soil. This causes particular problems with root crops such as carrots.

Clay soils do help hold water—a definite plus in the arid West. However, they also tend to drain poorly and easily become waterlogged. This results in oxygen starvation as the water drives out the soil oxygen required by roots to breathe and grow. Such conditions reduce the ability of the plant to absorb needed nutrients such as iron, producing deficiencies. Waterlogged soils also favor many fungi and bacteria, such as Phytophthora wilt, which can produce serious garden diseases.

Sandy soils are less common in the region, but do occur locally. Roots move easily through these soils and water drains rapidly. However, problems with drought and loss of nutrients can plague a sandy soil.

You can get a fairly good idea of your soil's texture by rolling some *slightly* moistened soil between your fingers. If it forms a firm ball, feels smooth and becomes sticky when moistened, it is too high in clay. If you cannot form a ball, the soil won't stay together and it feels somewhat grainy, the soil is of better texture. If, on the other hand, the soil feels very coarse, it may be too sandy and will not hold an adequate amount of water.

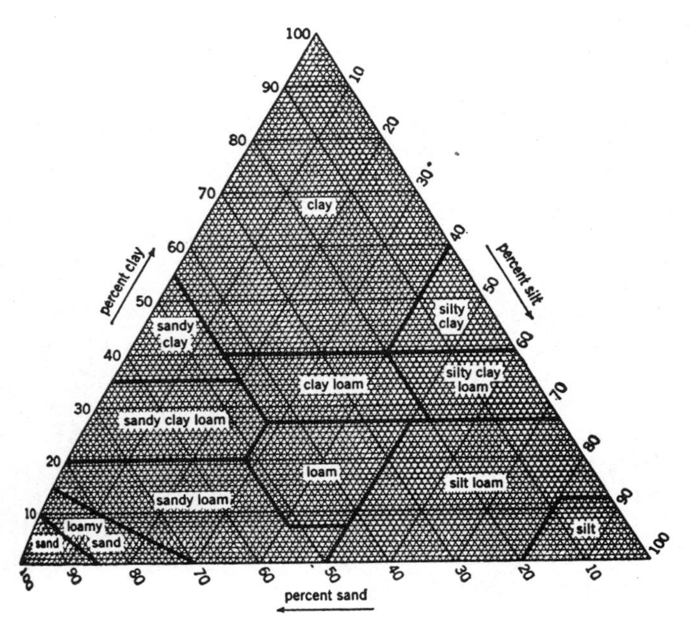


*Chart showing the percentages of clay (below 0.002 mm.), silt (0.002 to 0.05 mm.) and sand (0.05 to 2.0 mm.) in the basic soil textural classes*

## Subsoil Drainage

Poor drainage is one of the problems for growing plants in the West. This is a soil texture problem, usually occurring in the subsoil, below the area where most plant roots grow. Drainage problems, because they occur below ground, often go unnoticed by the gardener.

A simple test of soil drainage is done by digging a small hole about 12 inches deep and filling it with water. Return the next day and fill it again, this time taking note of how long it takes for the water to soak into the soil. If the water takes more than 1 hour to soak in, this is a symptom that your subsoil drainage might be poor. Water will tend to "pond" in your garden, increasing problems with salts and favoring root rotting problems. If the water soaks in within a few minutes, the subsoil drainage may be *too* good. Such soils do not hold water well and will tend to have problems with drought if not carefully tended. These soils will also easily lose some of the water-soluble plant nutrients, such as nitrogen, through leaching. If the water soaks in within 10 to 30 minutes, your subsoil drainage is probably pretty good. Get to work on other parts of your soil-building program.

## Correcting Soil Texture Problems

One common reaction to a problem with heavy clay soils is to consider adding sand. On the face of it, this seems logical, an attempt to balance the mixture of clay and sand particles. However, attempting to correct a clay soil in this way does little to increase the amount of air spaces in the soil (soil pores), providing little improvement. ("Sand on clay, money thrown away.") Unless *at least* one-third, and often more, of the soil volume is changed, there will be few effects on the texture of clay soils common to the region. In other words, a layer of 3 inches of coarse sand might be needed to successfully improve the texture in the top 6 inches of the garden soil using sand. Small amounts of sand sometimes even worsen drainage, creating "adobe" soils.

The situation is different where sandy soils are present. In this case, there can be some improvement in soil texture by amending soils with clay. ("Clay to sand, money in hand.") However, the added clay needs to be mixed thoroughly with the soil, a difficult chore to do well.

Addition of *organic matter* is the key to improving both clay and sandy soils. In clay soils, decomposition products of organic matter (polysaccharides) help to "glue" fine soil particles, creating secondary soil particles and allowing larger pores to form in the soil. Organic matter acts like a sponge to hold water in sandy soils. Although

## *Earthworms—Nature's Soil Builders*

After running the rototiller across the garden each year for several years to mix organic matter, gardeners will see dramatic improvement in soil texture. However, eons before tillers arrived on the market, soils were similarly improved by natural means, much of it the result of earthworms.

Earthworms are nature's most important soil builders. As they feed on dead plants and other organic materials, they act as *macrodecomposers*, organisms that break large particles into smaller pieces. During digestion by earthworms, some nutrients are released to the plants. More important, the smaller pieces left behind are more rapidly broken down by bacteria and other soil microbes, which completes the recycling process.

As they tunnel, feed and excrete their casts, earthworms thoroughly mix organic matter with soil minerals, improving soil structure. These tunnels help aerate the soil, providing an ideal environment for plant roots and improving drainage.

Several types of earthworms occur in regional gardens, many of them introduced species. Best known are the large nightcrawlers (species of *Lumbricus*), which make a permanent series of tunnels and forage on the soil surface at night. (Nightcrawlers also pile their casts on the soil surface, sometimes causing problems with lumpy lawns.) Much more abundant are the smaller garden worms (usually species of *Allolobophora*), which continuously tunnel through the soil. (Red wrigglers, often sold as fish bait, for worm farming or for some home compost methods require very moist conditions for survival. They are rarely found in garden soils and are not useful for building soil structure.)

Once established in a garden, earthworms usually thrive. Acidic soil, rarely a problem in the region, is one of the biggest factors limiting earthworm development. Earthworms are also tolerant of most pesticides, although some, such as carbaryl (Sevin), trifluralin (Treflan) and benomyl (Benlate), can temporarily reduce their numbers. In the absence of these stresses, earthworm populations increase as the amount of organic matter they use for food increases. Intolerant of heat and drying, earthworms normally are most active in spring and fall, but can be encouraged throughout the summer by the shading and watering a garden receives.

soil texture (the proportion of clay, silt and sand) is little affected by addition of organic matter, *soil structure* is improved.

Other practices that are sometimes recommended to improve soil structure but provide little benefit to improving soil structure are gypsum and liquid amendments. Gypsum (calcium sulfate) provides levels of calcium that are already in excess in most area soils. The only soils that can benefit by the addition of gypsum are certain uncommon types of salt-affected soils that are also high in sodium (*sodic soils*).

Various liquid products are also marketed to improve soils. These typically make claims to "break up clay soils" or to "condition the soil." However, most work to merely reduce the surface tension of the soil particles (as do detergents), allowing deeper water penetration. Modest amounts of organic acids may be added, but they have little effect on increasing the size of soil pores, fundamental to improving soil structure. The short-term benefit of such products is rarely worth their high cost.

When correcting soil texture problems, work

the soil deeply with the addition of organic matter or other materials. Problem subsoils may also need attention. Light spading or a once-over with the rototiller doesn't reach these lower areas, which require mixing soil a foot or more deep. Several gardening practices, such as double-digging and raised bed designs, provide this thorough tilling of the soil.

## *Soil Acidity (pH)*

One of the most fundamental properties of any soil is how acidic or basic it is. This is determined by the ratio of two types of charged particles in the soil: hydrogen ions (or $H^+$) and hydroxyl ions ($OH^-$). The hydrogen power (*pouvoir hydrogene* in French) is measured as *pH*. The pH scale runs from 0 to 14, with a neutral soil measured as pH 7.0. Acidic soils have higher levels of "hydrogen power" and a pH below 7.0; basic soils have lower numbers of hydrogen ions and a pH above 7.0. Basic soils are also known as alkaline soils. (However, they should not be confused with *alkali* soils, which result from an excess of soil salts.)

Soil pH affects the chemistry of soils, thereby affecting the availability of plant nutrients. Some pH conditions cause plant nutrients to chemically bind with other particles, making them unavailable for the plant to use. Alternatively, other nutrients may be released with changes in soil pH, allowing them to be picked up in the soil water around plant roots.

In areas of high rainfall, soils tend to be acidic, the result of leaching action by water, which removes the acidifying hydrogen ions. In the arid West, the opposite situation is the general rule. Also, the high level of calcium common to regional soils acts to maintain (buffer) the basic nature of our soils and bind certain minerals, such as iron, in a form so that plants cannot use them.

Most plants do best with a pH that is fairly neutral or slightly acidic. Within the mid-range of pH 6.0–8.0, there is often little difference in plant growth. Problems arise with extremes of acidity or alkalinity. Soils of pH 8.5 or higher can occur in parts of the region, which can seriously affect plant growth. Garden plants have a range of tolerance to these highly basic (alkaline) soils. Plants that are alkaline sensitive are often difficult to grow in these soils.

**SENSITIVITY OF GARDEN PLANTS TO SOIL ALKALINITY**

| | | |
|---|---|---|
| **Alkaline-tolerant** | | |
| Snap beans | Broccoli | Carrots |
| Cauliflower | Onions | Peaches |
| Nectarines | Rutabaga | |
| **Mid-range** | | |
| Apricots | Basil | Cherries |
| Chives | Melons | Parsley |
| Peas | | |
| **Alkaline-sensitive** | | |
| Brussels sprouts | Beets | Cabbage |
| Corn | Cucumbers | Dill |
| Eggplant | Horseradish | Lettuce |
| Peppers | Pumpkins | Radishes |
| Raspberries | Tomatoes | Turnips |
| Watermelons | | |
| **Extremely Alkaline-sensitive** | | |
| Blueberries | Rhododendron | Azalea |

### Correcting Problems with Soil Acidity

Problems with acidic soils are easily corrected by addition of materials that contain calcium. This includes limestone (calcium carbonate), gypsum (calcium sulfate) or dolomitic limestone. Wood ashes and the ground shells of shellfish can also be good materials to help "sweeten the soil," making it less acidic. Gardening col-

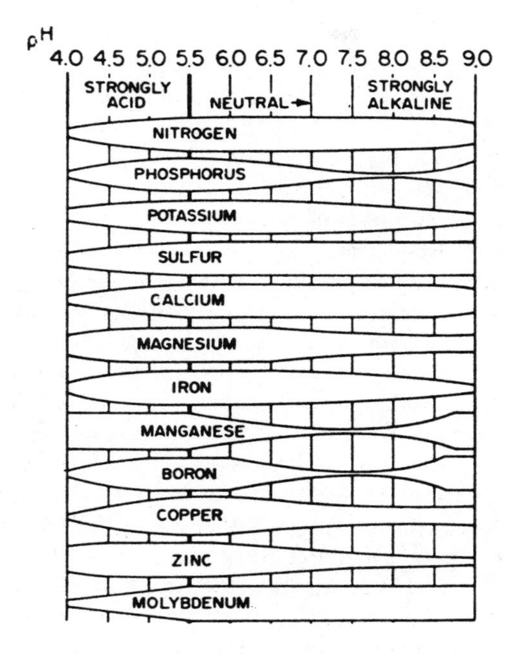


Courtesy of Kentucky Agricultural Experiment Station

*Theoretical relationship between soil pH and relative plant nutrient availability*

umns or books may recommend these practices, and limestone and gypsum are even sold in some local nurseries. These materials are suitable for many areas in the United States and Canada, but in the West, the already alkaline soils can be damaged by these products.

Reducing the pH in the typically basic (alkaline) soils of the region is considerably more difficult. Sulfur, either alone or in combination with acidic fertilizers such as ammonium sulfate, can decrease the pH of soils. However, sulfur added to the soil combines readily with calcium to form gypsum—a slightly basic (pH 7.4) compound. The high level of calcium present in most soils negates most of the positive effects of this treatment. Very high levels of sulfur may ultimately overcome this problem, but it should be added in small amounts over several seasons to prevent plant damage.

Addition of organic matter is the only consistently effective means of lowering the pH of basic soils when high calcium levels are present. Decomposition of organic matter produces acids as a by-product. Some of these organic materials may be highly acidic, such as sphagnum peat moss. Manures or compost tend to be more neutral but are also very valuable for lowering pH. Each year, mix about 1 inch of organic matter into the vegetable or flower garden (a cubic yard covers about 350 square feet), preferably in the fall, so it can begin to decompose and build the soil before spring planting.

## *Salt-affected Soils*

Salt is a common kitchen condiment and essential nutrient. However, high intake of salts produces human ailments, such as high blood pressure. Similarly, high levels of salts can be devastating to plants, as demonstrated by a brief visit to one of the western salt flat areas, which are largely devoid of plant life.

Salts affect plants by "pulling" water from the roots, making it difficult for plants to take up enough water or nutrients. Young plants and germinating seeds are particularly susceptible to injury by salts. Plants and plant varieties vary in sensitivity to salt-affected soils.

Salts are common problems in arid areas throughout the world. When moisture is lost through evaporation faster than it is replaced, salts move upward and concentrate in the root zone. When drainage is poor, salts can't be flushed (leached) out of the root zone. Salty soils have caused serious damage to agriculture, sometimes forcing crop production to be abandoned.

Sodium and/or water-soluble salts, such as sulfates or chlorides, are most common. These become very obvious as they are drawn to the soil surface by evaporation, forming white crusts on the soil surface. However, even when such soil crusts are not present, plant growth and seed germination can be greatly reduced.

Salts can also be introduced into a garden through fertilizers or manures. Many nitrogen fertilizers, such as ammonium nitrate, are high in salts. Cattle manure from feedlots and poultry manure can be high in salts, causing "fertilizer burn." Aging these manures, by storing them outdoors for a year or more, often allows the salts to leach out so that the manure can be more safely used. A soil test is recommended where large amounts of manure of suspect quality are to be added to the garden.

**RELATIVE SALT TOLERANCE OF VEGETABLE AND FRUIT CROPS**

| | | |
|---|---|---|
| **Most Tolerant** | | |
| Asparagus | Beets | Broccoli |
| Tomatoes | Cantaloupe | Spinach |
| Cabbage | Grapes | |
| **Intermediate Tolerance** | | |
| Cucumbers | Peppers | Potatoes |
| Sweet corn | Lettuce | Sweet potatoes |
| Apples | Pears | |
| **Least Tolerant** | | |
| Peaches | Plums | Onions |
| Carrots | Apricots | Beans |
| Raspberries | Strawberries | |

Sodium salts also produce other problems. High-sodium soils (*sodic soils* or *black alkali*) affect how soil particles clump, changing soil structure. Sodic soils are cloddy, drain poorly and tend to crust badly.

## Treatment of Salt-affected Soils

Problems with salt-affected garden soils can be difficult to correct. Attention should first be given to improving drainage conditions. This is best done by fixing soil structure problems (discussed earlier), usually by addition of organic matter. However, one important caution regarding fresh animal manures is needed. Since manures are often high in salts, especially those from feedlot cattle operations, manures should always be aged a year or more. Where precipitation is adequate, this allows some of the salts in the manure to be leached by rains and snow.

In compacted or very heavy clay soils, you may need to further work the soil to improve drainage. Use of raised beds can provide this extra drainage. Deep tillage, such as double-digging, can also improve drainage—after the addition of organic matter has improved soil structure. Deep tilling is particularly important in gardens where heavy equipment (such as large rototillers) or other practices have compacted the lower soil layers, creating a *hardpan* layer that prevents drainage.

Salts can also be leached out of soils by watering. About 6 inches of water will normally reduce the amount of salt in the soil by about 50 percent; 12 inches of water reduces the salts in the root zone by approximately 80 percent. However, some local water sources can be very high in salts and will not be suitable for leaching. Also, if soil drainage problems have not yet been corrected, salts will likely again accumulate in the root zone.

Where raised beds are used, salts tend to accumulate at the top of the hills as they are drawn up during evaporation. Plantings made off-center on the hills will lower salt levels in the root zone.

Problems with excess sodium can also be corrected by applying soil amendments that increase the amount of calcium in the soil. Calcium can replace the sodium in the soil, allowing it to be more easily leached out of the root area. However, most soils in the region already have high amounts of calcium. In these instances, the goal is release of the existing calcium, usually by lowering the alkalinity (pH). This is usually achieved by adding acidifying materials, such as sphagnum peat or other types of organic matter.

In those rare soils where calcium is deficient, calcium supplements can be useful. These act to convert sodium that is insoluble in water to forms that can be leached. Gypsum (calcium sulfate) is the most commonly available calcium supplement, typically applied at rates of 5 to 10 pounds

## *Wood Ashes and Garden Soils Don't Always Mix*

Some gardening publications recommend wood ashes as a method for providing soil nutrients and "sweetening" soils. Use of ashes in the garden allows recycling of a waste product generated by wood stoves.

Unfortunately, wood ashes are a poor mix for most western gardens. The ashes have a high pH, typically 10 to 11. Since most soils are already alkaline, the addition of wood ashes would only increase the high pH of the soil. This can affect how well plants grow and reduce the availability of nutrients, such as iron and zinc, to plant roots.

Wood ashes are also rich in salts. Excess salts are already common in western soils and the addition of wood ashes only exacerbates this problem. Potassium and calcium are also abundant in wood ashes. These two nutrients are rarely deficient in regional soils, however.

Wood ashes do have several benefits in eastern and southern gardens, where high rainfall has produced acidic soils. These areas also may have low levels of potassium and calcium. In the West, wood ashes are best used to cover icy patches on the sidewalk.

per 100 square feet. Calcium chloride is a faster-acting, but more expensive, calcium supplement. Use of these materials should be followed by leaching to remove the sodium. However, since calcium also can occur naturally in lower soil layers, a soil test for calcium (or gypsum) should be done before any calcium amendments. If such a calcium layer is detected, tilling the soil to mix it through the topsoil can solve this problem.

## *Soil Nutrients*

Garden and landscape plants, like the gardener, require many different nutrients to grow and remain healthy. Some of these, such as carbon, hydrogen and oxygen, the plant acquires from the air and water through the miracle of photosynthesis. The other remaining thirteen plus nutrients needed by the plant must be gleaned from the soil.

Nitrogen, phosphorus and potassium are the three nutrients most heavily drawn from the soil. These are found in almost all fertilizers and are called *macronutrients*. Another group of nutrients used for many plant processes is calcium, sulfur and magnesium. These secondary nutrients tend to be more common in soils and are less commonly found in fertilizers. Small amounts of other nutrients needed by plants include iron, chlorine, boron, manganese, zinc, copper and molybdenum; these are called *micronutrients* or *trace elements*.

Some soils are deficient in levels of one or more critical nutrients, causing problems with plant growth. For example, nitrogen is often deficient in native soils and is usually the nutrient that most limits plant growth. Garden soils can become depleted of nutrients as we harvest plants and do not replace the nutrients that are removed.

Other nutrients occur in adequate levels in soils, but not in a form that plants can use. Soil nutrients must be carried in water to be picked up in plant roots. Under various conditions, they become insoluble in water and unavailable. For example, our high-pH soils cause most of the iron, zinc and copper to become unusable to the plants. These are released more easily when soils are more acidic.

Organic matter in the soil also affects the availability of soil nutrients. Nutrients in manures and plant matter are not usable by plants until they decompose and are released by soil microbes. When decomposition is slowed, as in cool soils,

## SUMMARY OF COMMON PLANT NUTRIENTS AND DEFICIENCY SYMPTOMS

| Plant nutrient | Plant use | Deficiency symptoms |
| --- | --- | --- |
| Nitrogen | Constituent in chlorophyll, proteins and nucleic acids | Retarded plant growth and overall yield, yellow color (chlorosis) most pronounced on older leaves, very common |
| Phosphorous | Involved in energy storage and transfer, part of cell nuclei | Retarded growth, purple color of leaves, reduction in fruit production, commonly deficient in many soils |
| Potassium | Used in enzyme action and cell permeability, involved in cell division | Bronzing and brittleness of leaves, rarely deficient in regional soils |
| Sulfur | Constituent of proteins, found in several vitamins and plant oils | Reduced yield, rarely deficient in regional soils |
| Calcium | Important in production of cell walls. Also involved in formation of nucleus and mitochondria | Very rarely deficient in regional soils, although sometimes inadequately distributed within the plant. New growth may be retarded. Also blossom-end rot symptoms |
| Manganese | Involved in many enzyme systems | Speckling or spotting of younger leaves, infrequently occurs in region |
| Chlorine | Affects root growth | Rarely observed except experimentally |

| Plant nutrient | Plant use | Deficiency symptoms |
|---|---|---|
| Boron | Needed for growth of new cells | Plant terminal stops growing and may die, restricts flowering and fruit development, rarely observed |
| Copper | Involved in many enzyme systems | Yellowing, off-colors or small spotting, very uncommon in region |
| Iron | Involved in many chemical reactions (oxidation-reduction reactions, electron transfer) | General yellowing, concentrated on younger leaves, very common in region |
| Zinc | Involved in many enzyme systems | Areas between veins of younger leaves become yellow (chlorotic) and may die, leaves often smaller and thicker, very common in region |
| Molybdenum | Used in small amounts in certain nitrogen reactions | Reduced growth, very uncommon in regional soils |

nutrients may temporarily be deficient. Phosphorus and nitrogen deficiencies are very commonly associated with cool soil temperatures.

Bacteria and other microbes that decompose organic matter may also have "first grabs" on other soil nutrients, notably nitrogen. Addition to the garden of plant materials that are high in carbon but low in nitrogen, such as leaves or straw, causes the microbes to withdraw nitrogen from the soil for their own growth. This can lead to a temporary shortage of nitrogen to the plants. Ultimately the nitrogen is again released to the plant as decomposition progresses, but this may take months or even years. Organic matter also forms strong chemical bonds with copper, reducing the available copper in the soil.

Finally, the proportion of various nutrients in soil is important, since they compete for uptake by plant roots. For example, calcium, potassium and magnesium are all brought into the plant by the same chemical process. Excess levels of any one can produce plant deficiencies of the other nutrients. Problems with molybdenum deficiency in plants may be caused by high levels of competing manganese or copper. High levels of bicarbonates and phosphates lower iron availability. In addition, efforts to relieve iron deficiencies by using too much chelated iron can induce zinc deficiencies.

## Soil Sampling

Fundamental to any plan to correct soil problems is a soil test. Several laboratories are found throughout the region and can be reached through county offices of the Cooperative Extension system or state universities. The test results describe soil texture, pH and salt levels as well as various plant nutrients. Most soil testing labs also point out deficiencies or other problems and suggest methods of correcting them. Testing garden soils as you start, and every few years thereafter, can identify problem areas and mark your progress.

The key to an effective soil test of your garden is giving a proper soil sample to the testing lab. The sample must be representative of the garden area that you are developing so that you can correct soil problems. Soil testing labs usually provide guidelines for taking a soil sample, but general practices include testing sample areas that are fairly uniform. Often, garden soils around the yard have a checkered history due to past gardening practice or construction activities. Approach such different sites individually, sampling various areas separately. Otherwise the sample provided will be an average that does not truly represent any garden area.

Within the sample area, take several samples. A minimum of ten subsamples is often requested. These should be thoroughly mixed together and a portion (usually about 1 to 2 pints) should be submitted to the soil testing lab. Sampling should take cores of soil of uniform thickness to a depth of at least 6 inches.

When submitting soil samples, indicate what tests you would like to have run. Although in the beginning it may be useful to go for the works, later testing can be more selective. For example, the amounts of many soil nutrients, such as micronutrients, are removed very slowly from soils. If earlier tests indicated that these were in adequate quantities, you can spare the extra expense of later testing for these.

## Treating Soil Nutrient Deficiencies

Changing the pH is one of the best ways to increase soil nutrients, particularly in alkaline soils. As soils become more acidic, many of the nutrients, such as iron, are released by chemical changes so that they can be used by the plant.

Addition of fertilizer is the most obvious means of adding nutrients that are deficient in the soil. Most garden fertilizers contain various concentrations of the macronutrients nitrogen (N), phosphorus (P) and potassium (K), often marked boldly on the bag in a 3-number shorthand (e.g, 10-10-10 or 20-10-5). Try to select the fertilizer blend that best matches the soil needs.

*Caution:* Never use "weed and feed" types of lawn fertilizers in the vegetable or flower garden. These contain herbicides that can kill garden plants.

Micronutrients also are sold, usually in some of the smaller (and much more expensive) containers of fertilizer designed to be mixed with water. In alkaline soils, however, application of nutrients such as iron to the soil may do little good because they are rapidly changed to forms that plants cannot absorb. Use of special forms of these nutrients, called *chelates*, which resist these actions of the soil, can circumvent this problem. Plants may also be treated with fertilizers that can be absorbed through the leaves, called *foliar feeding*.

Organic matter is also an important source of nutrients and helps to improve soil texture and reduce pH. Plant and animal materials contain these nutrients in different proportions. As they are decomposed, the nutrients are released. This process is much slower than the release of nutrients found in most garden fertilizers. However, the steady release can help to provide nutrients to the plant more consistently and prevent imbalances.

## APPROXIMATE CARBON/NITROGEN (C/N) RATIO AND LEVELS OF NITROGEN (N), PHOSPHORUS (P) AND POTASSIUM (K) IN SOME ORGANIC MATERIALS

| Organic Fertilizer | C/N ratio* | % N | % P | % K |
|---|---|---|---|---|
| Sewage sludge | 5/1–10/1 | | | |
| (activated) | | 2–6 | 2–7 | 0–1 |
| (digested) | | 1–3 | 0.5–4 | 0–0.5 |
| Alfalfa | 13/1 | | | |
| Compost | | 1.5–3.5 | 0.5–1.0 | 1–2 |
| Sheep manure | 17/1 | 0.6–4 | 0.3–2.5 | 0.75–3 |
| Beef cattle manure | 17/1 | | | |
| Dairy cattle manure | 25/1 | 0.25–2 | 0.15–0.9 | 0.25–1.5 |
| Swine manure | 17/1 | 0.3 | 0.3 | 0.3 |
| Poultry manure | 18/1 | 1.1–2.8 | 0.5–2.8 | 0.5–1.5 |
| Horse manure | 50/1 | 0.3–2.5 | 0.15–2.5 | 0.5–3 |
| Dried blood | | 12 | 1.5 | 0.6 |
| Steamed bone meal | | 0.7–4.0 | 18–34 | |
| Small grain straw | 80/1 | | | |
| Wood chips | 400/1 | | | |
| Sawdust | | 0.2 | 0.1 | 2 |
| Wood ashes | | 0 | 1–2 | 3–7 |

* Carbon/nitrogen ratios of 20/1 or higher require microbes to acquire nitrogen from other soil sources and can induce soil nitrogen deficiencies.

*Sampling for a soil test*

CHAPTER 2

# Natural and Biological Controls

When hornworms are eating the tomatoes and spider mites are sapping the roses, it may seem that garden pests are doing very well indeed. But pause for a minute and think about the normal sorry lot of any insect.

For example, a mated female insect typically may lay 100 eggs. On average, 98 of these usually fail to reach the adult stage and reproduce, a miserable percentage by any measure. The rest suffer from starvation, losing bouts with severe weather or end up as breakfast for some other animal.

All garden pests suffer tremendous mortality, with populations greatly affected by *natural controls*. Natural controls, or "the balance of nature," often prevent pest species from causing loss. Since small shifts in this balance can determine whether a pest becomes abundant enough to damage crops, it is important for the gardener to recognize and work with existing natural controls.

## Abiotic (Environmental) Controls

Many of the most important natural controls are *abiotic controls*. These involve the effects of weather, such as temperature, moisture and wind. For example, rainfall, though not often abundant in the West, is very destructive to small insects and spider mites. Raindrops that we barely notice can bash a small aphid or mite, washing it from the plant to a certain death. Insects commonly drown in puddled water on plant leaves. Wetting of soil also determines when weed seeds germinate, and water on leaves is needed for the spread of fungal and bacterial diseases.

Strong winds, common in the West, are also very destructive to insects and mites. Leaves brushing against each other crush or dislodge small insects. Weak flying insects, such as aphids, are carried and often destroyed by strong winds (or at least blown into the next county, where you don't need to worry about them). Sand grains picked up on winds are like boulders to a hapless spider mite.

Temperature extremes are the most important control of overwintering insects. Several garden pests in this region, such as the corn earworm and aster leafhopper, can't survive western winters and must migrate each year to milder areas in Mexico and the southern United States. Very cold weather may kill some insects that are only marginally adapted to survive in the area, such as the tobacco budworm (geranium budworm). The effects of winter freezing are most important when snow cover is not present.

## Biological Controls

Perhaps more familiar to the gardener, and a lot more fun to watch, are the various types of biological (or *biotic*) controls of garden pests. These are the natural enemies of pest species—their *predators*, *parasites* and *diseases*.

Predators of insects and mites are organisms that move about and hunt, feeding on several prey before they become full-grown. Lady beetles (ladybugs), green lacewings, ground beetles, damsel

## *Seasonal Summer Visitors*

Many of the most important insect and disease problems in the region result from annual migrants returning to the area. These "snowbirds" spend the winter in the southern United States and Mexico, then fly or are blown by winds to the north. Among the more important migrants are the following:

- Potato (tomato) psyllid
- Corn earworm
- Spotted cucumber beetle
- Armyworm
- Aster leafhopper (transmits aster yellows disease)
- Powdery mildew of cucurbits
- Garden leafhoppers (species of *Empoasca*)

As a result, appearance of these insect and disease organisms is more irregular than other hometown garden visitors. They move through the area in different numbers each season, and local fallout is highly dependent on weather patterns during migrations. Outbreaks tend to be sporadic, with bad years often followed by seasons with relatively few problems. Since the insects either die out or fly back to the overwintering areas after each season, next year's problems cannot be predicted.

Some of the more benign garden visitors, such as the monarch butterfly and painted lady butterfly, are also annual migrants. You may see large migrations of these insects heading north in the spring and south in the late summer.

bugs and garter snakes are common predators of insects.

The term *predator* is also used to describe certain natural enemies of weeds. Insects and mites that feed upon the plant and the seeds produced by the weed are described as predators, although they would be classified as pest insects if they fed on desirable plants. For example, the Colorado potato beetle is welcome when it cleans up the nightshade weeds in the garden, but becomes a different animal when the eggplant and potatoes are later threatened.

There are also several groups of insect parasites. These are usually wasps or flies that grow in or on a host insect. The host insect is almost always killed by the developing parasite, and several parasites can develop in a single host insect.

Insects and mites are also commonly killed by diseases caused by infection with fungi, bacteria, viruses, protozoa and nematodes. Other fungi, viruses and nematodes may attack weeds; some even kill or displace bacteria and fungi that cause plant disease.

## Applied Biological Controls

Sometimes we may wish to get a little more help with biological controls than is provided by *laissez-faire* gardening. The term *applied biological control* is used to describe this manipulation of biotic controls for pest management. These applied biological controls can take many approaches, including:

- introducing natural enemies of pests into an area where the natural enemies did not previously exist;
- periodically releasing additional natural enemies to increase their numbers and effectiveness;

- conserving of existing natural enemies of pests; and
- improving the effectiveness of pest natural enemies by creating favorable environmental conditions for them.

Most use of applied biological control has involved management of insect and mite pests. However, some plant disease organisms and weeds have been controlled by biological means. In recent years, the study of biological controls has received increasing attention by researchers in many countries, with future benefits to the gardener.

## Recognizing Common Biological Control Organisms

Although there are many biological control organisms in our yards and gardens whose daily activities often escape us, the garden patch is a constant war zone for garden insects. (It's a jungle out there!) It is important that we recognize and appreciate these organisms so we can better use them in pest management. Plus, their activities add an interesting dimension for the observant gardener.

Although some are commonly recognized, such as the adult stage of the lady beetle, others are frequently overlooked, such as the parasitic wasps and tachinid flies. Furthermore, some insects change in appearance through metamorphosis, and we may not recognize all the stages of these beneficial organisms.

### *Common Predators of Insects and Mites*

**Lady beetles.** Often called ladybugs, lady beetles (Coccinellidae family) are the the most familiar insect predator to most gardeners. Although dozens of species exist, they are all typically a round or oval shape. Many are brightly colored and spotted.

Females periodically lay masses of orange-yellow eggs that hatch in about 5 days. The immature or larval stages look very different from the more familiar adults and often are overlooked or misidentified. Lady beetle larvae are elongated, usually dark-colored and flecked with orange or yellow. They can crawl rapidly over plants, searching for food.

Adult and larval lady beetles feed on large numbers of small, soft-bodied insects such as aphids. Lady beetles also eat the eggs of many insects. One group of very small, black lady beetles, aptly dubbed

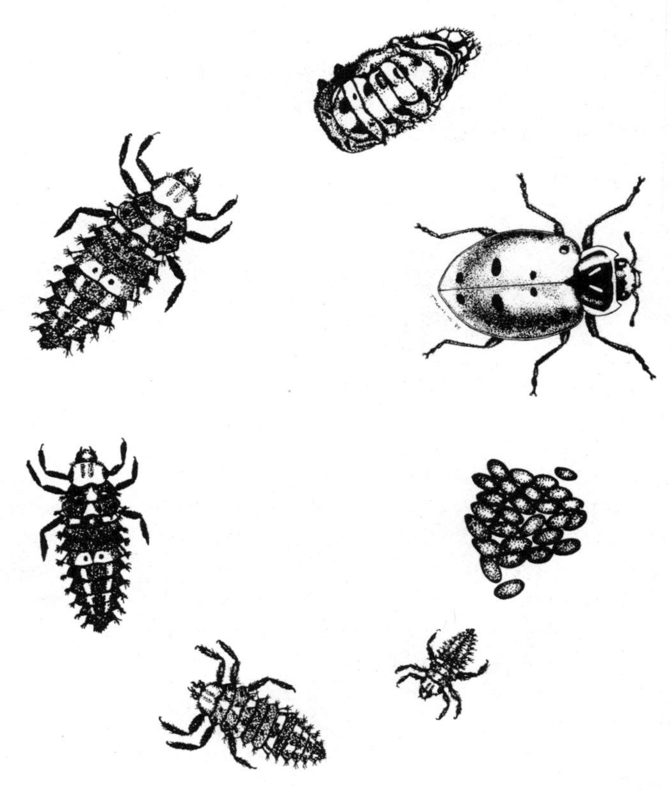

*Life cycle of a lady beetle*

the "spider mite destroyers" (species of *Stethorus*), are very important in controlling spider mites. Lady beetles reproduce rapidly during the summer and can complete a generation in less than four weeks under favorable conditions. As a result, they often overtake a pest outbreak, controlling many potential insect problems.

Unfortunately, lady beetles tend to be fair-weather insects that are slow to arrive in the spring and often leave the garden by late summer. (A few kinds even "head for the hills," spending the cool seasons at high elevations, protected under the snow.)

Over fifty species of lady beetles occur in the West. All are beneficial in that they feed on other insects and mites, with one exception. The lady beetle known as the Mexican bean beetle is the "bad apple" of the group, which has developed a fondness for bean leaves. Similar in shape but slightly larger in size than the insect-feeding lady beetles, the Mexican bean beetle is light brown and highly spotted. Larvae of the Mexican bean beetle are yellow and spiny.

**Ground beetles.** Ground beetles (Carabidae family) are very common around the yard and garden. They are general feeders with powerful jaws that prowl the soil and soil surface at night. Almost any garden insect that spends part of its life on the soil surface may be prey for these insects.

Adult ground beetles are usually black and 1/2 to 1 inch long. Most are poor climbers, although some (notably species of *Lebia*) crawl readily onto plants. Immature stages of ground beetles are less commonly encountered since they remain underground. Immature ground beetles are wormlike in shape, but have a pair of powerful jaws that allow them to feed on many soil insects.

**Rove beetles.** Rove beetles (Staphylinidae family) share the soil surface with ground beetles. The adult beetles are elongate with short wing covers, resembling an earwig without the pincers. Most rove beetles feed on the small insects (e.g., springtails) that in turn feed on decaying organic matter. They are often found prowling around compost piles and manure. Some rove beetles have more specialized habits and can be very important controls for fly larvae, such as seedcorn maggot and onion maggot.

**Green lacewings.** Several species of green lacewings (Chrysopidae family) are commonly found in gardens, particularly on shrubbery and fruit trees. The adult stage is a pale green insect with large, clear, highly veined wings that are held over

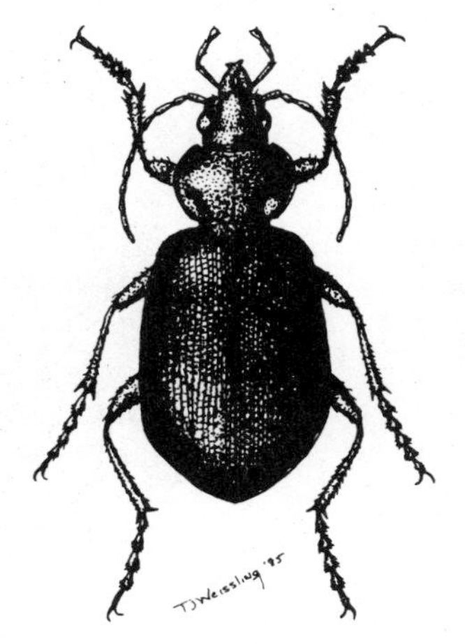

*Ground beetle*

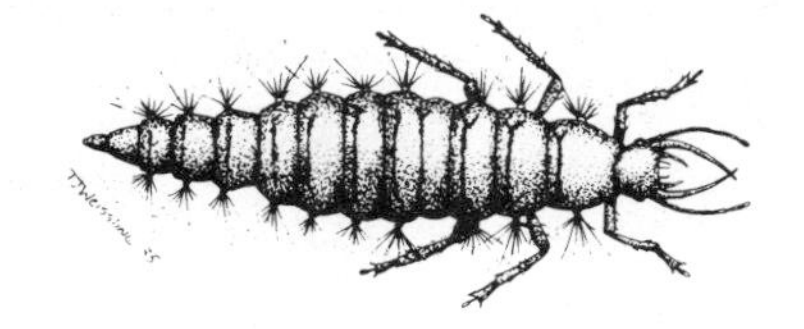

*Green lacewing (top), green lacewing larva (bottom)*

the body when at rest. They are delicate and attractive insects that feed primarily on nectar. The females lay distinctive stalked eggs, approximately 1/2 inch in height, in small groups or singly on leaves of plants throughout the yard.

Lacewing larvae emerge from the egg in about a week. These larvae, sometimes called aphid lions, are voracious predators capable of feeding on small caterpillars and beetles as well as aphids and other insects. In general shape and size, lacewing larvae are superficially similar to lady beetle larvae. However, immature lacewings usually are light brown and have a large pair of viciously hooked jaws projecting from the front of the head. Whereas lady beetles often limit their feeding to smaller insects such as aphids, green lacewings are capable hunters that can easily kill insects larger than themselves. Several generations of lacewings are born during the summer, and a green-brown, cold-tolerant species can be found late into the fall.

**Syrphid flies.** Syrphid flies (Syrphidae family), also called flower flies or hover flies, are common, brightly colored flies. Typical markings are yellow or orange with black, making them resemble bees or yellow jacket wasps. However, syrphid flies are harmless to humans. They are usually seen feeding on flowers.

## *It's a Bee, It's a Wasp—It's Syrphid Fly*

One of the most beneficial groups of insects in the average yard and garden is the syrphid fly (pronounced sir´fid). Adult stages of these insects are usually bright orange or yellow and black. Most resemble bees and wasps, and some even carry this act to convincing extremes by buzzing. However, adult syrphid flies (sometimes called flower flies or hover flies) are harmless and feed on nectar and other fluids.

The immature larva is the beneficial stage, as it feeds readily on aphids and other soft-bodied insects. It is a tapered, usually green or gray maggot that crawls over plants in search of prey.

Perhaps it is their appearance (it *is* hard to find good press about any maggot), but syrphid flies are one of the most important natural controls against aphids found in the garden. Many syrphid flies show up early in the season and retire late, existing in parts of the season when more highly touted predators, such as lady beetles, have vacated the yard. Syrphid flies also can squirm into tightly curled leaves that confound the bowlegged lady beetles and lacewings, providing biological control of leaf-curling aphids.

Syrphid fly maggots also, fortunately, don't get bothered by ants, which routinely drive off other predators that threaten aphids. Syrphid maggots appear able to feed undisturbed on aphid prey, perhaps because they often cover their body with disposed aphid carcasses. These "sheep in wolves' clothing" often get past the defense that ants erect against other more conspicuous predators, such as lady beetles.

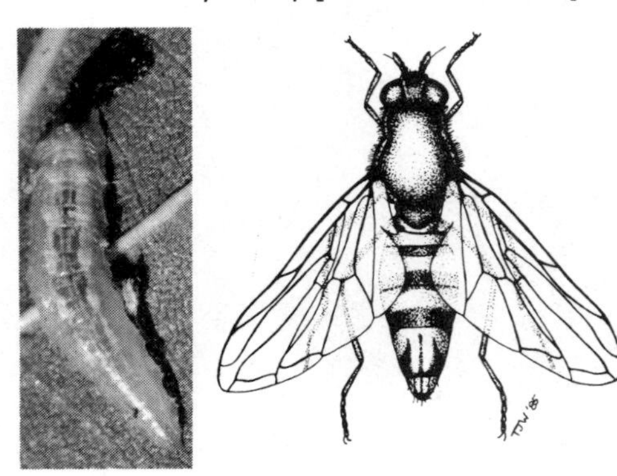

*Flower fly (larva)* *Flower fly (adult)*

It is the larval stage of the syrphid fly that is an insect predator. Variously colored, the tapered maggots crawl over foliage and down dozens of aphids daily. Syrphid flies are particularly important in controlling aphid infestations early in the season when it's still too cool for lady beetles and other predators.

A few species of syrphid flies, such as the narcissus bulb fly, develop by feeding on and tunneling plant tissues. These plant-feeding syrphid flies often develop into large, stout-bodied flies that may resemble bumblebees.

**Predatory bugs (Hemiptera order).** Several "bugs" are predators of insects and mites. All feed by piercing prey with their very narrow mouthparts and sucking out body fluids. A few may give the unsuspecting gardener a little nip if they are handled or get trapped in clothing.

*Stink bugs* (Pentatomidae family) are the largest of the predatory bugs and are characterized by their distinctive shieldlike body and ability to produce an unpleasant odor when disturbed. Most common in the vegetable garden is the twospotted stink bug, a red and black species with a distinctive horseshoe band on the back, a predator of beetle larvae such as the Colorado potato beetle. Several other stink bugs found within the region also feed on insects, although a few species feed on plants and are occasional pests.

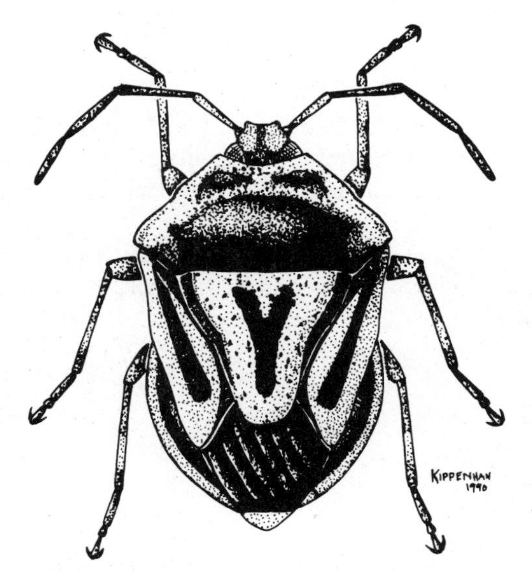


*Twospotted stink bug—a common predator of the Colorado potato beetle*

*Assassin bugs* (Reduviidae family) are capable predators that can subdue large insects such as caterpillars and beetles. Most assassin bugs are elongate, spiny and have a pronounced snout in front, which is the base for the stylet mouthparts. Despite their prodigious ability to dispatch most garden pests, they rarely become very abundant in a garden because they have too many enemies of their own (mostly egg parasites).

*Damsel bugs* (Nabidae family) are far more common but less frequently observed insect predators. The damsel bugs (also called nabid bugs) are light brown and may reach a size of 3/8 inch. Damsel bugs are found on the foliage of all plants, where they seek out aphids, insect eggs and small insect larvae.

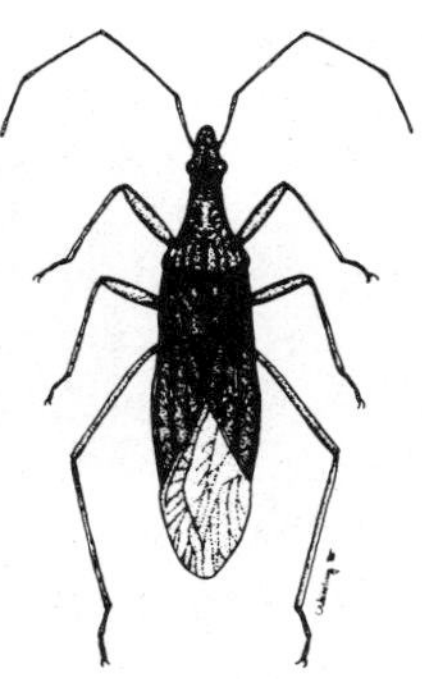

*Damsel bug*

*Minute pirate bugs* (Anthocoridae family), the most common of the predatory bugs, are very small (less than 1/8 inch). The adult stage is marked by black and white, but the pale orange nymphs may resemble hyperactive aphids. Minute pirate bugs are observed most frequently in flowers or in the crevices of a green plant, where they feed on thrips, spider mites and insect eggs.

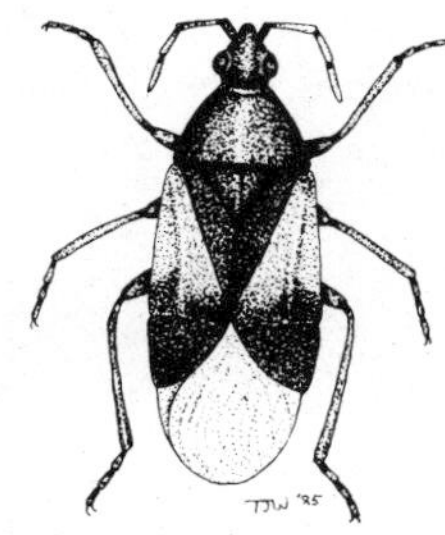

*Minute pirate bug*

**Predatory thrips.** The banded-winged thrips (Aleothripidae family) include some of the most important predators of spider mites and plant-feeding thrips. They are very small but may be observed upon close inspection of blossoms or other areas where thrips are common. The banded-winged thrips have distinctive black and white markings that distinguish them from other thrips. Although they are not common, they can demolish a budding spider mite colony in short order.

**Hunting wasps.** Many wasps prey on insect pests and feed them to their young. These hunting wasps build nests out of mud or paper. Other hunting wasps construct nests by tunneling into soil or pithy plant stems, such as rose or raspberry. The adult wasps then capture insect prey and take them back to the nest, whole or in pieces, to feed to the immature wasps.

Most hunting wasps are *solitary wasps.* The females construct the entire nests, working alone, while the males hang out on plants waiting for "action." The females then search for prey, which they immobilize with a paralyzing sting and carry back to the nest. The young wasps develop by eating the food provided by the mother wasp.

The solitary hunting wasps (Sphecidae family) have very specialized tastes that cause them to search only for selective types of prey. For example, some develop on leafhoppers, others attack caterpillars, and beetles are prey for some hunting wasps. The largest hunting wasp is the cicada killer, which resembles a giant yellow jacket and kills the dog-day cicada. Despite their often fearsome appearance, solitary hunting wasps rarely sting and do not contain the potent venom of the social wasps.

*Polistes wasp*

Other hunting wasps, such as paper wasps and yellow jackets (Vespidae family), are social species in which many individuals work together in a colony of specialized *castes* (queen, drone and worker). These make nests of a papery material formed by chewing wood, cardboard or other materials. The nests are constructed under eaves, in trees or underground in holes around building foundations or abandoned rodent burrows. Every year these *social wasps* create new colonies, which may be aggressively defended by stinging guard wasps, much to the dismay of the unwary gardener. However, most social wasps rear their young on a diet of caterpillar paste or other insects and are extremely useful for control of garden pests. At the end of the season, the nests of the social wasps are abandoned.

**Predatory mites.** Many types of mites are predators of plant-feeding spider mites (primarily members of the Phytoseiidae family). Typically, these predatory mites are a little larger than spider mites and faster moving than their prey. Predatory mites can provide good control of spider mites, but most are slowed down by dry weather—they like it hot and humid. Predatory mites are also more susceptible to insecticides than are plant-feeding species.

**Spiders.** Although hardly a favorite of most gardeners, spiders (Araneida order) are among their best allies. All spiders feed only on living insects or other small arthropods, protecting garden plants.

Although most gardeners observe the many web-making spiders, such as the yellow-and-black-banded garden argiope, there are many other spiders that do not build webs, such as wolf spiders,

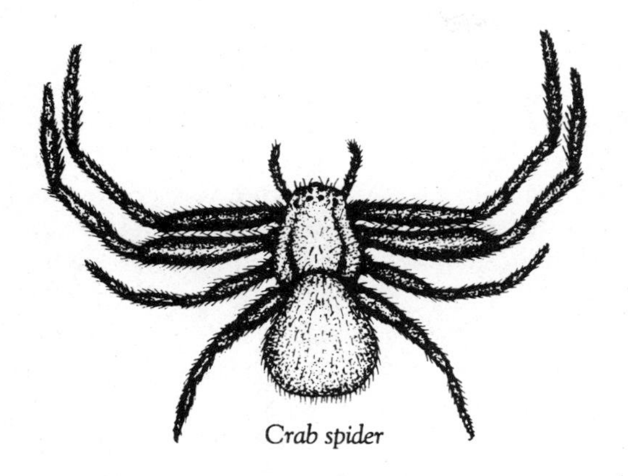
*Crab spider*

crab spiders and jumping spiders. These less conspicuous spiders move about the plants and hunt for their prey on soil or plants. They can be more important than their flashier, web-spinning relatives in controlling insect pests such as beetles, caterpillars, leafhoppers and aphids.

***Common Parasites of Insects***

**Tachinid flies.** Tachinid flies (Tachinidae family) develop as parasites inside other insects. Tachinids are about the size of a housefly, generally gray or brown and covered with dark bristles. They are rarely seen but often leave their "calling card": a white egg laid on various caterpillars, beetles and bugs, usually near the head. The eggs hatch within the day and the young fly maggots tunnel into their host. There they feed for about a week (carefully avoiding the vital organs until the end), eventually killing the host insect.

*Tachinid fly*

**Braconid and ichneumonid wasps.** The parasitic wasps, including the braconid and ichneumonid wasps (Braconidae and Ichneumonidae families), are a diverse group of wasps that develop as insect parasites. Some are very small and rarely observed, attacking small insects such as aphids. Others even live in the eggs of various pest insects. Larger parasitic wasps attack caterpillars or wood-boring beetles.

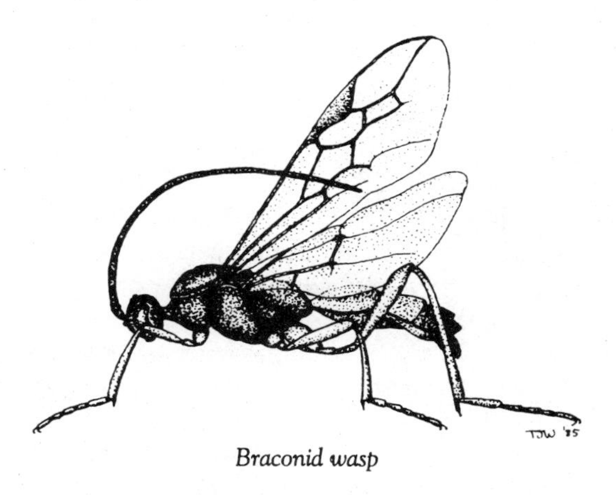
*Braconid wasp*

There is usually little evidence of parasitic wasps in the garden because the young wasps develop inside the host insect from eggs that were inserted by the mother wasp. However, parasitized insects may be somewhat different in form. For example, aphids that are parasitized by these wasps are typically small and discolored and are called aphid mummies. Other common braconid wasp species spin conspicuous yellow or white pupal cocoons after emerging from a host.

**Chalcid wasps.** There are hundreds of species of chalcid wasps (Chalcidoidea superfamily), which attack and kill other insects. However, most chal-

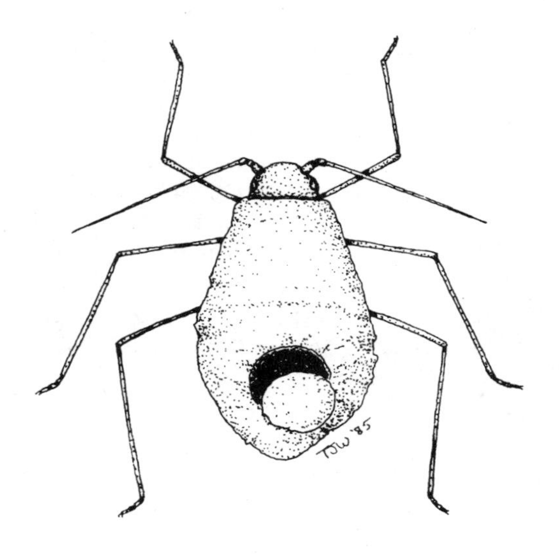

*Aphid "mummy"*

cid wasps are very small and are rarely observed by the gardener. Like the braconid and ichneumonid wasps, chalcid wasps do not sting and are harmless to humans.

Some of the more important chalcid wasps attack aphids and the caterpillar stages of cutworms, fall webworms and cabbage loopers. Some chalcid wasps are also available from suppliers of biological control organisms. For example, the trichogramma wasps, which develop within the eggs of various caterpillar pests, are widely sold in garden catalogs. A parasite of the greenhouse whitefly, *Encarsia formosa*, is also available from several sources. (See "Commercially Available Biological Controls," later in this chapter, for more details.)

***Insect Diseases***

Although not often observed, insects and mites can suffer from disease. Periodically, waves of disease caused by fungi, bacteria, protozoa or viruses may sweep through a population, making the Black Death of the Middle Ages (bubonic plague, a bacterial disease of humans and certain rodents) seem like a mild head cold. Although the classes of organisms that cause human disease are the same for many other animals and plants (e.g., viruses, bacteria and fungi), it is important to keep in mind that insect diseases are very specific in their effects. They do not infect mammals or birds, restricting their effects to the arthropods. Furthermore, most insect diseases are so specific in their effects that they can infect only a few insect species.

*Viruses* are most commonly found among the caterpillars and sawflies. One particularly gruesome group of these viruses (nuclear polyhedrosis viruses) cause the wilt disease. Caterpillars infected by these are killed rapidly, their virus-filled bodies hanging limply by their hind legs. At the slightest touch, the insects rupture, spilling the virus particles on the leaves below them to infect other insects.

Other types of viruses cause less spectacular infections. Most are slower-acting than the wilt viruses. External evidence of these viruses includes a chalky color and general listlessness of the insects. (Of course, making observations on the relative listlessness of insects is not of great interest to most gardeners.)

Although viral diseases of insects are widespread in nature, rarely have they been adapted for applied biological control. Much of this is due to problems in registering them as insecticides. Regulatory agencies have experienced difficulties in deciding how to ensure the safety of such mysterious particles as viruses. Manufacturers have also been leery of developing viruses because of problems in production (they need live cells to develop) and because the selectivity of viruses allows them to be used against only a few insects, limiting their marketability.

*Bacteria* have received more attention, due almost entirely to the successful adaptation of *Bacillus thuringiensis*. Several manufacturers have produced and marketed various strains of this famous bacteria. Many other types of bacterial disease occur naturally among insects, often producing similar symptoms.

One of the outstanding success stories for biological insect control, *Bacillus thuringiensis*, or Bt, works by disrupting the gut lining of susceptible insects, ultimately killing them by a type of starvation or blood poisoning. Infected insects shrivel and darken. This naturally occurring bacteria was first identified as a disease of insects some eighty years ago. Bt is now manufactured by several companies and used as a biological insecticide. It is sold under a variety of trade names, such as Dipel, Thuricide, Biobit, M-One and Caterpillar Attack.

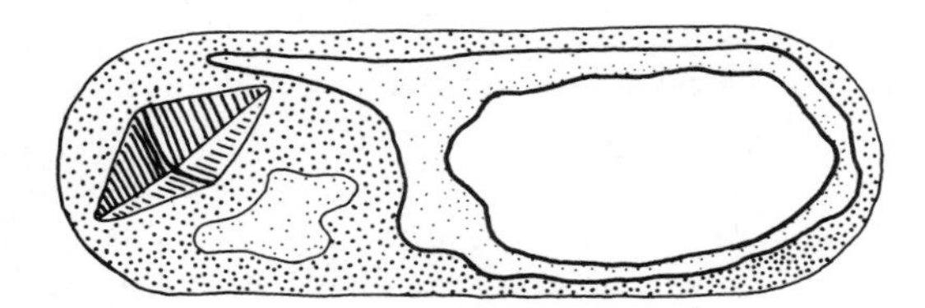

*Sporangium of* Bacillus thuringiensis; *note the protein crystal that has most of the insecticidal activity (greatly magnified).*

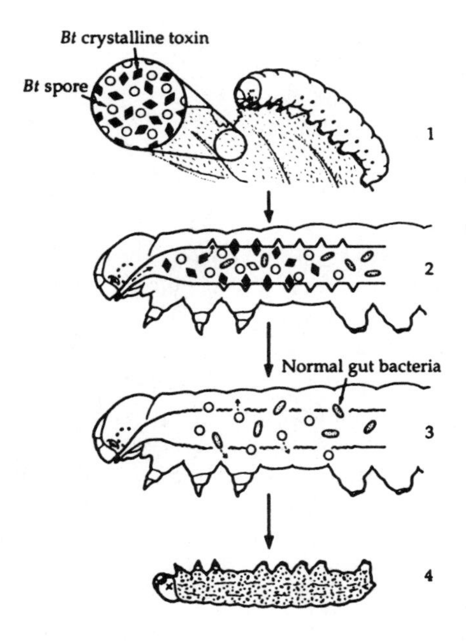


*Action of* Bacillus thuringiensis *var.* kurstaki *on caterpillars. (1) Caterpillar consumes foliage treated with Bt (spores and crystalline toxin). (2) Within minutes, the toxin binds to specific receptors in the gut wall, and the caterpillar stops feeding. (3) Within hours, the gut wall breaks down, allowing spores and normal gut bacteria to enter the body cavity; the toxin dissolves. (4) In 1-2 days, the caterpillar dies from septicemia as spores and gut bacteria proliferate in its blood.*

The outstanding advantages of Bt are its highly specific activity and associated safety. Individual strains of Bt affect only limited groups of insects, and must be eaten to provide control. Hazards to other desirable species, such as humans, pets or beneficial insects, are negligible or nonexistent. This safety makes Bt very valuable for use on garden crops, and it fits well in Integrated Pest Management programs because it spares naturally occurring biological controls.

Bt can be very toxic to some insects, however. The disease kills by producing a toxic protein crystal that is activated in the digestive tract of susceptible species. Very rapidly it paralyzes the gut of the insect, causing it to cease feeding. (However, it may take several days for it to die. In a sense, insects succumb to a sort of terminal "Montezuma's Revenge," often complicated by blood poisoning.)

Bt has generated tremendous interest for decades, which has accelerated recently with developments in biotechnology. Dozens of different strains have been identified, each effective against specific types of insects. Some of the older and most widely used strains (such as *kurstaki* and *thuringiensis*) are widely available for garden use and can control several caterpillars that damage the garden, including the following.

- Imported cabbageworm
- Diamondback moth
- Cabbage looper
- Tomato hornworm
- Parsleyworm
- Leafroller

- Climbing cutworms
- Tomato fruitworm
- Achemon sphinx
- European corn borer (in corn whorl)

More recently commercial development has involved the *israelensis* strain. This kills immature stages of mosquitoes and related flies such as fungus gnats. It is becoming a standard for community mosquito control programs.

Within the last four years, strains that control leaf beetles such as the Colorado potato beetle (e.g., *san diego* and *tenebrionis*) have begun to be offered for sale. Bt products that can be used for leaf beetle control are just now reaching the market, sold under trade names such as Foil, M-One, M-Trak and Trident.

*Fungi* produce some of the more spectacular diseases of insects. A wide variety of insects succumb to fungal disease around the yard and garden. Fungus-killed insects and mites become stiff and often are tightly attached to a leaf or stem. When conditions are right, they become covered with a white, light green or pink fuzz—the spores of the fungus.

Perhaps the most commonly observed fungal diseases involve species of *Entomophthora*. These induce an interesting behavior in the infected insect, causing it to crawl to the top of a high object before it dies. This is a useful characteristic for the fungus, since it allows the spores to be easily spread. How the invading fungus produces this behavior in the hapless insect is another intriguing question. In any case, be on the lookout for mass epidemics of fungus-killed root maggot flies following wet weather in spring. Also watch for the all-too-rare outbreak of fungal disease among grasshoppers, which causes them to stick, still and dead, to the waving grasses they ordinarily would be eating.

Infections by fungi were the earliest insect diseases observed, and attempts have been made for over a century to use them for applied insect control. Unfortunately, they have been difficult to use, resisting most human efforts to manipulate them consistently. Sensitivity to environmental conditions, notably a need for high moisture, has usually foiled their use. However, research in this area remains active, and several fungi now under development (including some for mosquitoes and greenhouse pests) show promise.

## Conserving and Enhancing Biological Controls

The most important need of any insect predator or parasite is enough food to allow reproduction and development. This means that there must always be some pest insects in the vicinity to maintain a population of their natural enemies. Biological controls have a rough time surviving and thriving in a garden where all pest insects are eliminated.

Sometimes alternative hosts of other species of insects are necessary to allow biological control organisms to thrive. For example, many predators, such as lady beetles, are general feeders on a wide variety of small insects. Nondamaging aphid populations on one plant may provide food that allows the lady beetles to later control aphids on another type of plant. Minute pirate bugs, an important predator of spider mites, may use thrips as an alternative host and thus build up their populations. Mixed plantings help provide a constant source of alternative hosts on which beneficial insects can feed.

Other types of food are used by insect predators and parasites. Many adult stages of the wasps and flies that develop inside pest insects feed upon nectar as a source of sugar-rich energy. Although many flowers contain nectar, most are not usable by small insects, since they cannot reach deeply into the flower. Shallow flowers of plants in the carrot family (Umbelliferae family), such as dill, are particularly good food plants for these beneficial insects. The well-fed wasps can live longer, lay more eggs and do a better job of insect control in the garden.

## *Ants and Aphids—a Benevolent Sisterhood*

Insects are not generally known for friendly, cooperative behavior. Certainly there are examples of truly social insects, such as honeybees, paper wasps and termites, that develop elaborate caste systems with different members of the colony given separate tasks. These are insects that truly cannot survive except as colony animals. However, even these insects rarely cooperate with other insect species.

Ants are an exception. Several of the ants common in the West seek out sweet materials as a major food source. Honeydew, the sugary by-product of aphids, and many other related insects, such as leafhoppers, soft scales and whiteflies, provides this favored food. Ants collect the honeydew from leaves or directly prod the insect to excrete honeydew for collection. Later, the honeydew is returned to the colony to feed the developing ants.

Ants that find rich sources of honeydew will often provide a "guard detail" that effectively protects their food provider. Ants are commonly associated with developing colonies of aphids and other honeydew producers. In return for the steady food source, ants drive away potential natural enemies, such as lady beetles and parasitic wasps. This commonly results in a temporary ant-induced disruption of biological controls. This benevolent sisterhood (males are rare in both aphids and ants) may cause problems in insect control. In many cases, the best control of aphids is control of their protectors.

*Cornfield ant—a common associate of aphids*

Artificial foods are another alternative. Sugar water or various sugar/yeast mixtures can increase the activity of lacewings and lady beetles in gardens. Applied as a thick paste on plants, these mixtures apparently simulate the honeydew produced by insects that are common prey, such as aphids. Sugar/yeast sprays can also allow adult predatory insects to live longer and mature more eggs.

Some insects need shelter to be most effective. Ground beetles can be important predators of cutworms and other insects found in the soil. But ground beetles need shelter during the day. Mulches or other ground covers can improve the performance of ground beetles. Earwigs commonly feed on pest insects and require specific shelter. Earwigs seek out cool, dark areas in which to hide during the day. Banding fruit trees with a burlap band provides suitable shelter and allows earwigs to better search trees at night for common prey such as aphids.

Biological controls are conserved most easily by use of chemical and cultural control practices that minimize damage to them. Insecticides vary widely in their ability to kill natural enemies of pest insects. Selective insecticides can conserve the effectiveness of these natural enemies. Pesticide applications may also be timed in a manner that makes them more selective; for example, the use of dormant oils on fruit trees to kill eggs of pest insects spares insect predators and parasites that do not overwinter on trees. The use of pesticides is discussed further in Chapter 4.

Cultural practices that can be important in biological control include reducing the amount of soil and dust on plants. Spider mite problems can be greatly aggravated in dusty areas, since the dust particles stick in the webbing, protecting them from predators.

## Introducing Biological Controls to the Yard and Garden

Several natural enemies of insect pests have been introduced into the region and become permanently established. For example, one of the most common natural enemies of the cabbageworm is the small parasitic wasp *Apanteles glomeratus*, which was introduced into the United States in 1872. Much more well known (and effective) was the Vedalia lady beetle, introduced in 1888 to control the cottony cushion scale in California.

This branch of biological control—introduction and permanent establishment of natural insect enemies—is called classic biological control. Over the past century, this technique has been used successfully many times in the United States, and continues to be developed by the USDA and many state agencies. Newly established natural enemies can provide ongoing control and greatly reduce the damage caused by pest insects.

Insects have also been introduced to help control weeds, particularly introduced rangeland or pasture weeds. For example, weevils that attack the seeds of musk thistle have been introduced and established in most of the West. Programs that introduce species for control of serious pest weeds, such as leafy spurge, continue to be developed.

## Commercially Available Biological Controls

Some natural enemies of pests can be mass-produced or collected in large numbers. These are then released in a yard or garden as a biological pesticide. These types of biological controls typically have to be released annually or repeatedly during a season to achieve effective pest control. Several companies produce biological control organisms, which are sold through mail-order catalogs or in garden centers.

The effectiveness of these "bugs for hire" is variable. Some appear to provide highly effective control under proper conditions. The track record of others is more mixed, while a few, such as the praying mantid, are probably best considered as "recreational pest controls." Biological control organisms available from commercial suppliers include the following.

***Bugs for Hire***

**Lady beetles**—Lady beetles (ladybugs) are one of the most common and beneficial kinds of insects found in the yard and garden. Commercially available lady beetles (*Hippodamia convergens*) are also available from many mail-order outlets and some area nurseries. These are collected from areas, usually in California, where lady beetles mass in large concentrations to pass periods when aphid prey are scarce. Similar aggregations occur during winter in the foothill areas of the Rocky Mountains. Lady beetles collected from these overwintering aggregations almost invariably disperse after they become active and rarely remain in the release area. Release during cool periods, particularly near existing aphid colonies, can improve their retention in a yard. In most cases, however, there is little additional benefit from introduction of these purchased lady beetles into a yard because they rapidly leave.

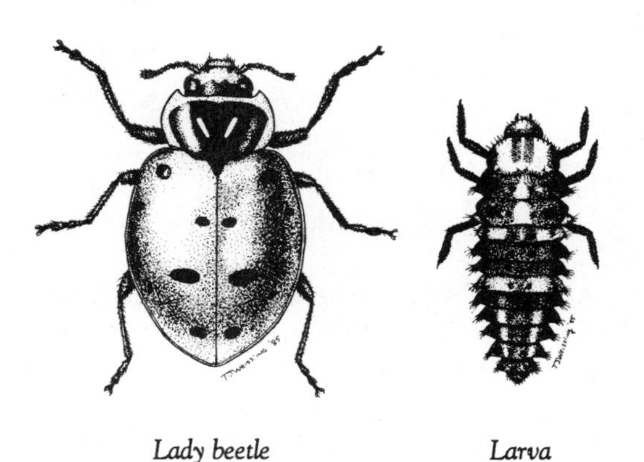

*Lady beetle* *Larva*

**Green lacewings** —Various species of green lacewings, which feed on a wide variety of insect pests, are native to the region. Several commercial sources of lacewing eggs are also available. Distribution of these eggs in a planting can provide some control of pest insects as the developing lacewing larvae feed and develop. Because of the wide host range of green lacewings and because the larvae cannot disperse until they become full-grown, green lacewings have been used as general-purpose predators. After lacewings reach the winged adult stage, they will disperse from an area unless large numbers of aphids and similar plant pests remain.

*The Chinese praying mantid,* Tenodera aridifolia sinensis. *A) Egg case with newly hatched nymphs; B) Adult.*

**Praying mantids**—The praying mantid is one of the largest insects that feed on other insects. Praying mantids are fascinating to watch, but provide little or no control of pest species. There are three reasons for this: They have a nonrestrictive feeding range that includes both pest and beneficial insects; they cannot rapidly increase in number if a pest increases in abundance; and they limit most of their feeding to only those insects that are relatively fast-moving.

Several small, native species of praying mantids occur in the shortgrass prairies of the West. However, purchased praying mantids (usually the Chinese mantid or European mantid) are not native to the area and often fail to overwinter except in areas of mild winters.

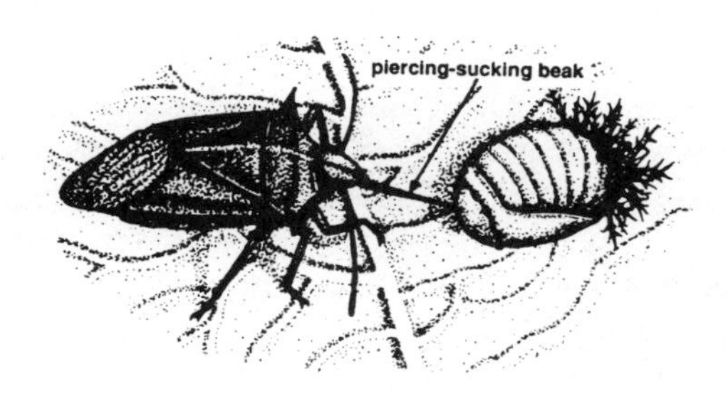


*An adult spined soldier bug, feeding on a Mexican bean beetle pupa*

**Spined soldier bug**—A type of predatory stink bug, the spined soldier bug (*Podisius maculiventris*) has recently been offered for sale as an insect predator. These large insects feed on caterpillars and beetles and can be useful predators. However, they are also native insects in western gardens and the value of supplemental releases is unproven.

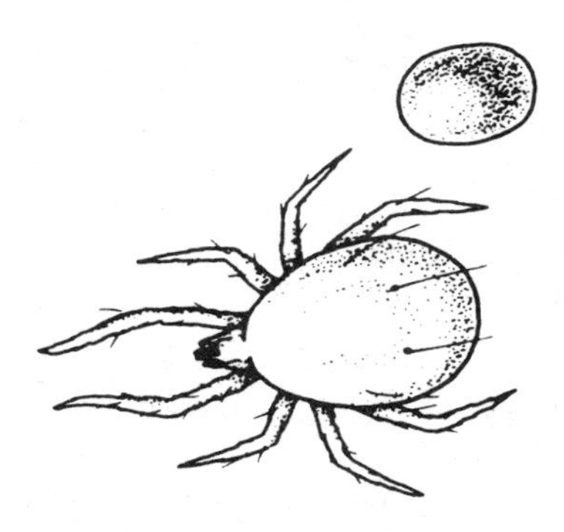

*Predatory mite*

**Predatory mites**—At least four species of mites that prey on plant-feeding mites are available from specialty suppliers. These predatory mites have provided effective control of spider mites on certain greenhouse crops. Use of these mites for control of mite problems on outdoor fruit or shade trees is still in the experimental stage.

Rearing predatory mites is more difficult than most other biological controls. They are also more perishable. It is best to discuss the purchase with your supplier to ensure that you are getting the best species of mite and to arrange for express mailing.

**Trichogramma wasps**—Several species of trichogramma wasps are sold. All of these tiny wasps attack and kill insect eggs, particularly eggs of caterpillar pests, such as cabbage loopers and corn earworms. Insect larvae already hatched are not susceptible to trichogramma attack.

Commercially available trichogramma wasps are usually used as a form of biological insecticide where they are expected to eliminate most of the developing eggs of the pests shortly after release. Multiple releases of trichogramma wasps are recommended, since persistence of the parasites may be short-term. The many trichogramma species differ in terms of the environment in which they are most effective. Some are better used in orchards, while others are more adapted for garden conditions.

**Whitefly parasite**—A small wasp, *Encarsia formosa*, attacks and develops within immature whitefly nymphs. Introduction of this parasitic wasp has proven useful for whitefly management in greenhouses where average (day and night) temperatures remain at 72°F or above. In cool greenhouses, this parasite will not be effective for whitefly control. Sticky yellow traps, which are also used to control whiteflies, catch few of the parasites and can be used compatibly for whitefly control.

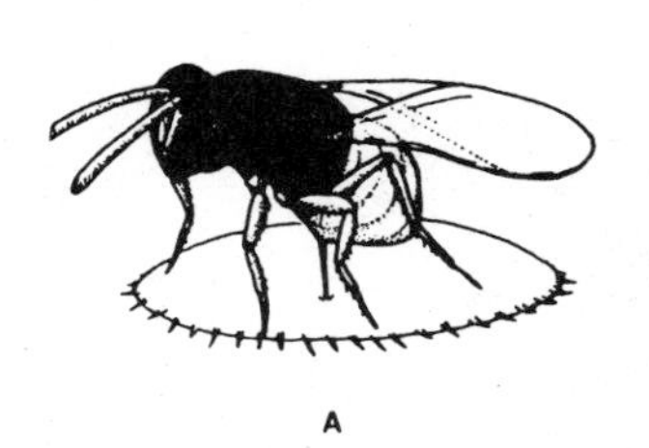

*A) An adult female* Encarsia formosa *depositing an egg into the scalelike fourth nymphal stage of the greenhouse whitefly; B) An Adult trichogramma wasp.*

**Mexican bean beetle parasite**—*Pediobius foveolatus* is a small parasitic wasp that develops within immature stages of the Mexican bean beetle. This parasite can control the Mexican bean beetle if it is released when beetle larvae are first observed. Releases of the parasite must be made annually, since it cannot survive winter conditions in the West.

**Bacillus thuringiensis**—*Bacillus thuringiensis* (Bt) is a bacterial disease organism that has been formulated into a number of microbial insecticides. Most Bt products are highly effective for control of leaf-feeding Lepidoptera (webworms, cabbageworms, leafrollers, tussock moths, cutworms, etc.). Trade names include Dipel, Thuricide, Caterpillar Attack and Biological Worm Spray. The exceptional safety of Bt products and their specific activity, which conserves beneficial organisms, make them very useful for control of many insect pests.

More recently, new strains of Bt have been identified and developed for control of other pests. The *israelensis* strain is effective against fungus gnats in soil and mosquito larvae in water. The *tenebrionis* and *san diego* strains are capable of controlling leaf beetles such as the Colorado potato beetle and the elm leaf beetle.

**Nosema locustae**—*Nosema locustae* is a protozoan disease that can infect many species of grasshoppers. Common trade names include Nolo bait, Grasshopper spore, Semaspore and Grasshopper Attack. As a typical protozoan disease, infection tends to debilitate rather than kill most insects. Young grasshoppers are more likely to be killed than older grasshoppers. More common symptoms of infection are sluggishness, reduced appetite and reduced egg-laying. In rangeland areas, where immediate suppression of grasshoppers is a less critical concern, *N. locustae* has had greater effect.

*N. locustae* also occurs naturally and can be spread via the egg or by cannibalism of infected insects. However, different species of grasshoppers vary greatly in susceptibility to infection. Although most of the common pest species of grasshoppers are fairly susceptible to the disease, some are not.

Proper timing and use of quality strains of the disease organism are very important in the successful application of *N. locustae*. Much greater infection rates occur when younger stages of grasshoppers are exposed to the disease organism. Fresh, well-stored preparations are also important in maintaining maximum viability of the spore. *N. locustae* is perishable, so poorly prepared or stored formulations may be relatively ineffective for control.

**Parasitic (predatory) nematodes**—Several species of nematodes develop within and kill insects. These insect-parasitic (or predatory) nematodes have become commercially available in the past few years and are now sold in many garden catalogs under such trade names as BioSafe and ScanMask. This is one of the most active areas of applied biological control, and a great deal of research on insect parasitic nematodes is in progress. Because they require conditions of high moisture, insect-parasitic nematodes appear particularly useful for controlling insects that occur

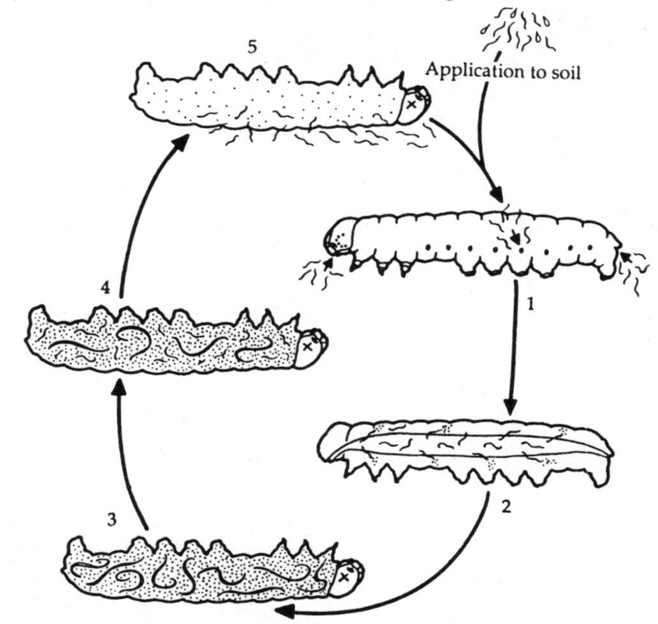


*Insect parasite nematodes. (1) Entering natural openings of an insect host and moving into the body (2) Releasing a bacteria that kills the insect (3–4) Reproducing within the dead insect and (5) Leaving to infect other insects.*

on or in the soil, such as white grubs, root weevils, sod webworms, cutworms and raspberry crown borer.

The most widely available species of parasitic nematode, *Steinernema carpocapsae*, is particularly effective against Lepidoptera (e.g., crown borers, sod webworms, cutworms). *S. carpocapsae* also has the advantage of being fairly easy to rear and store. Recently, techniques have been developed to allow nematodes to be stored for longer periods, so they are becoming more available through garden catalogs and nurseries.

Species of *Heterorhabditis* have been the most promising parasitic nematodes for control of various types of beetle grubs in the soil, such as root weevils and white grubs. Additional strains and species are under development.

Parasitic nematodes typically kill insects within a few days following infection. Some reproduction of nematodes occurs in many insects that are infected, such as caterpillars and grubs. However, the persistence of nematodes in a western yard or garden is unknown. Extremely high temperatures (greater than 95°F), rapid drying and the presence of certain pesticides (e.g., mercurial fungicides) will kill or inactivate most parasitic nematodes.

CHAPTER 3

# Cultural and Mechanical Controls

Aside from natural and biological controls, gardeners have a wide variety of options for managing garden pests. Gardeners can manipulate the garden environment to avoid problems by making conditions unfavorable for pest species. Other practices can be employed specifically to destroy garden pests, such as use of traps, sprays or mulches. Pest problems are most effectively managed if a combination of techniques is used.

## Cultural Controls

The techniques we use to grow plants affect many of the problems we may have with insects, diseases and other pests. What we plant, where we put it in the garden, when it gets planted and how we do it all may greatly affect susceptibility to pest problems. When consciously employed, these *cultural controls* are among the most important means of managing pest problems in the garden.

### Grow a Healthy Plant

Fundamental to garden pest management is the use of cultural practices that promote healthy plant growth. Growing conditions can greatly influence plant tolerance to injuries, allowing plants to "outgrow" attacks. Healthy plants also prevent populations of many pests from reaching damaging levels.

Plants stressed by poor nutrition and inadequate water are more susceptible to several of the leaf-spotting fungal diseases, such as early blight. Many "weak" pathogens, such as Cytospora canker, are more damaging to weakened plants. Some insects also prefer to feed on plants that are deficient in certain nutrients. For example, the squash bug has been shown to prefer squash plants deficient in sulfur, magnesium and potassium, since these plants contain higher levels of the amino acids desired by squash bugs. Other insects, such as shothole borers, spider mites and wood borers, thrive on plants stressed by drought and in poor health.

On the other hand, many insects, such as aphids, typically grow best on healthy plants. However, even with these pest problems, healthy plant growth allows plants to better outgrow the damage. Furthermore, many seedling problems, such as injury by flea beetles or slugs, can be avoided if the plants can grow past the susceptible seedling stage. Healthy growth thus allows plants to better tolerate damage that does occur, reducing the need for other control measures.

Proper plant nutrition and culture generally involve providing for plant growth needs throughout the season. Fertilization and watering should be moderate but consistent, thereby avoiding sudden changes in growing conditions.

### Use of Pest-free Planting Materials

Sometimes you get more than you paid for when making garden purchases. Many pest problems are inadvertently brought into the garden through infested soil or plants. Later problems with these pests can be avoided by purchasing clean, healthy plants.

One of the most avoidable garden insect problems is the greenhouse whitefly. In the West, all whiteflies die outdoors during the winter, but they may remain indoors on infested houseplants and in greenhouses. Use of whitefly-free transplants during the spring can prevent serious problems during the summer. Always be careful to look under leaves for the pale, scalelike, immature stages of whiteflies. (Just say no! to whitefly problems by purchasing insect-free plants.)

Several viral diseases can be introduced into the yard and garden and spread to healthy plants. Viral diseases may occur on plants that are reproduced by cuttings, tubers or other vegetative plant parts, such as potatoes, strawberries and raspberries. Another problem in recent years has been tomato spotted wilt virus, which flourishes in transplants of greenhouse-grown tomatoes, peppers and many garden flowers. Prevent these problems by buying plants from a reputable nursery and carefully checking plants before placing them in the garden.

It is particularly important to avoid introducing pests that can become permanently established in a garden. Slugs are easily introduced into a yard as clusters of pearl-like eggs carried inside containers of nursery or bedding plants. Fungal diseases such as Fusarium wilt of tomatoes, Verticillium wilt or white mold can be severe permanent problems once they have been brought into a garden on infested soil or plant parts. Cheap deals on nursery plants often are no bargain. (I learned this lesson the hard way after unintentionally purchasing narcissus bulb flies via some inexpensive mail-order bulbs. Annual replacement of daffodils and hyacinth has since become a major garden expense.)

Weeds can also be accidentally introduced into a yard. Animal manures and straw mulches are commonly infested with weeds. Soil or borrowed garden equipment may spread weed seeds. It is important to inspect all plants carefully prior to purchase. Discard or return plants that appear diseased or infested with insects. Sterilization or high-temperature composting can help avoid transferring weed seeds.

## Adapted, Pest-resistant Varieties

Some plants have been bred specifically to resist plant diseases (and rarely insects). Most common are the many VFN or VFTN tomato varieties, which are resistant to such diseases as Verticillium wilt (V), Fusarium wilt (F), Tobacco mosaic virus (T) and root knot Nematodes (N).

Plants may also be bred to better tolerate effects of weather or garden conditions. For example, some varieties of broccoli and spinach resist premature flowering (bolting) under hot conditions, an important characteristic for the warmer areas of the West. Similarly, cold tolerance or shortened maturity are characteristics found in other plant varieties, which make them useful for gardening in the high country.

It is always important to choose types of plants that are adapted to local conditions. In the West, problems such as soils high in salts or pH and late spring frosts often damage plants that are poorly adapted to these conditions, making them more susceptible to pests. Unfortunately, most garden plants were developed in areas of the country with very different growing conditions, and seed companies often distribute standard product lines to nurseries. As a result, plants offered for sale may not be locally adapted. Consult local gardeners and Extension offices and draw from your own experience to find the best plants to grow in your yard and garden.

## Planting Site Preparation

Proper site preparation can help limit or avoid many pest problems. In vegetable and flower gardens, good soil preparation promotes rapid, uniform germination and emergence. This is very useful in avoiding injury by seedling pests, such as seedcorn maggots, cutworms and flea beetles. Be

## *The International Yard and Garden*

Most of the garden plants we now take for granted are actually exotic species that were introduced from other areas of the world. Similarly, many of our garden visitors also originated from other areas, including the common honeybee. Indeed, a gardener in western North America 100 years ago would have faced an entirely different set of garden pests. Many have only recently colonized the region.

### SOME REGIONAL GARDEN PESTS NATIVE TO WESTERN NORTH AMERICA

| | | |
|---|---|---|
| Cabbage looper | Corn earworm/tomato fruitworm | Colorado potato beetle |
| Mexican bean beetle | Western corn rootworm | Migratory grasshopper |
| Redlegged grasshopper | Potato (tomato) psyllid | Tomato hornworm |
| Aster leafhopper | White grubs | Western cabbage flea beetle |
| Wireworms | Peach tree borer | Raspberry crown borer |
| Lygus bugs | Hairy nightshade | |

### SOME REGIONAL GARDEN PESTS INTRODUCED INTO WESTERN NORTH AMERICA

| | | |
|---|---|---|
| Imported cabbageworm | Green peach aphid | European earwig |
| Diamondback moth | European corn borer | Asparagus beetle |
| Asparagus aphid | Gray garden slug | Codling moth |
| Pear psylla | Pillbug (roly-poly) | San Jose scale |
| Fusarium wilt | Tomato spotted wilt | Beet curly top |
| Redroot pigweed | Lambsquarters | Purslane |
| Field bindweed | Canada thistle | Dandelion |

sure that the soil is adequately warm before planting, particularly when growing warm-season plants such as beans, corn and squash, which require soil temperatures of at least 50°F to 60°F. You may need to break up soil crusts and keep the soil moist to allow emergence of small seeded plants like onions, carrots and many flowers.

Preparation for proper planting is particularly important for perennial plants, such as fruit trees and berries, that will remain in place for many years. Poor preparation often prevents plantings from ever becoming well established and thriving.

Fruit trees and shrubs should be planted in a hole wide enough to encourage future root growth. Amending the existing soil with about one-third quality topsoil or peat moss is useful where heavy clay predominates. Scoring or breaking the sides of the planting hole can also help root soil penetration and healthy growth. Plantings should be made at the proper depth. Planting too deep is a common problem, which causes suffocation of the roots and a slow decline of the plant. All containers, bags, string and wire should be removed at planting. Guy wires should be removed as soon as possible so they do not eventually grow into, girdle and kill the plant.

## Fertilization

Many pest problems can be reduced in more vigorously growing plantings, and the proper balance of available nutrients is essential for healthy growth. As the garden is developed, have the soil tested periodically to detect deficiencies or excesses. Correct soil problems by increasing organic matter, adjusting pH or adding deficient nutrients. (Soils and soil testing are discussed in Chapter 1.)

Some pest problems may worsen with excess nutrients. For example, excess nitrogen produces succulent growth, which is favorable to many leaf diseases and some insects. Overly stimulated top growth can produce imbalances because good root growth is needed to sustain a healthy plant. Common root diseases of vegetables, such as Fusarium wilt, are aggravated when plants are excessively fertilized.

Fruit trees may need to be fertilized if characteristic annual growth is lagging. Nitrogen fertilizers should not be used during the first year of a planting since they may stimulate excessive top growth. Late-season nitrogen applications during any season should be avoided to prevent succulent growth that may not be frost-hardy.

## Crop Rotation

Crop rotation is fundamental to managing problems with many of the most important diseases of the vegetable garden, such as Verticillium and Fusarium wilt. Disease problems tend to become more serious when related plants are planted repeatedly on the same soil, allowing steady growth of the disease organisms. These survive in the plant debris, and some produce resistant stages that can persist for years in the soil. Rotation to nonsusceptible crops can stop development of the disease organism and, ultimately, allow it to die out. Crop rotation is also good for control of insects that overwinter in gardens in egg stages, such as the corn rootworm.

Rotation of plants within a vegetable and flower garden can greatly reduce disease problems. When possible, set up a garden plan that involves annual movement of related plants to new areas. This can include rotations of the most common families of vegetable plants, such as the following.

- Cucurbit family (cucumbers, squash, melons, pumpkins)
- Nightshade family (tomatoes, peppers, potatoes, eggplant)
- Sweet corn
- Onions, garlic, leeks, shallots
- Crucifer family (broccoli, cauliflower, Brussels sprouts, cabbage, turnips, radishes, kohlrabi, kale, etc.)
- Legume family (beans, peas)

Other plants, such as lettuce, beets and carrots, can also be used in crop rotations.

A rotation of at least three to four years should adequately reduce most soil disease problems. However, the benefits of crop rotation can be reduced if infected soil is moved about the garden during tillage or when infected plants are introduced into the garden.

## Sanitation

Sanitation can help with several common yard and garden pest problems. Pruning is useful for managing problems on trees and shrubs. For example, fire blight can be controlled by removing diseased limbs, preferably during the dormant season. Control of cane borers in raspberries or roses is best done by removing canes when the insect is still inside. Pruned wood should be disposed of promptly since disease organisms and insects can continue to develop in the prunings.

## FAMILIES OF COMMON YARD AND GARDEN PLANTS

**Chenopodiaceae**

| | | | |
|---|---|---|---|
| Beets | Lambsquarters | Pigweed | Spinach |
| Swiss chard | | | |

**Compositae**

| | | | |
|---|---|---|---|
| Aster | Chrysanthemum | Cosmos | Dandelion |
| Dusty Miller | Jerusalem artichoke | Lettuce | Marigold |
| Strawflower | Sunflower | Tarragon | Zinnia |

**Convolvulaceae**

| | | | |
|---|---|---|---|
| Bindweed | Morning glory | Sweet potatoes | |

**Cruciferae**

| | | | |
|---|---|---|---|
| Broccoli | Brussels sprouts | Cabbage | Cauliflower |
| Chinese cabbage | Collard | Horseradish | Kale |
| Kohlrabi | Mustard | Bok choy | Parsnip |
| Radishes | Rutabagas | Turnips | |

**Cucurbitaceae**

| | | | |
|---|---|---|---|
| Cantaloupe | Cucumbers | Gourds | Muskmelons |
| Pumpkins | Summer squash | Watermelons | Winter squash |

**Gramineae**

| | | | |
|---|---|---|---|
| Foxtails | Kentucky bluegrass | Quackgrass | Perennial ryegrass |
| Popcorn | Sweet corn | | |

**Labiatae**

| | | | |
|---|---|---|---|
| Basil | Catnip | Mint | Oregano |
| Rosemary | Thyme | | |

**Leguminosae (Legumes)**

| | | | |
|---|---|---|---|
| Broad beans | Dry beans | Lima beans | Peas |
| Snap beans | Soybeans | | |

**Lilliaceae**

| | | | |
|---|---|---|---|
| Artichokes (globe) | Asparagus | Chives | Garlic |
| Leeks | Onions | Scallions | Shallots |

**Malvaceae**

Okra

**Portulacaceae**

| | | | |
|---|---|---|---|
| Moss rose (portulaca) | Purslane | | |

**Solanaceae**

| | | | |
|---|---|---|---|
| Nicotiana | Nightshade | Potatoes | Tomatoes |
| Tomatillos | Peppers | Eggplant | |

**Umbelliferae**

| | | | |
|---|---|---|---|
| Carrots | Celery | Dill | Parsley |
| Fennel | Coriander/cilantro | | |

It is also a good idea to remove and dispose of infested plants that die out in fall but still harbor pest species. Several of the fungal diseases common in gardens, such as Alternaria leaf blight, survive on crop debris that has not yet decomposed. Some insects, such as the aphids attacking asparagus and columbine, remain attached to the old growth of the previous season. Problems with these pests can be reduced by removing old plant parts or tilling them into the soil so that they are physically covered and can later decompose.

"Off" type plants that are discolored or show unusual growth changes may indicate infection with viral disease. Removing these plants (roguing) as they are observed can keep other plants from becoming infected.

Sanitation is particularly important in managing weed problems in the yard and garden. A vigorous program of weed control, before weeds reseed, can greatly reduce problems next year.

Many gardeners compost organic materials to help recycle their nutrients and produce a very useful soil amendment. Properly constructed and maintained, compost piles can be an important part of any garden operation and help destroy garden pests. However, compost piles that decompose slowly at low temperatures allow weed seeds and disease organisms to survive. In addition, slugs, millipedes and other pests that thrive in moist organic matter can build up to damaging populations around compost piles. Proper compost management is an important sanitation practice.

## Tillage

Tillage in the flower border or vegetable garden, either through hand turning of the soil or rototilling, is usually done to prepare a more uniform seedbed. But tillage can also limit many plant disease problems. By incorporating old plant materials, the soil cover prevents many fungi and bacteria from spreading. Crop residues and the disease organisms in them also decompose more rapidly in the soil. This is important because many fungi that cause plant disease survive on intact crop debris but are killed when it decomposes. In much of the region, this breakdown of plant matter occurs primarily in the following season, since rapid soil cooling (and drying) in fall inhibits activity of soil microorganisms.

Tillage can also affect some insect pests. Grasshopper eggs can be exposed and killed during tillage. Cutworms that overwinter in the garden can be exposed or crushed during tillage. However, most garden insect pests are highly mobile and overwinter outside of the garden or resist the effects of tillage.

## Watering

For control of many fungal and bacterial diseases of vegetables and flowers, plantings should be watered in a way that allows the leaves to remain wet for a short time. Several things can influence leaf wetting, especially *how* and *when* you water.

Several methods of watering the garden reduce leaf wetting. These include drip irrigation, use of soaker hoses or construction of basins or furrows to hold the water. An additional benefit of spot watering systems, such as drip or basin irrigation and soaker hoses, is that water is conserved and weed germination is reduced in drier areas. However, these types of watering can contribute to a problem with thrips, spider mites and cabbageworms, which is decreased by hosing or overhead irrigation.

Watering during the day usually allows rapid drying, unless overcast or rainy conditions exist. Watering during midday, however, often results in substantial water loss through evaporation, an important consideration in the water-short West. Watering in early morning usually solves this problem.

Early watering also has other benefits for pest management. Dew formation on plants is often the result of *guttation fluids*, which plants exude as droplets from leaf pores. These beads of water, often seen in early morning along leaf edges, contain sugars and other nutrients that provide a particularly good environment for fungus spores to germinate. Watering in early morning washes off these sugary droplets.

# *Mechanical and Physical Controls*

Mechanical and physical controls are practices or devices used specifically to kill or exclude pests. Many of these controls are fairly labor-intensive but may work well in small gardens. Furthermore, they can be satisfying. While cultural controls work unseen, you get immediate satisfaction as you pick off a potato beetle or watch a weed shrivel after hoeing.

## Barriers

Several important pests of vegetables and fruits can be excluded from plantings by using barriers. *Collars* of cardboard around seedlings can prevent injury by cutworms and other insects that move to transplants from other areas of the garden.

*Floating row covers*, used to extend the growing season, also provide insect control. Row covers are particularly useful for insects that are highly mobile in the adult stage and do not spend the winter next to susceptible crop plants. Cabbage worms, Colorado potato beetles, most aphids, Mexican bean beetles, flea beetles, squash bugs and tomato hornworms are some of the insects that can be excluded by row covers.

Sticky materials applied to plant stems prevent ants from climbing plants and interfering with biological controls. Although not commonly found in area garden centers, specially designed sticky substances used to trap insects (Tanglefoot, Stick 'em Special, etc.) are sold in many garden catalogs.

Some control of slugs is possible by surrounding susceptible plants with abrasive materials, such as diatomaceous earth. Slugs also will not cross barriers of copper foil, which may be placed around new garden beds.

## Mulching

Mulches are barriers placed on soil and they have a variety of uses in gardens and landscape plantings. Weed control is the big reason for mulching, since weed seedlings are smothered by an effectively applied mulch. Organic mulches, such as straw or wood chips, have long been used in gardens or in landscape plantings. Straw also helps repel many insects. Recently, various fabric weed barriers have been marketed, particularly for weed control around shrubs and trees. (Further details on the uses of mulch for weed control can be found in Chapter 8.)

Mulching around the base of trees keeps mowing and weed control activities farther away from the plant. This avoids trunk injuries by lawn mowers and "weed eaters," one of the biggest threats to new plantings.

Aluminum foil and similar *reflective mulches* can repel many flying insects. Aluminum foil has been used extensively to prevent aphid transmission of viral diseases to vegetables such as squash and peppers.

Mulches also modify the growing environment of the plant. Plastic mulches warm the seedbed, allowing warm-season plants such as peppers and tomatoes to grow more rapidly. Mulching also reduces evaporation of water, reducing plant stress caused by fluctuating soil moisture.

One problem with the use of organic mulches, such as straw, is that weed seeds may be carried in the mulch. If possible, buy straw from known sources that is free of weeds.

## Solar Heating (Solarization)

The high temperatures and intense solar radiation common to the West can sometimes be used to kill garden pests. One technique involves intensified soil heating by solar energy, a process known as *solarization* or *solar pasteurization*. Weeds, weed seeds and many disease-producing fungi that produce plant disease can be controlled by solarization under the proper conditions.

Solar heating is intensified by use of clear plastic sheeting. First, prepare the garden site by loosening the soil with a tiller or spade. Moisten the soil thoroughly because wet soil better conducts heat and makes many fungi and weed seeds more susceptible to heating. Although a single deep watering may provide lasting moisture throughout a solarization treatment, consider installing soaker hoses or drip irrigation to maintain soil moisture. Cover the area with the thinnest polyethylene sheeting available, taking care to cover the edges and produce a tight seal.

Solarization is most effective when local solar radiation and temperatures are greatest, usually in early summer. The effects are also increased with increased length of treatment. A month-long treatment is considered the minimum in most areas. If garden areas can't be taken out for an entire season, solarization may be squeezed in between cool-season crops, such as spinach, lettuce and cabbage family plants.

Soil treated by solarization is very susceptible to reinfestation with weed seeds and fungal pathogens. Contaminated soil and running water are two ways that garden soils can be reinfested. Take steps to avoid recontamination problems.

## Hosing/Syringing

Many insect and mite pests can be controlled by forcefully hosing (*syringing*) plants with water. Small, soft-bodied insects and mites can be destroyed by these sprays, including aphids, mites and spittlebugs. The force of the water crushes or flings susceptible insects from the plant. Some scale insects can also be dislodged by forceful jets of water. (Plus, it's fun!)

Other pests, such as small caterpillars, may be drowned during hosings. Cabbageworms are particularly susceptible, since the waxy leaves and structure of the plants cause small pools of water to collect at the base of leaves.

## Handpicking/Crushing

Handpicking or destruction of insects and insect eggs is an option in small plantings. Several important vegetable garden insects, such as the Colorado potato beetle and the Mexican bean beetle, produce conspicuous egg masses. Large beetles and caterpillars, such as the tomato hornworm, can be easily collected or killed by hand. Handpicking is most effective if it is done frequently, at least once a week, since the insects may develop rapidly and cause injury in a short time.

Hard scales, such as oystershell scale and San Jose scale, can be rubbed off the plants with a plastic scrub pad. The eggs, which overwinter under the old scale coverings, dry out and die soon after this treatment. Woolly apple aphids and colonies of other soft-bodied insects are sometimes easily destroyed by light scrubbing.

Some insects can be collected by shaking the plants. Weevils, such as those that attack fruit trees and roses, are easy to collect in this manner, since most weevils play dead in response to disturbance and readily drop from plants.

## Hoeing

Hoeing, handpulling or well-timed tillage treatments can be very effective for weed control.

Weeds are most easily controlled when they are young. Hoeing early in the day can be particularly effective, since it allows the sun to dry and kill the plants rapidly. Irrigation should be suspended for at least a day following weeding to prevent possible rerooting.

Large weeds more easily resist weeding, since an underground root system may be in place that allows regrowth. Even when pulled, these plants may reroot, drawing on reserves of stored energy. Large plants that have been pulled, and succulent weeds such as purslane, should be removed from the garden area to prevent rerooting.

Well-established perennial weeds, such as bindweed and Canada thistle, can be very difficult to eradicate. Repeated hoeing is necessary to eliminate these weeds. It is more effective to hoe these plants within two weeks after they reemerge, since the plants begin to store new food reserves in the roots after this time.

Regardless of the type of weed, it is very important to prevent the plants from setting seed that will establish future infestations. Flowering and seeding can occur in only a couple of weeks with some weeds, particularly late in the season. Periodic weeding is a necessary chore in all yards and gardens and should not be neglected at the end of the summer.

## Traps

A wide variety of traps are available to aid in control of insects and slugs. Typically, these traps involve visual or chemical attractants that lure the pest and then destroy it.

Color traps are attractive to many flying insects. Yellow sticky traps are widely available and can be used to capture whiteflies, thrips, psyllids and some aphids. White is attractive to whiteflies and plant bugs; light blue is most effective for capturing flower thrips. Sticky coatings, such as Tanglefoot, Stick 'em Special or Tack-Trap, cover the traps to ensnare insects. Heavy-grade oil or grease can be used to trap insects, although these may liquify and drip in high temperatures. Several sticky traps are now being offered in garden catalogs, although homemade ones may be equally effective. (*Note*: If you make your own color traps, cover them with a plastic bag, applying the sticky adhesive to the inside of the bag and inverting it when covering the trap. This reduces the mess of covering the traps with the adhesive and makes the covers easy to replace.)

Fermenting materials, such as beer or sugar water and yeast, are well-known attractants to slugs. Slugs will travel several feet to visit saucers containing these baits and drown. Millipedes and sowbugs also fall for this trap.

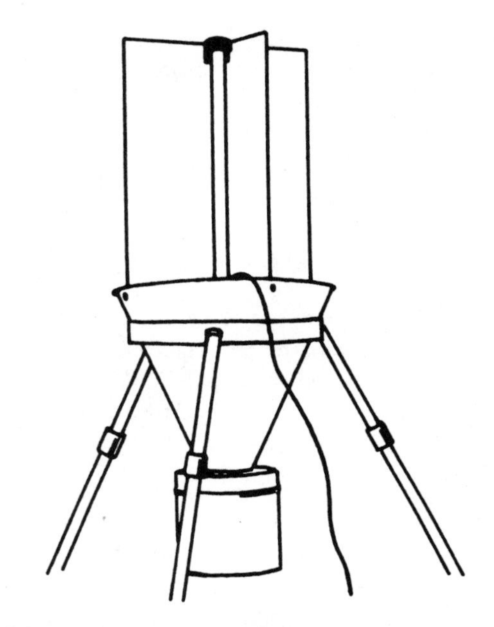

*A light trap used to survey night-flying insects. Most light traps use ultraviolet lamps and capture a wide range of moths, beetles and other insects.*

Molasses, or molasses and vinegar baits, have been used to attract many kinds of moths. Fermenting molasses baits have long been used for controlling adult stages of the codling moth, a common pest of apples.

Light traps, or variations on the ubiquitous "bug zappers," are widely sold for insect control. Light traps can help control some insects in confined situations. Fly control in dairy barns is one such use of these traps. However, light traps have repeatedly been shown to be ineffective for controlling garden pests. Moreover, some studies show *increases* in some pest problems, such as white grubs, in the vicinity of light traps. Although many insects are killed by light traps, most of them are nonpest insects or are beneficial species, such as lacewings. (Sometimes there is a lessening of insects 100 feet or so from a light trap, at the periphery of its attractive effects. Therefore, the best use of a "bug zapper" to protect *your* yard may be to give it to your neighbor to use.)

Other traps provide shelter sought by insects. Dark, tight-fitting, moist areas are favored by earwigs and slugs. Moistened, rolled-up newspaper can provide an ideal site for these types of pests to collect. They can then be destroyed during the early morning.

Caterpillars of several insect pests, such as the codling moth and the tent caterpillar, search for protected locations to pupate and will concentrate under burlap bands or corrugated cardboard placed on the trees on which they feed. The insect pupae can then be collected and destroyed.

*Trap crops* involve the purposeful use of certain types of plants that are particularly favored by the pest. By growing these plants, you can divert some of the feeding damage from the desired main crop. For example, radishes and Chinese cabbage are highly preferred by the western cabbage flea beetle, which may then spare the cabbage or broccoli. For getting a new seeding past the slugs, lettuce or beans are favored foods to use as trap crops. Colorado potato beetles usually home in on the eggplant, potatoes—and particularly the garden weed hairy nightshade—before touching the tomatoes.

## Those Fantastic Pheromones

Although their vision is so poor they are considered legally blind and they cannot speak (some produce sounds), insects often communicate by odors. Chemicals that animals of the same species use to communicate are called *pheromones*. These signals are used in a number of ways. Ants lay down a *trail-marking pheromone* as they forage for food, allowing them to return to the nest and direct others to the food. Honeybees produce an *alarm pheromone* that allows them to launch a mass attack to defend the colony against an intruder. *Aggregation pheromones* are used by bark beetles to concentrate egg-laying on a single tree to help overcome the plant's defenses.

It is the *sex pheromones*, used to attract and court mates, that have received the most attention. Use of sex pheromones appears to be widespread in the insect world; these are almost always chemicals produced by the female to attract a male. Among the moths, sex pheromones of incredible potency have been discovered. Often only a few molecules of the "eau de female" can be perceived by a receptive male moth, making insect pheromones among the most biologically active chemicals known.

During the past twenty-five years, tremendous progress has been made in our understanding of insect pheromones. A great deal of research has been conducted on how we might use these fantastic chemicals for taming insect pests.

The greatest use of pheromones at present is in baiting traps. These traps have been a great boon for government agencies in detecting when new pests, such as gypsy moths, have entered an area. Pheromones also are used by farmers and some gardeners to detect when the adult stage of a pest is active. This allows them to better time the use of sprays or other controls when they will be most effective, such as when eggs are being laid.

Typical pheromones used for traps are produced by female insects to attract males for mating. Pheromone-baited traps can be extremely powerful and can attract large numbers of insects, often from considerable distances. Since it is the males, rather than the egg-laying females, that most often respond to these traps, the most effective use of pheromone traps has

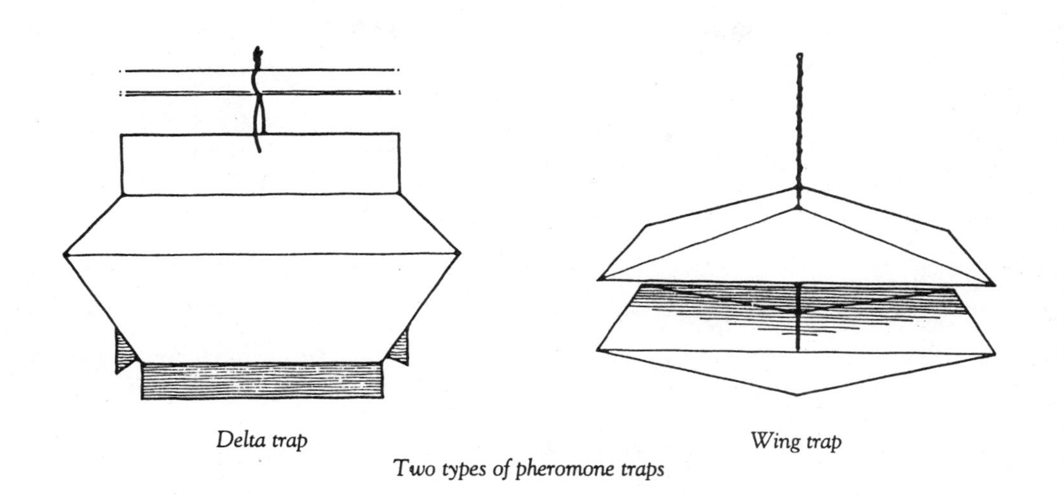
Delta trap

Wing trap

*Two types of pheromone traps*

been to identify when flights of the moths occur. This can be helpful in timing controls such as sprays or releases of biological controls.

Unfortunately, the main limitation of pheromones for insect control is that the potent ones attract only males. It is the females that are responsible for bearing the young, which later chew up the garden, and the trapping of a few males has little effect. Insects are not models of fidelity, and the loss of a few male moths to a pheromone trap is easily made up by the increased mating of the remaining moths. (However, some dealers will try to sell pheromone traps to gardeners for insect control.)

A relatively new approach for the use of sex pheromones for insect control, the *male confusion method*, or *mating disruption*, shows more promise. This involves spreading numerous pheromone sources throughout the crop to permeate the air with the sex pheromone. The male moths exhaust themselves visiting all the "false" females or simply become so used to the chemical that the female is no longer attractive.

Success with the male confusion method has been limited mostly to insects that attack a single crop. Prominent successes have included the control of artichoke plume moths on artichokes, pink bollworms on cotton and grape berry moths on grapes. However, researchers continue to expand this work, with such garden notables as the peach tree borer and the codling moth—the common "apple worm."

*Sam Cranshaw checks an insect trap.*

CHAPTER 4

# Chemical Controls in the Garden

Chemical controls are considered by most gardeners as a backup control for garden pests. They are usually employed when other types of controls have not been sufficient or have been too slow-acting to solve an imminent problem. These controls, however, should never be used as a substitute for good garden culture.

## Pesticides

Managing pest problems can also involve use of chemical controls, or *pesticides*.

Proper use of pesticides is more complex and involves hazards not associated with other pest management techniques. It is very important to understand their characteristics and how they should be used. A wide variety of different chemicals are used for the control of weeds, insects, disease organisms or other pests. Based on their use, garden pesticides fall into one of the following classes.

- *Herbicide*—a chemical substance sold to kill undesirable plants (also known as weed killers)
- *Insecticide*—a chemical substance sold to kill undesirable insects
- *Fungicide*—a chemical substance sold to kill undesirable fungi
- *Miticide*—a chemical substance sold to kill mites
- *Bactericide*—a chemical substance sold to kill bacteria (includes many antibiotics)
- *Molluscicide*—a chemical substance sold to kill pest mollusks such as slugs and snails
- *Nematicide*—a chemical substance sold to kill nematodes

Some pesticides may be discussed and classified in a number of ways, based on their chemistry, origin or mode of action and use. (Characteristics of specific garden pesticides are discussed in Appendix II.)

### Botanical Pesticides

Some pesticides, particularly insecticides, are described by how they are produced. This is because several are obtained from natural sources, often derived from plants. Pyrethrum, rotenone

*Stages of the pyrethrum daisy, the source of natural pyrethrins*

| SOURCES OF BOTANICAL INSECTICIDES | | |
|---|---|---|
| **Insecticide** | **Active ingredient** | **Source** |
| Nicotine sulfate | Nicotine | Extract from leaves of tobacco and related species of *Nicotiana* |
| Neem | Azadirachtin | Extracted from seed pods of the Neem tree, grown primarily in southern Asia. |
| Pyrethrum (pyrethrins) | Several compounds known as pyrethrins | Crushed flowers (pyrethrum) or extracts of active ingredients (pyrethrins) of the young flowers of the daisy *Chrysanthemum cinaeriofolium,* grown primarily in West Africa. |
| Rotenone | Rotenone and other resinous rotenoids | Found in dozens of species of legumnous plants. Commercially produced rotenone is derived from either species of *Derris*, found in Malaysia and the East Indies, or *Lonchocarpus* (cubé), from South America. Presently, almost all rotenone sold in the U.S. is from the latter area. |
| Ryania | Ryanodine | Ryania is the powdered extract from the roots and stems of the plant*Ryania speciosa,* native to South America. Ryania is sold primarily as a minimally processed extract from ground stems, formulated on a wettable powder. |
| Sabadilla | Veratrine alkaloids | Ground seeds of a South and Central American plant in the lily family, sabadilla (*Schoenocaulon officinale*). |

and nicotine are examples of naturally derived insecticides.

## Mineral-based or Inorganic Pesticides

Other pesticides are derived from minerals, such as sulfur and copper. These include the oldest pesticides in use. For example, Homer of ancient Greece extolled "pest-averting sulfur," and sulfur remains the basic material in several fungicides and miticides. Frequently, these mineral-based pesticides are combined with other materials to produce Bordeaux mixture (a mixture of lime, copper sulfate and water) or lime sulfur (sulfur salts produced by boiling water, lime and sulfur).

## "Organic" Pesticides

The term "organic" is widely used in reference to various gardening practices, including the use of certain pesticides. Because this term has various scientific, popular and legal definitions, however, its meaning has become muddled.

Technically, all chemical compounds are classified as having *inorganic* or *organic* chemistry. According to this definition, organic compounds are those that contain molecules of carbon. Inorganic substances do not contain carbon. Under the strictly chemical definition, inorganic pesticides would include most of the older pesticides, such as Bordeaux mixture, sulfur, lime sulfur, copper fungicides and boric acid used for cockroach control. Since essentially all widely used pesticides have a carbon chemistry, most are technically organic.

However, use of the term "organic" in reference to garden pesticides is more often applied to the pesticide source or history of use. Materials derived from natural sources, such as the botanical insecticides pyrethrum, sabadilla and rotenone, as well as certain mineral-based pesticides, such as sulfur and Bordeaux mixture, are among the pesticides commonly considered to be acceptable for organic gardening. Selective insecticides, such as *Bacillus thuringiensis* and soaps, may also be included in such listings. The underlying assumption for pesticides described as organic in this manner is that being naturally derived or having a long history of use, they involve minimal hazards.

Because of the tremendous interest in commercial production of organic produce, several states and grower organizations have recently established legal systems to certify produce as being grown organically. This has involved defining what materials are acceptable in production of "Certified Organic" fruits or vegetables. Although regulations among the various states vary, most define pesticides by their source. Naturally derived pesticides are generally acceptable; materials that have involved some synthetic alteration are usually prohibited.

## Insecticidal Soaps and Horticultural Oils

These types of insecticides are receiving increased attention by gardeners because of their safety, effectiveness and selective action. These advantages, as well as their associated limitations, are based on their action as short-residual *contact insecticides*. Both soaps and oils must be applied directly to the body of the insect, and effects are of very short duration. With these materials, "what you spray is what you get," and they are not forgiving of poor coverage. However, soaps and oils usually cause little harm to desirable organisms such as birds, pets, beneficial insects—and you.

The soaps being touted for pest control are *just* soaps. (Chemically, most insecticidal soaps are the product of 8- to 18-carbon-chain fatty acids that have undergone reaction to form a water-soluble potassium salt.) The insecticidal activity of soaps has been known and exploited for over 200 years. Their use has always been limited by their potential for damaging plants, a problem further exacerbated by uneven manufacture. In recent years, several companies have developed these products (and the price!) by more carefully selecting the soaps that both kill insects and are safe for plants. During this process, highly plant-destructive soaps were also developed, some of which are being sold as herbicidal soaps.

Similarly, household detergents can have insecticidal action. Many liquid dishwashing detergents are very effective insecticides, and when used as a dilute spray (1 to 2 percent concentration or about 3 to 5 tablespoons per gallon), can be safe for most plants. They are also much cheaper. Liquid detergents are often not of consistent manufacture, however, and can be more irregular in ensuring effectiveness and plant safety than the insecticidal soaps sold especially for garden use.

Oils used for insect and mite control are highly refined specialty oils designed for plants. They generally distill at low temperatures (412°F to 438°F) and have many of the sulfur impurities removed to decrease their risk to plants. Oils are usually sold for application as dormant oils to fruit trees and other woody plants before buds break in spring. Scales and spider mite eggs, which winter on such trees and plants, are the most common targets of these treatments. However, many of the newer spray oils have been refined so that they can be safely used on plants even after leaves emerge. Recently, their use has expanded to include whiteflies and other small insects on vegetable and greenhouse crops as well as for summer use on trees. Spray oils also contain a small amount of emulsifier, which allows them to be mixed with water. (Some vegetable oils can also kill insects when mixed with a little soap or detergent as an emulsifier and diluted in water.)

Both soaps and spray oils are designed to be used as sprays, diluted to a concentration of 1 to 3 percent. Soaps apparently work by disrupting the cell membranes of susceptible insects, mites or plants. They are fast-acting, usually showing results within a few hours. Oils are generally regarded to act as suffocants that plug the small spiracles through which insects breathe. They can also kill insect eggs. Both oils and sprays are primarily effective against small, soft-bodied insects or mites. Larger insects (and other animals) are little affected by these treatments.

## Microbial Insecticides

Some pesticides are derived from various microorganisms. The most common of these *microbial* pesticides use naturally occurring microbes to kill specific types of susceptible insects. Although microbial diseases of pests are widespread

SOAPS AND OILS—THE PROS AND CONS FOR PEST CONTROL

| Insecticidal Soaps | Horticultural Oils |
|---|---|
| **Advantages** | |
| Safety to humans | Safety to humans |
| Safety to wildlife, pets | Safety to wildlife, pets |
| Conserve insect natural enemies | Conserve insect natural enemies |
| Effective against many small insects | Effective against some difficult insects to control with other methods (leafcurling aphids, scales) |
| **Limitations** | |
| No residual activity | No residual activity |
| Purely contact action/high coverage requirement | Purely contact action/high coverage requirement |
| Effective only against small insects, mites | Limited range of effectiveness |
| Some potential plant injury (phytotoxicity) | Some potential plant injury (phytotoxicity) |
| Works best when environment allows slow drying | |
| Less effective in hard water | |

in nature and are important naturally occurring controls, relatively few are manufactured and sold for pest management. Most important are the *Bacillus thuringiensis* products (see page 22). Several strains of these bacteria have been developed primarily for control of caterpillars (Thuricide, Biobit, Dipel, MVP and Caterpillar Attack). Newer strains have recently appeared on the market for control of leaf beetles or mosquito larvae in water.

A protozoan (Microsporidia) disease of grasshoppers, *Nosema locustae*, is sold under various trade names such as NoLo Bait or Semaspore. This causes a slow-acting disease in infected grasshoppers that tends to debilitate rather than kill. Unfortunately, several of the grasshopper species found in the region appear not to be very susceptible to this organism.

Insect parasitic nematodes (species of *Steinernema* or *Heterorhabditis*), also called predator nematodes, are among the newest of the microbial insecticides available for garden use. These have been most effective for control of soil insects, such as cutworms, crown borers and root weevils, since the nematodes require moist conditions and protection from sunlight.

## Synthetic Organic Pesticides

Almost all pesticides now used are synthetically produced compounds with a carbon-based chemistry. Synthetic organic pesticides include such compounds as carbaryl (Sevin), diazinon, malathion, glyphosate (Roundup) and triforine (Funginex). Some of these are based on naturally occurring chemicals. For example, the insecticides known as pyrethrins derived from the pyrethrum daisy are the chemical godparents of the widely used class of insecticides known as the *pyrethroids*. (A discussion of the characteristics and uses of all the common garden pesticides is included in Appendix II.)

## Systemic Pesticides

Pesticides may also be classified as *systemic* or nonsystemic compounds. Systemic pesticides used on plants are those that can be picked up through leaves and/or roots of plants and moved within the plant. Systemic pesticides vary in how readily they move in the plant. However, some systemic pesticides move easily in plants, often in specific directions. For example, the herbicide glyphosate (Roundup, Kleenup), when applied to leaves, may move into and kill the plant roots. Alternatively, the insecticide DiSyston, applied to roots, moves upward in the plant and concentrates in actively growing foliage.

Systemic activity often improves the effectiveness of a pesticide by providing better plant coverage. However, the systemic pesticides often come with some special hazards. For example, most systemic insecticides, such as Orthene, should not be used on food crops. Accidental use on these plants results in pesticide residues that are illegal and potentially unsafe. Most systemic insecticides are also highly toxic to humans and other mammals, particularly DiSyston, the active ingredient found in various systemic rose and flower care products. Systemic herbicides such as glyphosate (Roundup, Kleenup) have great potential to damage desirable plants that they accidentally contact.

## Eradicant versus Protectant Fungicides

Fungicides are often classified as either *protectants* or *eradicants*. Protectants, such as maneb and chlorothalonil (Daconil 2787), must be in place on the surface of the plant before fungal infection of the plant occurs. Eradicants, such as benomyl (Benlate, Tersan 1991), can move within the plant and kill developing fungi.

## Preemergence and Postemergence Herbicides

Many herbicides available for use in vegetable and flower gardens are *preemergence herbicides*. These must be applied before seeds germinate, since they act to prevent early seedling development. Dacthal (DCPA) and Treflan are examples. After seeds have germinated and the weeds have grown a supporting root system, these herbicides are no longer effective for weed control. Conversely, other herbicides are applied to plants after they emerge. These *postemergence* herbicides include 2, 4-D and glyphosate. (For more discussion on herbicides, see Chapter 8.)

## Mode of Entry

Special terms are used to describe insecticides by the means through which they enter the insect. *Stomach poisons*, such as *B. thuringiensis*, must be ingested by the insect to be effective. *Contact insecticides* can penetrate through the external skeleton (cuticle) of the insect. (This term is also used to describe herbicides that kill plant cells at the point of contact, such as herbicidal soaps.) Insecticides that become a gas and enter through the breathing openings (spiracles) are known as *fumigants*.

## Pesticide Selectivity

Pesticides are also described as *selective* or *nonselective*. Selective herbicides are those that can be used to control one type of plant but cause little or no injury to another plant. For example, fluazifop-butyl (Grass-B-Gon) is generally considered to be selective, since it is much more active against grasses than against broadleaf plants. Several selective insecticides are also available. For example *B. thuringiensis* only kills caterpillars that eat the pesticide; beneficial insects such as lady beetles—and nonsusceptible pests such as aphids—are not affected by *B. thuringiensis*. Insecticidal soaps are fairly selective because they primarily control small, soft-bodied insects and mites that are contacted with the soap spray during application. Many larger insects are not susceptible to soap sprays, and insects that visit the plant after the soap has dried are not killed.

Selectivity is also determined by how a pesticide is used. Glyphosate is a nonselective herbicide that can be used selectively to kill a thistle in the flower garden if it is carefully placed on the weed. A spot treatment of Sevin may be useful for controlling a patch of blister beetles on a potato plant, and is increasingly selective in its effects as the treated area is more limited. Dormant oils could kill beneficial insects on a fruit tree, but they are selective, since these insects are not often on the tree at the time oils are sprayed.

If used at the improper time or mixed in high concentrations, all pesticides can become nonselective and can injure plants, insects or other organisms that otherwise could tolerate the treatment.

Nonselective herbicides are compounds that are toxic to both crop and weed species of plants. Glyphosate (Roundup, Kleenup) is an example of a nonselective herbicide that will kill or weaken almost any sprayed plant. Because of this potential to cause injury, they must be used with extreme care around desirable plants.

Often the term *broad spectrum* is applied to nonselective insecticides. This indicates that a wide range of insects—and sometimes other organisms—are susceptible to their effects. For example, chlorpyrifos (Dursban) and diazinon are insecticides that can kill most species of insects as well as many mites. In some situations, such as when several pests commonly attack the plant, broad-spectrum activity in an insecticide can be useful. However, broad-spectrum insecticides also commonly kill beneficial insects, such as the natural enemies of insects and mites, and thus do not integrate well with existing biological controls.

*Soil sterilants* are nonselective herbicides applied to soil that can kill actively growing plants and prevent growth of new plants for as long as the compound remains active, often several years. Soil sterilants are used primarily in maintaining rights-of-way or in areas where no vegetation is desired, such as driveways and sidewalks. Serious inadvertent injuries to desirable plants, including trees and shrubs, have occurred from use of soil sterilants such as pramitol (Triox) near landscape plantings. Soil sterilants have no place in and around garden areas. Some states have even moved to make these products "Restricted Use Pesticides," available only to certified pesticide applicators who have passed an exam administered by the state Department of Agriculture or the Environmental Protection Agency.

## Integrated Pest Management (IPM)

A variety of control techniques may be required to effectively manage pest problems. For example, management of cutworm problems in a garden can include rototilling to kill overwintering larvae (cultural control), use of cardboard collars to protect plants (mechanical control) and selective use of insecticides (chemical control). None of these techniques alone may be sufficiently effective.

Care must be taken to coordinate these various control techniques so that their effects don't conflict in overall pest management. For example, certain uses of insecticides can seriously reduce the effectiveness of biological insect controls. Mulching used for weed control can increase problems with slugs, particularly in high-moisture conditions. Tillage and harvest practices may affect beneficial biological control organisms as well as pests.

Maintaining effective pest management is also often a problem in areas where single control approaches are repeatedly used. This has most commonly been documented with the use of pesticides. Many insects, mites, fungi and even a few weeds have become resistant to pesticides that formerly were effective. Pesticide use may also cause shifts in pest problems. For example, nightshade weed problems have developed in many fields following the use of several common herbicides that were ineffective against nightshade. Mite problems in fruit and field corn are often aggravated by early-season insecticide use directed against other pests.

Finally, other serious nontarget effects may result from pest management practices. Many pesticides are highly toxic to the applicator or to wildlife around a treated area. Pesticides may move into and pollute ground or surface waters. Pesticide residues can contaminate food or feed, and plants may be inadvertently damaged by pesticides. These hidden costs from pest control practices are not always fully considered during pest control decision-making.

*Integrated Pest Management* (IPM) is a system for making optimal pest management decisions based on both ecological and economic considerations. It is not a particularly new idea, since sound pest management practice has always involved IPM approaches.

Fundamental to Integrated Pest Management are a number of assumptions.

- Optimal pest management is achieved by a variety of techniques.
- Pest management techniques should be used in a coordinated (integrated) manner so that their effects complement each other.
- Since eradication of many pests is not often achievable or desirable, controls should be applied only when pest populations are sufficient to justify their action. Scouting or monitoring pest populations is usually required.
- Pest management techniques should consider the long-term costs of the practice, including social and environmental costs.
- The performance of any pest management practice will be continually re-

evaluated to determine if it is achieving the desired result in the most effective manner.

# Common Mistakes Using Yard and Garden Pesticides

## "Spray First, Read Labels Later (If At All)"

The information that appears on a pesticide label is the result of extensive research and experience and includes the effective uses and potential hazards of the pesticide. Although often difficult to read, the label contains information essential to proper use of the pesticide. Furthermore, the label instructions are a legal document, so use of the pesticide in a manner that is not indicated is a violation of federal law. Failure to thoroughly read label instructions can result in several problems.

- Plants that are sensitive to the effects of the pesticide may be injured.
- The pesticide may be applied to a food crop for which its use is not permitted. This results in illegal residues appearing on the food.
- Preharvest intervals required between the time of application and harvest may be neglected, resulting in excessive pesticide residues on food.
- Pesticides may be applied in a manner that poses high risk to wildlife, pets or other humans.
- The applicator may receive excessive exposure to the pesticide.
- The applicator will be legally liable for any damage resulting from an illegal pesticide application.

## "Got A Problem? Spray It"

No single pesticide can control all pest problems. In addition, few yard and garden problems are simple enough to be solved just by spraying pesticides. It is always important to first diagnose the problem so that the most effective controls can be used. Inappropriate use of pesticides as a first-line defense may result in the treatment being ineffective. The underlying problem may persist and continue to cause damage. Pesticide treatments may even aggravate the problem. For example, problems with spider mites are increased with the use of insecticides that kill their natural enemies. In either case, the money and effort involved in controlling the pest are wasted.

## "If a Little Is Good, More Is Better"

Using pesticides in a casual manner may result in many problems. One very common habit is to use a stronger rate of pesticide for insurance. If pesticides are used at rates exceeding those recommended in the instructions, they can cause various plant injuries. The rates of pesticides listed on the label were established, in part, to avoid plant injury, and even insecticides and selective herbicides, such as 2,4-D, can injure normally tolerant plants when applied at high rates. The pesticide label is a legal document. Use of any pesticide in a manner that is not specified on the label is illegal.

The rates of pesticides allowed on the label are also designed to ensure that pesticide residues on the harvested crop do not exceed safe levels. Using higher rates of pesticides can result in amounts of residues that exceed known safety levels. Increasing the rates of pesticides applied also increases the risk of injuring beneficial insects, wildlife and other "nontarget" organisms.

Using too much of a pesticide is usually the result of "guesstimating" the amount needed. Typically, these estimates are off by 25 to 50 percent or more. Always read and follow the label directions.

Courtesy of University of Illinois Cooperative Extension Service

*Proper precautions should be taken whenever applying pesticides of any sort. Use of water resistant gloves is particularly important since much exposure to pesticides during application occurs through the hands. Disposable dust masks, goggles for eye protection, and use of long pants and shirt sleeves are also minimal precautions that should be taken.*

*Measure* the amount of pesticide to be used. *Never* guess.

## "Protective Clothing Is for 'the Other Guy'"

The use of pesticides assumes that basic protective measures will be followed by the applicator. It cannot compensate for the "worst case" applicator who may expose himself or herself to very high rates of the pesticide by careless handling. Potential for exposure to pesticides is especially great during mixing, since concentrated pesticides are being handled. Exposure to the pesticide through inhalation or skin contact is very common during application.

Gloves are the most important piece of protective equipment because the great majority of pesticide exposure occurs through the hands. Gloves used for mixing and applying pesticides should be made of neoprene rubber or some other material that is not absorbent. Cloth or leather gloves should not be used since they are absorbent and can concentrate pesticides next to the skin.

A dust mask is also very useful as protection when applying pesticides. Fine droplets produced by sprayers are easily inhaled. Pesticides used as dusts present even greater risks of being inhaled. Dust masks also help prevent irritation from materials such as diatomaceous earth or sabadilla.

Finally, some sort of eye protection is always a good idea. Several pesticides, such as sulfur, can be very painful if they get into the eyes. Drift into eyes is also very common when spraying upward into a fruit tree. A pair of goggles or safety glasses can protect eyes from pesticides at these times.

## "Herbicides Only Kill Weeds"

Herbicides are pesticides designed to kill plants. Most herbicides have a particular action that makes them more lethal to certain plants than to others—the basis of their selectivity. However, used inappropriately, no herbicide distinguishes between "good plants" and weeds. Proper use of herbicides is very much up to the applicator.

One of the most devastating injuries caused by herbicides around the yard involves the *soil sterilant* herbicides. These are used to control weeds in driveways, sidewalks and waste areas for long periods of time. Unfortunately, they often produce unwanted effects as tree roots grow under the driveway or water carries the herbicides downhill. Established trees may be killed by careless use of soil sterilant herbicides.

Herbicides used in gardens, such as Dacthal or trifluralin (Treflan), are usually designed to prevent germination of weed seeds. However, some weeds are close relatives of our cultivated plants, and careful reading of the label instructions will indicate that many of these are also susceptible.

The nonselective herbicide glyphosate (Roundup, Kleenup) has also been used increasingly around gardens. Most plants are susceptible to this herbicide, as gardeners are generally aware. However, unless used with great care, fine droplets may drift or sprayer nozzles may drip so that desirable plants are accidentally treated. Glyphosate also must be used carefully around the base of trees and shrubs because it can be absorbed by green wood.

Most injury by herbicides and other pesticides can be prevented by reading the label instructions. "The label is the law," but the label is also your best guide to using pesticides without harm.

CHAPTER 5

# Common Disorders of Garden Plants

This chapter will help you troubleshoot and diagnose the cause of your particular garden problem. Under each crop, a list of common disorders will refer you to subsequent chapters and sections for details on the control of specific problems. Since the cultural requirements of the crop may also be important in diagnosing problems, a brief description prefaces each section. Vegetables are discussed first, followed by fruit crops, flowers and, finally, roses.

## Vegetable Crops

### Asparagus

Asparagus is one of the few perennial vegetable plants common in the vegetable gardens of the West. The delicious shoots, which emerge each spring from the crown of the plant, are one of the first plants ready to harvest. If the shoots are not harvested, they become ferns that produce energy for the next season's crop.

Asparagus is generally planted as roots that are one to two years old; seeds are also usually available through catalogs. Because the plants are not moved after planting, extra care needs to be taken to prepare the site before planting. Asparagus is very tolerant of salts, but high levels of organic matter and a soil that drains well are important prerequisites for asparagus culture. The plants also require phosphorus and potassium, which are best applied before planting. Protection from frost and weed control should also be considerations when planning an asparagus bed.

Because asparagus plants need to build root reserves, they are usually not harvested for two to three years after planting. Peak production usually follows five to ten years later. Application of a nitrogen-containing fertilizer is required to maintain plant growth.

Asparagus produces either male or seed-bearing female plants. Yields from female plants are lower because of energy used in seed production. Recently, high yielding, all-male varieties have been introduced.

Asparagus can have some serious pest problems. Asparagus beetles and asparagus aphids damage plants in some areas, and asparagus rust is an occasional problem. Removing old ferns helps reduce problems with aphids and rust.

*Asparagus*

***Common Disorders of Asparagus***

Affecting spears
- New spears die back, shrivel
  - Frost injury (199)
- New spears chewed
  - Asparagus beetle (93)
  - Spotted asparagus beetle (93)
  - Cutworms (115, 116, 117)
- Black eggs on spears
  - Asparagus beetle (93)
- Spears tough, pithy
  - Spears too old
  - Lack of fertility

Affecting ferns
- Ferns chewed, appear bleached
  - Asparagus beetle (93)
- Stunted, bushy ferns
  - Asparagus aphid (155)
- Fern growth slow
  - Excessive picking during establishment
  - Cultural problems (e.g., poor fertility or watering, insufficient sunlight, over-crowding)
- Fern growth slow, yellow/reddish/black areas on ferns
  - Asparagus rust (190)
- Dieback of ferns, crown death
  - Fusarium wilt (178)
  - Phytophthora root rot (176)

## Beans

Beans are an herbaceous annual plant grown for seeds (dry beans) or immature pods (snap beans). Several different species of beans are grown, but most are varieties of the common bean, *Phaseolus vulgaris*.

Bean seeds require warm temperatures and good soil moisture to germinate and grow. Minimum soil temperature for seedling emergence is 60°F. The plants are susceptible to frost and optimal temperatures for growth are in the 70°F–80°F range.

Maturity of most garden beans is not affected by day length. Numerous varieties of beans are available, with bush types generally maturing more rapidly than vine varieties. Other types of beans, such as limas, chickpeas, black-eyed peas and asparagus beans, require warmer temperatures and a longer growing season than snap beans.

Beans are legumes, capable of extracting nitrogen from the air through bacteria that grow on root nodules. As such, they do not require additional nitrogen but can require moderately high levels of potassium and phosphorus for best growth. Beans are also susceptible to iron and zinc deficiencies related to high-pH conditions in many regional soils.

Garden-grown beans usually suffer few problems with pests, although the overeager gardener risks problems with seedcorn maggot and damping off when planting too early in cool, wet soils. The Mexican bean beetle, the "bad apple" of the lady beetle family, can be quite damaging in some locations.

***Common Disorders of Beans***

Seedling problems
- Poor seedling emergence
  - Cool soil temperatures
  - Uneven watering
  - Seedcorn maggot (130)
  - Damping off disease (187)
- Seedlings chewed irregularly, slime trails may be present
  - Slugs (166)
- Seedlings emerge but growing point killed ("snakehead" beans)
  - Seedcorn maggot (130)
- New leaves curled
  - Wind injury
  - Onion thrips (136)
- Seedlings appear cut
  - Rabbits (226)
  - Cutworms (115, 116, 117)

Affecting foliage
- New leaves curled
  - Wind injury
  - Onion thrips (136)

Herbicide (2,4-D, etc.) injury (200)
Bean aphids (154)

Aphids
Bean aphids (154)

Small pits chewed in upper leaf surface
Palestriped flea beetle (92)

Leaves clipped from stems, cut at sharp angle
Rabbits (226)

Leaves chewed, ragged
Grasshoppers (139)
Slugs (166)
Mexican bean beetle (89)
Western bean cutworm (climbing cutworm) (117)
Blister beetles (102)

Leaves with small veins remaining, lacy appearance
Mexican bean beetle (89)
Slugs (166)

Upper leaf surface with dark or yellow spotting
Halo blight (192)
Common bacterial blight
Bacterial brown spot
Bean rust (190)

Leaves with meandering tunnels
Vegetable leafminer (133)

Leaves with yellow, green or blue-green mosaic patterning
Common bean mosaic virus (171)

Leaves yellowed and puckered, plants stunted
Beet curly top (174)

Leaves yellowing between veins
Iron deficiency (iron chlorosis) (202)
Zinc deficiency (203)

Leaves with small white flecks, retarded growth
Spider mites (163)
Onion thrips (136)

Top leaves "scorched" along edges
Wind damage during hot, dry weather

Affecting pods, seeds
Flowers fail to "set," abort
High temperatures
Western flower thrips (136)

Tunneling into pods
European corn borer (112)
Lima bean pod borer (limas only) (114)
Mexican bean beetle (89)
Western bean cutworm (climbing cutworm) (117)
Slugs (166)

Twisted bean pods
Herbicide injury (200)

Poor seed set
Drought stress during blossom set

Seeds hard
Overmaturity

## Broccoli and Cauliflower

The edible broccoli head or cauliflower curd is a tightly packed mass of flower buds. In typical varieties, all the flower heads have merged to form a single head. Sprouting types of broccoli, such as Romanesco, can form several distinct heads. Intermediate types exist, such as the purple cauliflowers, which are classified as broccoli. Broccoli and cauliflower are biennial plants grown as annu-

*Broccoli*

als. However, many broccoli varieties can flower during a single season, particularly when extended cold weather follows transplanting.

Like all members of the cabbage family, broccoli and cauliflower are cool-season plants, with optimal temperatures for growth around 60°F to 70°F. High temperatures can produce a variety of growth disorders, such as poor head quality and off-flavors. Heading is also delayed by high temperatures.

Although they can be seeded, transplants are most commonly used. Broccoli can easily be transplanted. However, cauliflower is more sensitive to checks on growth, and transplanting is more difficult. Plants with about four true leaves that do not show the blue coloration associated with stress are best for transplanting.

Both broccoli and cauliflower require good soil fertility and structure for best production. However, watering is usually the most critical cultural practice since both crops are heavy users of water. The quality of broccoli and cauliflower can be greatly affected by periods of drought stress, so even amounts of water should remain available throughout growth.

Early-maturing "snowball" types of cauliflower usually require blanching of the head by tying the inner wrapper leaves. Many later-season varieties have infolding leaves that self-blanch the head.

Broccoli and cauliflower may be attacked by several insect pests, including western cabbage flea beetles and various cabbageworms.

### *Common Disorders of Broccoli and Cauliflower*

- Affecting whole plant
  - Plant growth slow, stem shriveled at soil line
    - Wirestem (Rhizoctonia) (188)
    - Wind injury
    - Root pruning (cultivator blight)
  - Seedlings cut
    - Cutworms (115, 116, 117)
- Affecting leaves
  - Small circular holes or pits in leaves
    - Western cabbage flea beetle (91)
  - Larger, irregular holes chewed in leaves
    - Imported cabbageworm/southern cabbageworm (106)
    - Cabbage looper (109)
    - Grasshoppers (139)
    - Diamondback moth (123)
    - Zebra caterpillar
    - Slugs (166)
  - Leaves with warty, brown lumps (cauliflower)
    - Edema (198)
  - Leaves with meandering tunnels
    - Vegetable leafminer (133)
  - White spotting on leaves, often distorted
    - Harlequin bug (144)
- Yellow angular spots progressing inward from leaf edges, wounds
    - Black rot
- Dieback of leaves from edges, wilting and rot of roots (cauliflower)
    - Phytophthora root rot (176)
- Affecting stem
  - Hollow stem
    - Hollow stem disorder (produced by fertilization with high rates of nitrogen and rapid growth)
- Affecting head
  - Premature setting of heads, head size very small
    - Exposure of transplants to shifting warm/cool temperatures
  - Bolting (flowering) of broccoli
    - Excessive maturity
    - Poorly adapted variety
    - High-temperature stress
  - Curd of cauliflower appears uneven and fuzzy (riceyness)
    - Warm temperatures during curd development
  - Flower buds of broccoli turn brown
    - Brown bud disorder (physiological problem associated with rapid growth)
  - Cauliflower head yellow
    - Exposure of head to direct sun

## Cabbage, Brussels Sprouts, Collards and Kale

Cabbage, Brussels sprouts, collards and kale are closely related members of the cabbage family, the latter two being in the headless group, which most closely resemble their wild ancestors. Cabbage produces a single large head of densely packed leaves, which form after the plants have produced a rosette of outside wrapper leaves. Brussels sprouts produce a large stalk that forms numerous smaller heads at the base of the leaves. Cabbage, Brussels sprouts, collards and kale are biennial plants that form a flower stalk in their second season.

Like all members of the cabbage family, they are cool-season plants, with optimal temperatures for growth around 60°F to 70°F. They are also frost-tolerant and usually survive early fall freezes that kill most other garden plants. Freezing temperatures can actually improve the flavor of these plants, particularly Brussels sprouts. High temperatures slow growth and reduce the quality and flavor of leaves. Kale and collards are more tolerant of high temperatures than cabbage.

All can be grown by directly seeding in the garden, and seeds germinate with soil temperatures above 40°F. However, transplants are most commonly used, particularly for cabbage. Young cabbage has good frost tolerance and is easily transplanted. Plants with about four true leaves that do not show the blue coloration associated with stress are best for transplanting.

All the cole crops are heavy users of water due to their large leaf surface area. Therefore, they grow best in heavier soils, which better retain water.They do not tolerate poor drainage, however. Cabbage and Brussels sprouts are among the vegetables that show the best growth response to fertilization. Kale and collards are more tolerant of poor growing conditions.

Disease problems with these vegetables are uncommon in the West. Cabbage is attacked by several insect pests, especially the western cabbage flea beetles and various cabbageworms. Brussels sprouts often develop large populations of cabbage aphids late in the season. Collards and kale are relatively pest-free.

***Common Disorders of Cabbage, Brussels Sprouts, Collards and Kale***

Seedling problems
- Seedlings wilt, die
  - Damping off/seedling blights (187)
  - Wirestem (Rhizoctonia) (188)
- Seedlings cut
  - Cutworms (115, 116, 117)
  - Rabbits (226)

Affecting whole plant
- Plant growth slow, stem shriveled near soil line
  - Wirestem (Rhizoctonia) (188)
  - Wind injury
- Plant growth slow, roots chewed and tunneled
  - Cabbage maggot (132)

Affecting leaves
- Small circular holes or pits in leaves (*Note*: a small purple halo often develops around these wounds on mustard greens)
  - Flea beetles (90, 91)
- Large holes chewed in leaves
  - Imported cabbageworm/southern cabbageworm (106)
  - Cabbage looper (109)
  - Diamondback moth (123)
  - Grasshoppers (139)
  - Slugs (166)
- White spotting on leaves, some distortion
  - Harlequin bug (144)

Yellow, angular spots moving inward from wounds and leaf margin
  - Black rot
- Rot progressing from lower leaves
  - Bottom rot (Rhizoctonia) (188)
- Dieback of outer leaves from edges, wilting and root rot
  - Phytophthora root rot (176)

Affecting head
    Head distorted, aphids present
        Cabbage aphids, turnip aphids (157)
    Raised brown warts on inner leaves
        Edema (198)
        Onion thrips (136)
    Dieback of inner leaves (cabbage, Brussels sprouts)
        Tipburn (physiological problem related to high temperature and high fertility)
    Head size small
        Overcrowding
        Poor culture (fertility, water)
    Cracking of cabbage heads
        Excess water following dry period
        Overmaturity

*Carrots and parsley*

## Carrots and Parsley

Carrots and parsley are biennial plants grown as annuals. They are very tolerant of cool temperatures and can overwinter in the garden if given some winter protection. If left in the garden over winter, they produce a flower stalk the following season.

Carrots and parsley prefer moderately warm soil temperatures. Germination can occur in soils as low as 40°F, but optimal sprouting is at 60°F to 65°F. The seeds are small and should be planted shallowly and kept moist. Seedlings are weak and may have difficulty growing through crusted soils. Carrots are almost never transplanted because the taproot and subsequent growth are easily distorted. However, parsley transplants readily.

Carrots require deep, loose soils, since soil compaction inhibits root growth. In heavy soils, varieties that develop short taproots are most adapted. Carrots respond well to nitrogen fertilization, but are less demanding of phosphorus and potassium.

The plants have few pest problems. East of the Rockies, aster yellows is the most common disease of carrots. Parsleyworm, the larva of the black swallowtail butterfly, is common on parsley and related plants, such as dill and fennel.

***Common Disorders of Carrots and Parsley***

Seedling problems
    Poor seedling emergence
        Crusted soil
        Slugs (166)
        Millipedes (165)
        High temperatures
Affecting leaves
    Top growth chewed
        Rabbits (226)
        Climbing cutworms (116, 117)
        Alfalfa webworms
        Parsleyworm (107)
        Slugs (166)
    Leaves show bronze discoloration, bushy top growth
        Aster yellows (196)
        Carrot aphid (155)
Flowering
    Flower stalk is produced
        Normal flowering by two-year-old plant
Affecting roots
    Roots hairy, top growth excessively bunchy
        Aster yellows (196)

Forked roots
Damage to root tip
Stony or heavy clay soils
Root knot nematode (172)
Overmaturity
Twisted roots
Overcrowding
Roots chewed
Slugs (166)
Lesser bulb fly (136)
Roots cracked
Overmaturity
Roots rotted
Phytophthora root rot (176)
Lesser bulb fly (136)

## Cucumbers

Cucumbers are warm-weather plants that do not tolerate frost. Germination requires minimum soil temperatures of 60°F with optimal temperatures somewhat higher. Optimal average temperatures for growth are in the 70°F to 75°F range. Many varieties mature 50 to 60 days after germination.

Cucumbers require pollination by bees or other insects. All garden cucumbers bear both male and female flowers on the same plant. Unlike many other squash family plants, most flowers are female, capable of producing fruit.

Cucumbers require large amounts of water, and the fruits have the highest water content of all garden vegetables. Some wilting during midday is common and unavoidable during high temperatures. However, wilting during early morning indicates drought stress. Inadequate water contributes to problems such as poor fruit development and off-flavors. Cucumbers are among the vegetables that respond well to fertilization.

In much of the region, cucumbers have few insect problems. Angular leaf spot and other fungal diseases can be a problem where leaf wetting allows infection.

***Common Disorders of Cucumbers***

Seedling problems
Poor seedling emergence
Seedcorn maggot (130)
Damping off disease (187)
Cool soils
Small pits chewed in seedlings
Striped cucumber beetle (95)
Palestriped flea beetle (92)
New leaves curled
Wind injury
Onion thrips (136)
Herbicide (2,4-D, etc.) injury (200)
Affecting whole plant
Sudden wilting of plant
Bacterial wilt (194)
Fusarium wilt (178)
Affecting leaves
Leaves with small white flecks, retarded growth
Spider mites (163)
Onion thrips (136)
Mottling/yellowing of leaves
Cucumber mosaic (172)
Watermelon mosaic (173)
Squash mosaic
Greenhouse whitefly (152)
Dark spotting and death of tissues between veins
Angular leaf spot (191)
Anthracnose (178)
Phytophthora wilt (176)
Meandering tunnels in leaves
Vegetable leafminer (133)
White covering on leaves
Powdery mildew (182)
New leaves puckered, distorted
Herbicide injury (200)
Viral disease (various mosaic viruses) (171)
Onion thrips (136)
Affecting fruit
Failure to set fruit
Lack of female flowers

Excessive heat, rain during flowering
Poor pollination
Knobby or misshapen fruit
Lack of pollination
Stress during fruit development
Cucumber mosaic (172)
Watermelon mosaic (173)
Fruit with blackened end
Blossom-end rot (197)
Fruit pitted
Angular leaf spot (191)
Anthracnose (178)
Cold damage (199)
Off-flavor
Water stress
Temperature fluctuations

*Eggplant*

## Eggplant

Eggplant is a warm-weather plant that grows best with optimal temperatures of 70°F to 85°F. Seed germination requires a minimum of 60°F. Because of the need for an extended growing season, eggplant is almost always grown from transplants started six to ten weeks before the last expected frost. Eggplant is highly susceptible to frost injury and grows poorly during periods of cool temperatures (below 50°F). Fruit will not set when temperatures drop below 60°F to 65°F.

Plants transplant with some difficulty. Short, stocky transplants with a large root mass do much better than spindly plants. Transplants are also susceptible to wind injury and drying.

Eggplant is a fairly heavy feeder of soil nutrients that needs a regular supply of nitrogen and water for best growth. During heavy fruit set, plants are susceptible to breakage in high winds.

The Colorado potato beetle is the most important pest of eggplant east of the Rockies.

***Common Disorders of Eggplant***

Affecting whole plant
Plants wilt, roots appear brown, dark streaks inside stem
Fusarium wilt (178)
Plants wilt, dead areas on stem around soil line
Phytophthora wilt (176)
Affecting leaves
Small holes chewed in leaves
Flea beetles (90)
Large areas chewed in newer leaves
Colorado potato beetle (96)
Edges of leaves chewed
Grasshoppers (139)
Colorado potato beetle (96)
Leaves with small white flecks, retarded growth
Spider mites (163)
Leaves die back, yellow from margins, dark streaks inside stem
Verticillium wilt (177)
Affecting fruit
Pitting, discoloration of fruit
Cold damage (199)
Verticillium wilt (177)
Poor fruit set
Low light
Cool night temperatures

## Lettuce

Lettuce prefers cool weather and, unlike many other vegetables, thrives in partial shade. It requires moist conditions during growth and may produce bitter flavors if grown in hot, dry weather. The plants tend to produce a seed stalk (bolt) when temperatures are too high, although heat-tolerant varieties are available. Spring and fall crops are most commonly grown in the region, with summer plantings more restricted to the cooler high-altitude gardens.

*Lettuce*

Lettuce is almost always grown from seed. During germination, moisture must be fairly high and temperatures must be favorable. Germination can occur in cool soils as low as 35°F, although the optimal temperature is somewhat higher. However, germination is inhibited by temperatures above 85°F.

Lettuce can be transplanted easily. It is quite cold-tolerant and can usually be grown as long as temperatures stay above 25°F.

Slugs show a particular fondness for lettuce and are often the worst pest of this crop. Aphids may be abundant, particularly in fall, and aster yellows is a common problem with varieties that form heads.

***Common Disorders of Lettuce***

- Seedling problems
  - Poor emergence
    - High temperatures during planting
    - Planting too deep
    - Soil crusting
    - Slugs (166)
- Affecting whole plant
  - Formation of seed stalk (bolting)
    - High temperatures
    - Cold weather during seedling stage
  - Wilting
    - Tomato spotted wilt (174)
    - Lettuce root aphid
- Affecting leaves
  - Twisting of head, yellowing, dark latex spots
    - Aster yellows (196)
  - Yellowing of foliage, no twisting or latex spotting
    - Yellows viruses
  - Chewing of foliage
    - Cabbage looper/celery looper (109)
    - Slugs (166)
    - Palestriped flea beetle (92)
  - Aphids on leaves
    - Lettuce aphid (154)
    - Potato aphid (154)
    - Green peach aphid (156)
  - Dieback of inner leaves
    - Tip burn (physiological problem related to high temperature and rapid growth)
- Affecting roots
  - White, cottony insects on roots
    - Lettuce root aphid
- Affecting heads
  - Soft heads
    - Immaturity
  - Head decay
    - Bottom rot (Rhizoctonia) (188)
    - Tomato spotted wilt (174)
  - Bitter taste
    - Hot, dry weather

## Muskmelon (Cantaloupe) and Watermelon

Muskmelons and watermelons are long-season, warm-weather plants that do not tolerate frost. Growth, yield and quality of these crops are best where warm, sunny weather prevails; melons are difficult to grow at higher elevations. Optimal temperatures are 70°F to 85°F for watermelon, somewhat lower for muskmelons. Varieties vary in maturity, ranging from 75 to 100 days.

Seeds of muskmelon and watermelon require a minimum of 60°F to germinate but are very tolerant of dry soil conditions during sprouting. To force in the garden, plants can be started early. However, like all cucurbits, they do not tolerate disruption of the root system, so transplanting must be done with extra care to avoid breaking roots.

Muskmelons and watermelons are dioecious (have separate male and female flowers) and require pollination. New seedless watermelon varieties require interplanting with a fertile pollinator.

Both muskmelon and watermelon require fairly high levels of watering, particularly during emergence and pollination. However, the root system of muskmelon does not go as deep as that of watermelon and thus requires more frequent applications of moisture. Plants should not be moist during the harvest period or sugar content will decrease.

Developing watermelons and muskmelons have few pest problems, primarily wilt and mosaic diseases. However, as fruits ripen, coyotes—and neighborhood children—develop a particular fondness for these garden favorites.

***Common Disorders of Muskmelon and Watermelon***

Seedling problems
- Poor seedling emergence
  - Cool temperatures
  - Seedcorn maggot (130)
  - Damping off disease (187)
  - Uneven watering
- New leaves cupped
  - Wind injury
  - Onion thrips (136)
- Small pits chewed in seedlings
  - Striped cucumber beetle (95)
  - Palestriped flea beetle (92)

Affecting whole plant
- Sudden wilting of plant
  - Bacterial wilt (muskmelon only) (194)
  - Fusarium wilt (178)
  - Phytophthora wilt (176)

Affecting leaves
- Leaves with small white flecks, retarded growth
  - Spider mites (163)
  - Onion thrips (136)
- New leaves curled, distorted
  - Wind injury
  - Onion thrips (136)
  - Herbicide (2,4-D, etc.) injury (200)
- Meandering tunnels in leaves
  - Vegetable leafminer (133)
- White covering on leaves
  - Powdery mildew (182)
- Mottling/yellowing of leaves
  - Watermelon mosaic (173)
  - Cucumber mosaic (172)
  - Greenhouse whitefly (152)
- Irregular dead areas on leaves
  - Anthracnose (178)
  - Phytophthora wilt (176)

Affecting fruit
- Failure to mature fruit
  - Cool temperatures
- Worms scarring bottom of fruit
  - Striped cucumber beetle (larvae) (95)
- Fruit pitting
  - Anthracnose (178)
  - Cold injury (199)
- Fruits chewed by large animal
  - Coyotes
  - Raccoons (225)

## Onions, Leeks and Garlic

Onions and related species of *Allium*, such as garlic, shallots and leeks, are biennial plants generally grown as annuals. Some *Allium*, such as bunching onions and chives, are also grown as spreading perennials.

*Onions and leeks*

Seeding should take place early. Germination requires a minimum of about 35°F, but optimal temperatures are above 50°F. They do not require large amounts of soil moisture to germinate, but seedlings are weak and have difficulty penetrating soil crusts. The onset of bulbing is determined by day length and soil temperature, so crops planted late may bulb early and be small.

Onions grown from bulbs generally produce larger, earlier crops. However, if cool weather after transplanting is followed by an early summer, many of the larger bulbs may prematurely bolt and produce a seed stalk. Early removal of these flowering spikes may force the plants to renew bulb growth.

Since onions are shallow rooted, they require frequent watering, particularly in sandy soils. When grown for bulbs, onions should be allowed to mature naturally. Top growth will fall over and the neck area will dry, reducing problems with diseases that might enter the bulb. As the plants mature, water should be reduced.

Onions generally have few pest problems in the home garden, except in areas where onion maggots have become established. Thrips, bulb rots and leaf diseases are occasional problems.

***Common Disorders of Onions, Leeks and Garlic***

- Seedling problems
  - Poor emergence
    - Soil crusting
- Affecting leaves
  - Leaves scarred, flecked with light areas
    - Onion thrips (136)
  - Elongate brown-purplish dead areas on foliage
    - Purple blotch (180)
    - Botrytis leaf blight (181)
    - Downy mildew (188)
- Affecting bulbs
  - Plants produce flowers (bolting), bulbs woody
    - Biennial bearing (overwintered bulbs)
    - Cool, wet weather following transplanting
  - Bulb size small
    - Late planting
    - Onion thrips (136)
    - Poor culture (watering, fertilization, etc.)
  - Bulbs rot
    - Fusarium basal rot (178)
    - Pink root
  - Bulb tunneled
    - Seedcorn maggot (130)
    - Onion maggot (130)
    - Lesser bulb fly (131)

*Garlic*

## *What about Inoculants?*

Peas, beans, alfalfa and related plants are legumes, members of the Leguminosae family. One of the most remarkable characteristics of the legumes is their ability to provide for their nitrogen needs by "fixing" the nitrogen gas that is such a common element of the atmosphere but often deficient in soils.

Nitrogen fixation is done with the aid of specialized rhizobia bacteria, which live on the roots of legumes. The bacteria and the plants have a symbiotic relationship. The plants derive the benefit of nitrogen in the form of usable ammonia, a necessary nutrient that is often limited in soils. The bacteria invade the roots of the plant and derive energy from it. Bacteria that have successfully established themselves in plant roots cause a noticeable small swelling, or nodule, to form on the root, a process called *nodulation*.

With few exceptions, each species of legume is associated with a specific type of rhizobia bacteria. For example, the nitrogen-fixing bacteria that produce colonies on the roots of peas cannot establish themselves on green beans or soybeans. To ensure that the bacteria are present around the roots, *inoculants* containing the bacteria are often sold to be used as seed treatments or applied to the soil. Sometimes these commercial inoculants contain only a single species of rhizobia bacteria. However, most inoculants sold through garden catalogs contain a blend of several species of *Rhizobia*, typically those used by peas, lima beans and snap beans. (Inoculants should not be confused with *seed treatments*, usually a coating of fungicide or insecticide, which are used to protect germinating seeds and seedlings.)

How valuable are these inoculants for the gardener? In most cases they are of marginal use and the gardener will see little or no increase in plant growth. This is because the rhizobia bacteria spread readily, so the garden is likely to already contain some if the crops have been previously grown in the garden or neighborhood. Furthermore, once established in the garden by growing the crop, the bacteria normally persist for a long time.

In some cases, inoculants may benefit the gardener. If a garden is being established in a fairly isolated location, inoculants may be needed to establish the bacteria during the first year. In gardens grown in very acidic soils, the bacteria may need to be periodically reintroduced, since they do not survive well in these soils.

## Peas

Peas are a cool-season crop generally planted in early spring and again in late summer, although areas with cool summer temperatures can sustain cropping throughout the summer. Seed germination requires a minimum soil temperature of 40°F, and crop growth is optimal at 60°F to 65°F. Growth is slowed and plants become stressed by high temperatures, although there is a wide range in tolerance of high temperatures among varieties. Soil moisture should be fairly high while seeds germinate, but seeds should not be soaked before planting or seeds can crack and rot.

Peas are moderate users of water during growth. They are legumes that do not require supplemental nitrogen fertilization.

Peas have few pest problems. Powdery mildew commonly develops in early summer.

*Common Disorders of Peas*
Seedling problems
  Poor seedling emergence
    Seedcorn maggot (136)
    Damping off disease (187)
  Seedlings chewed
    Slugs (166)
    Rabbits (226)
Affecting whole plant
  Wilting
    Pea aphids (158)
    Fusarium wilt (178)
  Retarded growth
    High temperatures
Affecting leaves
  Entire leaves chewed, leaving stems
    Rabbits (226)
  Leaves chewed, ragged
    Loopers (109)
    Climbing cutworms (116, 117)
    Grasshoppers (139)
    Slugs (166)
  Leaves with small white flecks, retarded growth
    Spider mites (163)
  White covering on leaves
    Powdery mildew (182)
Affecting pods, peas
  Pods with scarring
    Flower thrips (136)
    Onion thrips (136)
  Peas hard
    Overmaturity
    High temperatures

## Peppers

Peppers are a warm-weather vegetable that cannot tolerate frost. In the West, the length of the growing season is not long enough to allow direct seeding in the garden, so peppers are grown from transplants started about six to eight weeks before planting in the garden. Germination requires a minimum soil temperature of 60°F.

Peppers grow best when daily temperatures average 70°F to 85°F, with somewhat cooler night temperatures. Hot pepper varieties can tolerate and thrive at higher temperatures than sweet peppers. Pepper growth and ripening cease and blossoms may drop when night temperatures fall below 55°F. Peppers also need even watering, particularly when flowers are being pollinated and fruits begin to form.

Peppers have few serious pest problems in the West. Most garden problems with pepper production are related to culture, such as uneven watering, or temperature and sunlight.

*Common Disorders of Peppers*
Affecting whole plant
  Plants wilt, roots show browning, decay
    Verticillium wilt (177)
    Fusarium wilt (178)
    Phytophthora wilt (176)
  Plants wilt, roots appear healthy
    Tomato spotted wilt (174)
  Growth retarded
    Cool temperatures
    Viral disease (common mosaic, potato virus Y) (171)
Affecting leaves
  Small holes chewed in leaves
    Flea beetles (90, 91)
  Large areas chewed in leaves
    Tomato hornworm (115)
  New growth small, discolored
    Beet curly top (174)
    Tomato spotted wilt (174)
    Viral disease (common mosaic, potato virus Y) (171)
    Herbicide (2,4-D, etc.) injury (200)
  Leaves thick, leathery
    Viral disease (common mosaic, potato virus Y) (171)
  Leaves with irregular dead areas
    Phytophthora wilt (176)
  Sticky honeydew present
    Green peach aphid (156)

Greenhouse whitefly (152)
Affecting fruit
Fruit tunneled
Corn earworm (tomato fruitworm) (110)
European corn borer (112)
Fruit shows discoloring in ring pattern
Tomato spotted wilt (174)
Fruit shows blotchy, dead patches
Sunscald (198)
Phytophthora wilt (176)
Blossom end of fruit dead
Blossom-end rot (197)
Ripe fruit shows circular, slightly sunken patches
Anthracnose (178)
Fruit color dull
Cold injury (199)
Fruit fails to ripen
Cool temperatures (199)
Poor fruit set
Low nighttime temperatures
Low light
Fruit distorted
Uneven growing conditions (temperature, water)
Low temperature during blossoming

## Potatoes

Unlike most garden vegetables, potatoes are not grown with seeds. Instead, they are vegetatively propagated using seed pieces of cut or whole potatoes, and sprouts develop from the dormant eyes. Because true seeds are not used, potatoes are much more likely to carry over diseases between seasons. As a result, it is usually a good practice to plant certified seed potatoes, which are grown to meet strict standards of disease control.

Potatoes grow best in moderately cool weather, with optimal growth at 60°F–65°F. However, the plants are very sensitive to frost and generally should not be planted earlier than two to three weeks before the last spring frost is expected. Good tuber growth is dependent on even water and fertilization conditions. Plant needs are especially important when their vines are growing most vigorously.

The base of the growing plants is commonly covered with soil (hilling) to protect the tubers from light, which produces tuber greening. Because the plants grow shallow, spreading roots, weeding within 6 to 8 inches of the plants should be done carefully early in the season, to avoid root damage.

*Potatoes*

***Common Disorders of Potatoes***
Stand establishment problems
Poor germination
"Blind" seed pieces (no eyes)
Seed piece decay (Fusarium) (178)
Blackleg (soft rot/Erwinia) (195)
Affecting whole plant
Plants wilt frequently
Lack of water
Roots pruned by cultivation
Plants wilt, roots appear brown
Verticillium wilt (177)
Plants wilt, stems black and soft
Blackleg (soft rot/Erwinia) (195)
Affecting leaves
Small holes chewed in leaves
Flea beetles (90)
Large areas chewed in leaves
Colorado potato beetle (96)
Grasshoppers (139)
Blister beetles (102)
Cabbage looper (109)
Climbing cutworms (116, 117)

White flecking of leaves
Garden leafhoppers (151)
New growth small, discolored, curled
Potato (tomato) psyllid (161)
Aster yellows (196)
Viral disease (potato leafroll, potato virus Y, etc.) (171)
New growth may die back
Frost injury (199)
Lygus bugs (145)
European corn borer (112)
Affecting tubers
Tubers chewed
Cutworms (115, 116, 117)
Field mice
Slugs (166)
Tubers with dark streaking
Tuber flea beetle (tunnels irregular) (90)
Potato leafroll virus (phloem areas of tuber dark) (171)
Verticillium (on stem end of tuber only) (177)
Tubers appear punctured
Quackgrass (rhizomes) (211)
Wireworms
Tubers with hollow center
Hollow heart (physiological problem related to uneven growth)
Tuber size small, tuber skin rough
Potato (tomato) psyllid (161)
Knobby tubers
Potato (tomato) psyllid (161)
Uneven growing conditions during tuber set
Tubers show greening (virescence)
Exposure to light
Tubers sprout prematurely
Potato (tomato) psyllid (161)
Tuber skin with dark spotting ("dirt" won't wash off)
Rhizoctonia (188)
Tuber skin with small light, scabby patches
Enlarged lenticels (related to cool, wet soils)
Tubers form on stem above ground
Potato (tomato) psyllid (161)
Aster yellows (196)
Rhizoctonia (188)

## Radishes, Turnips and Rutabagas

Radishes, turnips and rutabagas are the commonly grown root crops of the cabbage family (Cruciferae). They grow best when temperatures are cool, and seeds can germinate in soils with a temperature as low as 40°F. Optimal growth is 60°F to 65°F, and growth is slowed when temperatures exceed 75°F. They can tolerate frosts; the flavor of turnips and rutabagas is actually improved by light frost.

These vegetables are usually planted in early spring and late summer, although they are more adaptable to varying growing conditions than are other garden plants of the cabbage family. Turnips and rutabagas are biennials, flowering the second year after planting. Radishes will often set a seed stalk a few months after sprouting. Turnips and radishes should be picked when small or they will develop a strong flavor and poor texture.

Radishes, turnips and rutabagas have relatively few pests. The western cabbage flea beetle can be severe in some areas, killing seedlings. The cabbage maggot tunnels roots in localized areas of the West.

***Common Disorders of Radishes, Turnips and Rutabagas***

Seedling problems
Seedlings die
Damping-off fungi (Rhizoctonia, others) (187)
Affecting leaves
Small shotholes in leaves
Western cabbage flea beetle (91)
Ragged holes in leaves
Diamondback moth (123)
Imported cabbageworm (106)

Affecting roots
    Tunneling of root
        Cabbage maggot (132)
        Slugs (166)
    Roots fibrous, with strong flavor (turnips, radishes)
        Overmaturity

## Spinach, Swiss Chard and Beets

Spinach is a cool-season crop that grows best when temperatures average 60°F to 65°F. Plants produce seed stalks (bolt) during periods of long days, a process accelerated by high temperatures. Bolt-resistant varieties are increasingly available. Spinach is usually grown as a spring and fall crop. It is extremely cold-tolerant and can overwinter in the garden. Seed germination can occur in cool soils (about 40°F) and is inhibited in warm soils. Spinach should be planted as early as possible to deter bolting.

Swiss chard is related to spinach, another member of the Chenopodiaceae family. Originally cultivated for its root (now grown as a separate vegetable—the garden beet), the outer leaves and petioles are used in cooking. Chard can resist bolting in high temperatures, a distinct advantage for many gardeners. If stressed by high temperatures and drought, it may produce a seed stalk. By pinching this off as soon as possible, leaf growth can again be stimulated. Like spinach, chard is grown by seeding and is fairly tolerant of a wide range of garden conditions.

New Zealand spinach, sometimes sold as a heat-tolerant spinach, is actually a very different species in a different plant family (Tetragoniaceae).

Beet seeds are usually clusters of four or five seeds. As a result, some thinning may be needed. Beets generally require less nitrogen and more phosphorus than do nonroot crops. They prefer lighter, looser soils. Beets can tolerate highly alkaline soils.

Spinach, beets and chard are attacked primarily by the spinach leafminer, which produces dark blotches in the older leaves.

***Common Disorders of Spinach, Swiss Chard and Beets***

Seedling problems
    Poor seed germination
        High soil temperatures
Affecting whole plant
    Formation of flower stalk (bolting)
        Long day length (accelerated by high temperatures)
    Roots rot
        Rhizoctonia (188)
    Plant wilts, roots tunnelled
        Sugarbeet root maggot (130)
    Plants stunted, may die
        Blue mold (downy mildew) (188)
        Sugarbeet root maggot (130)
    Beets fail to bulb
        Overcrowding
Affecting leaves
    Dark hollowed blotches in leaves
        Spinach leafminer (133)
    Holes in leaves
        Flea beetles (90, 91)
        Webworms
    Aphids on leaves
        Green peach aphid (156)
        Lettuce aphid
    White or red blisters on lower leaf surface
        White rust (190)
        Red rust (spinach)
    Leaves yellow, often rotted
        Blue mold (downy mildew) (188)
    Red spotting of leaves
        Leafhoppers (beets) (147)
    Small leaves, growth slow
        High temperatures

## Summer Squash and Zucchini

Summer squash are grown for their immature fruit. Almost all summer squashes have been bred for bush-type growth, unlike the vine growth of most hard or winter squashes. They have a relatively short growing season, producing fruit about 40 to 50 days after planting.

Summer squash are dioecious, producing separate male and female flowers. Generally, far more male flowers occur, which do not produce fruit. They require cross-pollination by bees or other insects.

Summer squash are warm-weather plants. They are fairly difficult to transplant and are usually seeded in the garden. Seed germination requires a minimum soil temperature of 60°F and growing plants prefer temperatures in the 65°F to 80°F range. Summer squash are intolerant of near-freezing conditions, and growth ceases below 50°F.

The plants produce an extensive but shallow root system. They are heavy users of water. However, fertilization needs are fewer than with most vegetable crops.

Summer squash have few pest problems. Plantings in cool soils are frequently attacked by seedcorn maggots. Viral diseases transmitted by aphids can be serious in some areas.

***Common Disorders of Summer Squash and Zucchini***

- Seedling problems
    - Poor seedling emergence
        - Seedcorn maggot (130)
        - Damping off (187)
        - Cool soils
    - Small pits chewed in seedlings
        - Striped cucumber beetle (95)
    - New leaves cupped
        - Wind injury
        - Onion thrips (136)
- Affecting whole plant
    - Sudden wilting and death of plant
        - Bacterial wilt (194)
        - Fusarium wilt (178)
        - Squash bug (143)
        - Squash vine borer
    - Plants wilt frequently
        - Salt injury
        - Root pruning
- Affecting leaves
    - New leaves curled
        - Wind injury
        - Onion thrips (136)
        - Herbicide (2,4-D, etc.) injury (200)
    - Leaves with small white flecks, retarded growth
        - Spider mites (163)
        - Garden leafhoppers (151)
        - Onion thrips (136)
    - Mottling/yellowing of leaves
        - Mosaic (cucumber, watermelon) viruses (172, 173)
        - Angular leaf spot (191)
        - Varietal character of some yellow squash
        - Greenhouse whitefly (152)
    - White covering on leaves
        - Powdery mildew (182)
    - Small gray bugs on leaves
        - Squash bug (143)
    - Dead areas on leaves
        - Angular leaf spot (191)
        - Squash bug (143)
    - Dark spots on leaves, often with concentric pattern
        - Early blight (179)
- Affecting flowers
    - Insects feeding on flowers, pollen
        - Western corn rootworm (94)
        - Striped cucumber beetle (95)
        - Spotted cucumber beetle (94)
    - Plants flower but don't set fruit
        - Male flowers only
- Affecting fruit
    - Gray-brown bugs on fruit
        - Squash bug (143)
    - Fruit mottled, misshapen
        - Cucumber mosaic (172)

Watermelon mosaic (173)
Poor pollination
Fruit chewed
Western corn rootworm (94)
Striped cucumber beetle (95)
Skin tough, thick
Overmaturity
Dark pitting of fruit
Cold damage (199)

## Sweet Corn

Sweet corn does not tolerate freezing. Seed germination requires a minimum soil temperature of 50°F, with the optimum above 60°F. Difficulties with germination are greater with the new "supersweet" varieties.

Sweet corn is wind-pollinated and should be planted in blocks to allow for optimal pollination. Plantings should be isolated from field corn, since cross-pollination will produce kernels that lack sweetness.

Sweet corn is a heavy user of nitrogen and water. Watering is particularly critical during pollination.

Raccoons can be the biggest pest problem in the sweet corn patch. Corn earworms are a common insect pest in southern areas of the West. The western corn rootworm often clips the silk of later plantings, causing poor pollination.

*Sweet corn*

***Common Disorders of Sweet Corn***

Seedling problems
Poor seedling emergence
Seedling blights (Penicillium, damping off) (187)
Seedcorn maggot (130)
Seedlings chewed at soil line
Cutworms (115, 116, 117)
Seedlings purple
Phosphorus deficiency
Cool soils
Affecting leaves
Lower leaves die back and yellow
Nitrogen deficiency (201)
Spider mites (163)
Lower leaves with streaks of chewing
Slugs (166)
Leaves with regular rows of holes
European corn borer (112)
Leaves in midsummer with streaks of chewing
Western corn rootworm (94)
Small red-brown raised area on leaves
Rust (190)
Affecting roots
Roots chewed
Western corn rootworm (94)
Affecting tassel
Aphids covering tassel
Corn leaf aphid (154)
Tassel chewed
House finch (231)
Affecting silk
Silk clipped
Western corn rootworm (94)
European earwig (141)
Affecting ear
Poor kernel set
Western corn rootworm (94)
Poor pollination (drought, lack of pollen)
Kernels lack sweetness
Overmaturity
Cross-pollination with field corn

Enlarged gray growths on stems, tassel or ear
    Common smut (185)
Insects in ear
    Corn earworm (110)
    Western bean cutworm (117)
    Dusky sap beetle (103)
    European corn borer (112)
    Bumble flower beetle (104)
    Western corn rootworm (94)
    European earwig (141)
Worms tunneling into ear, stalk
    European corn borer (112)
    Dusky sap beetle (103)
Ear tips chewed off by large animal
    Deer (228)
Ear pulled from plant and chewed with husk removed
    Raccoons (225)

*Tomatoes*

## Tomatoes

Perhaps the most popular garden vegetable, tomatoes provide more challenge to grow than most other vegetables. Tomatoes are warm-season plants that require a long growing season to produce. As such, they are almost always grown as transplants, started six to ten weeks before planting outdoors. Transplanting should be done in warm soils after frost. Growth is greatly slowed in cool soils, and plants are susceptible to frost injury.

Most tomatoes offered for sale are hybrids, although open-pollinated varieties, such as Rutgers, Beefsteak and Roma, are also widely available. Among the many tomato varieties is a great range in maturation time, with many newer varieties being developed for more rapid growth.

Tomatoes have either a determinate or indeterminate growing habit. Determinate varieties stop growing at a certain point, are compact and usually bear early. Indeterminate varieties continue to grow as long as conditions permit. They have a more sprawling growth and benefit from staking.

Pollination can be disrupted by either excessively warm or cool temperatures. Warm night temperatures are also important to ripen fruit.

Tomatoes require high fertility. However, excessive nitrogen produces top growth at the expense of flowering. They require steady moisture and become susceptible to blossom-end rot if watering is irregular.

Tomatoes are susceptible to hornworms and potat (tomato) psyllids throughout most of the West. Common diseases include the leaf-spotting fungi, such as early blight, and Fusarium wilt.

***Common Disorders of Tomatoes***

Seedling problems
    Seedlings cut near soil line
        Cutworms (115, 116, 117)
        Blowing soil
Affecting whole plant
    Growth retarded; leaves thick, leathery
        Common mosaic virus (175)
    Growth retarded; new leaves small and discolored
        Potato (tomato) psyllid (161)
        Beet curly top (174)
    Plants wilt, roots appear brown
        Fusarium wilt (note browning of vascular system) (178)
        Phytophthora wilt (176)

Plants wilt and may die, roots appear healthy
Tomato spotted wilt (174)
Affecting leaves
Small holes chewed in leaves
Flea beetles (90)
Large areas chewed in leaves
Tomato and tobacco hornworm (115)
Spotting, dieback of leaves
Early blight (179)
Septoria leaf spot
Lower leaves first yellow, then die
Fusarium wilt (178)
Leaves with brown, warty lumps
Edema (198)
Leaves thickened, curled
Beet curly top (174)
Potato leafroll virus (171)
Common mosaic (175)
Physiological leaf curl
Potato (tomato) psyllid (little thickening) (161)
New growth small, discolored
Potato (tomato) psyllid (161)
Beet curly top (174)
Cucumber mosaic (172)
Tomato spotted wilt (174)
Herbicide (2,4-D, etc.) injury (200)
Sticky honeydew present on leaves
Potato aphid
Greenhouse whitefly (152)
White, sugarlike pellets on leaves and soil
Potato (tomato) psyllid (161)
Affecting fruit
Fruit chewed
Tomato and tobacco hornworm (115)
Corn earworm (tomato fruitworm) (110)
Slugs (166)
Fruit shows discoloring in ring pattern
Tomato spotted wilt (174)
Fruit with brown spots
Early blight (179)
Overripe fruit with circular, sunken spots
Anthracnose (178)
Fruit with yellow spots
Plant bugs (147)
Blotchy white area on upper surface of fruit
Sunscald (198)
Blossom end of fruit dies
Blossom-end rot (197)
Blossom end of fruit distorted (catfacing)
Cool temperatures during early fruit development
Fruit cracks
Irregular growth due to environmental conditions (water, temperature, etc.)
Fruit size small
Potato (tomato) psyllid (161)
Flowering delayed
Long-season variety
Excessive nitrogen fertilization
Fruit fails to set
Excessive heat (at least 90°F) during pollination
Cool temperatures (nighttime temperatures below 50°F)
Fruit fails to ripen
Short growing season
Cool night temperatures
Skin becomes discolored, russeted
Cold injury (199)

## Winter Squash, Pumpkins and Gourds

Winter (or hard) squash and pumpkins are grown as long-season annuals, the fruit picked when mature. Typical varieties require 90 to 110 days, sometimes more, to become mature.

Winter squash and pumpkins require a minimum of 60°F for seeds to germinate. However, seeds are very tolerant of dry soil conditions during sprouting. Transplanting is difficult because the plants do not easily recover from disturbance of the roots.

Winter squash and pumpkins are warm season crops with growth optimal in the 70°F to 95°F range. The plants grow extensive, shallow root systems. Most have a sprawling growth habit and can be trained on vines. Bush varieties are also available for smaller gardens, although fruit quality is not as good.

Pumpkins and squash are heavy users of water. Growth is favored by soils of good texture that can provide water needs more evenly.

As with other cucurbits, the plants require pollination. Cross-pollination between different types of squash is often adequate but will produce fruit with intermediate characteristics if seeds are planted the following year.

Pest problems can be severe in some areas, especially in the warmer regions of the West. Squash bugs can be a particularly difficult insect to control, and striped cucumber beetles may damage seedlings. Root-rotting fungi, such as *Phytophthora*, can be a serious problem in heavy soils.

*Pumpkins and gourds*

***Common Disorders of Winter Squash, Pumpkins and Gourds***

- Seedling problems
  - Poor seedling emergence
    - Seedcorn maggot (130)
    - Damping off disease (187)
    - Cool soils
  - Small pits chewed in seedlings
    - Striped cucumber beetle (95)
    - Palestriped flea beetle (92)
- Affecting whole plant
  - Sudden wilting and death of plant
    - Squash bug (143)
    - Bacterial wilt (194)
    - Fusarium wilt (175)
    - Phytophthora root rot (176)
    - Verticillium wilt (177)
  - Periodic wilting, recovery
    - Lack of water
    - Root pruning
- Affecting leaves
  - New leaves cupped
    - Wind injury
    - Onion thrips (136)
    - Herbicide (2,4-D, etc.) injury (200)
  - Leaves with small white flecks, retarded growth
    - Spider mites (163)
    - Garden leafhoppers (151)
    - Onion thrips (136)
  - Leaves with irregular dead areas
    - Phytophthora wilt (176)
  - Leaves covered with white powder
    - Powdery mildew (182)
  - Mottling/yellowing of leaves
    - Cucumber mosaic (172)
    - Watermelon mosaic (173)
    - Angular leaf spot (191)
    - Varietal character of some yellow fruited varieties
  - Browning, dieback of leaves
    - Squash bug (143)
    - Garden leafhoppers (151)
    - Angular leaf spot (191)
- Affecting fruit
  - Pitting of fruit
    - Squash bug (143)
    - Striped cucumber beetle (95)
    - Western corn rootworm (94)
    - Gummy stem blight/black spot
    - Cold injury (199)
  - Irregular shape of fruit
    - Poor pollination
    - Cucumber mosaic (172)
    - Irregular watering
    - Watermelon mosaic (173)
  - Russeting of fruit
    - Cold injury (199)

# Fruit Crops

## Apples

Apples are the most widely adapted tree fruit in the West. However, as with most tree fruits, cold, dry, desiccating winds, sunscald and late spring frosts can be limiting. Thus, they are best grown in more sheltered areas.

Commercially available apples are not grown from seed, which can produce highly variable plant types. Instead, they are propagated by grafting from desirable varieties. Most apple trees sold are dwarf or semidwarf varieties, produced by grafting the fruit-bearing variety atop special rootstocks that produce a dwarfing growth. Since these rootstocks vary in winter-hardiness and anchorage in the soil, the choice of rootstock can be almost as important as the fruit-bearing variety in determining suitability. At higher elevations and in northern areas, only the most winter-hardy and early-maturing varieties are adapted.

Apples require pollination by insects to set fruit. Many varieties are not self-fruitful and further require cross-pollination, involving transfer of pollen from a separate variety, known as a pollinizer or pollinator. Fruit is produced on short fruiting spurs. The rosette of leaves surrounding the fruit is very important in determining fruit quality. It is useful to thin fruit to one per cluster to increase fruit size and to prevent alternate year, or biennial, bearing.

Apples need some pruning to remove dead or diseased wood and nonfruit-bearing sucker growth. Pruning is also used to train the growth of the tree. The most common training involves production of a central leader and strong, lower scaffold limbs. Pruning is best done in early spring, before growth starts, to avoid spread of fire blight.

The codling moth (appleworm) is a serious insect pest throughout the West. Fire blight is the most important disease problem that can kill trees.

*Apples*

***Common Disorders of Apples***

Affecting twigs, branches

- Twigs or branches die back, leaves wilt suddenly and do not fall
  - Fire blight (193)
- White woolly material on branches, trunk
  - Woolly apple aphid (159)
- Branches covered with small circular scales
  - San Jose scale (160)
- Twigs with series of splinter wounds, often dying back
  - Cicada egg-laying damage

Affecting lower trunk, roots

- Trees decline in vigor, trunk tunneled
  - Flatheaded apple tree borer/Pacific flathead borer (101)
  - Roundheaded apple tree borer
- Sawdust being pushed from holes in lower trunk
  - Roundheaded apple tree borer (102)
- Trees decline in vigor, root and crown area shows decay
  - Phytophthora root and crown rot (176)

Affecting leaves

- White speckling or streaking of leaves
  - White apple leafhopper (147)
  - Tentiform leafminer (133)

Leaves covered with white powder
Powdery mildew (183)
Leaves bronze-colored, may drop prematurely
Twospotted spider mite, McDaniel spider mite (163)
Leaves with brown, scabby patches
Eriophyid mites (164)
Cedar-apple rust (190)
Leaves chewed
Eastern tent caterpillar (119)
Red-humped caterpillar
Speckled green fruitworm (118)
Apple flea beetle (91)
Leaves curled
Rosy apple aphid, green apple aphid (154)
Leafrollers (121)

Affecting fruit
Fruit tunneled
Codling moth (120)
Apple maggot/western cherry fruit fly (134)
European earwig (enters previous wounds) (141)
Fruit deformed, with scabby wounds on surface
Leafrollers (121)
Oriental fruit moth
Speckled green fruitworm (118)
Hail injury
Fruit deformed, dimpled
Plant bugs (147)
Stink bugs (147)
Apple curculio (98)
Box elder bugs
Light spotting of fruit
San Jose scale (160)
Flower thrips (136)
Dark spotting of fruit
White apple leafhopper (147)
Hail injury
Fruit surface with general brown coloring (russeting)
Spray injury
Powdery mildew (183)

## Cherries

Commonly grown backyard cherries are either sweet cherries (*Prunus avium*) or sour (pie) cherries (*Prunus cerasus*). The latter are much more winter-hardy and widely adapted in the West. Hardiness is also related to the rootstock on which the plants are grafted. Mehalab rootstocks tend to produce hardier and earlier-bearing, but shorter-lived trees than cherries grafted to Mazzard rootstock. Recently, G.M. series dwarfing rootstocks have been introduced, which have the advantage of tolerating heavy clay soils.

Like winter freezing injury, blossom damage by spring frosts is a common weather-related problem with cherries. To delay blossoming, trees should be placed in more shaded areas of the yard, which warm later.

Cherries generally need less pruning and training than other tree fruits. Pruning is needed primarily to remove dead or damaged branches and crossing limbs, and to allow additional light into the tree.

Cherries have moderate water needs. Sweet cherries are much more sensitive to drought than sour cherries. Fertilization needs are modest for most sites. Cherries require pollination by insects. Sour cherries are self-fruitful, but most sweet cherries require a nearby pollinizer for cross-pollination.

Fruit damage by birds is usually the most conspicuous pest injury to cherries. Cherries, like all stone fruits, are susceptible to peach tree borer. As trees age, they commonly succumb to Cytospora cankers.

***Common Disorders of Cherries***

Affecting whole plant
- Trees decline in vigor
  - Peach tree borer (123)
  - Stem pitting virus
  - Phytophthora root rot (176)
  - X-disease (196)
  - Cytospora canker (186)

Affecting leaves
- Leaves curled, undersides often with dark aphids
  - Black cherry aphid (154)
- Leaves yellow
  - Iron chlorosis (202)
  - X-disease (196)
- Upper surface of leaves chewed, often leaving main veins
  - Pearslug (129)
- Leaves chewed and tent of webbing produced
  - Eastern tent caterpillar (119)
  - Fall webworm (119)
- White powdery covering on leaves
  - Powdery mildew (183)
- Small red bumps on leaves (finger galls)
  - Eriophyid mites (164)

Affecting branches, twigs
- Clear, amber ooze from bark
  - Cytospora canker (186)
- Branches die back
  - Cytospora canker (186)
  - Shothole borer (99)
- Small holes in bark of twigs, branches
  - Shothole borer (99)
- Tips of twigs die back, twigs tunneled
  - Peach twig borer (122)

Affecting lower trunk
- Clear, amber ooze from bark
  - Cytospora canker (186)
- Amber ooze mixed with sawdust, often near or below soil line
  - Peach tree (crown) borer (123)
- Pitting at graft union
  - Stem pitting virus
- Roots discolored and rotted
  - Phytophthora root rot (176)

Affecting fruit
- Fruit eaten but not tunneled
  - Bird damage (231)
  - Cherry curculio (97)
- Fruit small and conical with pebbly skin (sweet cherry only)
  - X-disease (buckskin) (196)
- Fruit burst, but not eaten (fruit cracking)
  - Heavy rainfall prior to harvest
- Fruit pit infested
  - Cherry curculio (97)
- Fruit infested with white worms
  - Western cherry fruit fly (134)
- Fruit swollen and hollow (chokecherry only)
  - Chokecherry gall midge

## Currants and Gooseberries

Currants and gooseberries are one of the most easily grown fruit crops in a garden. Although they require continuous moisture for best production, they can adapt to drier and more exposed sites than can other fruits. They may be planted in spring or fall and are tolerant of moderately alkaline soils but will develop problems with iron chlorosis in high-pH soils.

Regular pruning is needed for best fruit production, since most fruit is borne on young, year-old wood. Currants and gooseberries are self-fruitful and do not require cross-pollination with a separate pollinizer.

Several insects are common on currants, including the imported currantworm, currant borer and currant aphid. Gooseberries are less commonly damaged by these insects, although the gooseberry maggot is locally a problem on later-maturing varieties in some areas of the West.

***Common Disorders of Currants and Gooseberries***

Affecting leaves

- Leaves pucker and may show reddening
  - Currant aphid (154)
- Leaves lightly speckled or bronze-colored
  - Spider mites (163)
- New growth generally yellow
  - Iron chlorosis (202)
- White flecking on leaves
  - Leafhopper (147)
- White powdery coating on leaves
  - Powdery mildew (181)
- Dark spotting on leaves
  - Leaf spot (anthracnose) (178)
  - Rust (190)
- Leaves chewed
  - Imported currantworm (128)
  - Currant spanworm (128)
  - Currant flea beetle (91)
- Center of leaves dry out
  - Drought injury
  - Currant borer (124)
  - Bronze cane borer (100)

Affecting canes

- Canes die back during midsummer
  - Currant borer (124)
  - Bronze cane borer (100)

Affecting fruit

- Fruit soft, with pale spotting
  - Sunscald (198)
- Maggots in berries, berries discolored
  - Currant fruit fly (gooseberry maggot)

## Grapes

Two species of grapes are commonly grown, the European wine grape (*Vinis vinifera*) and the native North American species (*Vinis labrusca*). The native types, and the many hybrids between the two species, are far more winter-hardy and adapted to the West. Wine grapes are usually limited to only the mildest locations.

Grapes should be planted in sites that receive full sun and no extreme frosts. Since they use less water and fertilization than most plants, including turf, they should be placed where these cultural needs are met. Winter survival is improved by providing water to promote early-season growth, but allowing the soil to dry late in the season. Vines should be planted deeply, about 1 foot, to promote rooting and limit winter killing.

Fruit is produced on shoots that develop from buds produced during the previous season. Pruning and vine training are important in grape culture to maintain productive buds.

Native grapes have few pest problems—other than raccoons and neighborhood children. However, wine grapes commonly suffer from powdery mildew, grape leafhoppers and mites.

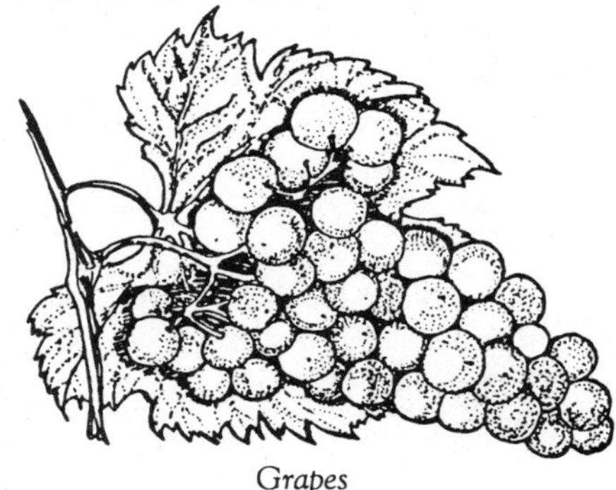

*Grapes*

***Common Disorders of Grapes***

Affecting vines

- Irregular woody growth on stem, usually at soil line
  - Crown gall (192)

Affecting leaves

- Buds tunneled
  - Grape flea beetle (91)
  - Climbing cutworms (116)
- Powdery covering on leaves, stems
  - Powdery mildew (184)
- Leaves with white spotting
  - Grape leafhopper (149)
- Leaves with grayish color, may drop prematurely

Twospotted spider mite (163)
Leaves with meandering pale streaks
Leafminer (133)
New leaf growth distorted, small
Herbicide injury (200)
Eriophyid mites (164)
Large areas of leaves chewed
Eightspotted forester
Achemon sphinx (115)
Grape flea beetle (91)
Affecting berries
Berries shrunken, rotten
Sour rot
Botrytis (180)
Berries split
Powdery mildew (184)
Berries covered with small dark spots
Grape leafhopper (149)
Berries chewed, damaged
Bird injury (231)
Grape berry moth
Berries eaten rapidly (overnight)
Raccoon (225)

## Peaches

Peaches are one of the most difficult fruit trees to grow in the West. Spring frosts commonly kill blossoms, and freeze damage often kills branches or twigs. Relatively few varieties and rootstocks are well adapted to the region. Trees should be planted in more slowly warming, well-protected areas of the yard to minimize this cool-season injury, although they need full sun during the growing season.

Peaches are most often pruned by the open vase method. This involves cutting out much of the center of the tree to promote spreading growth of the lower scaffold limbs. The open center allows for better light and air circulation. Pruning can occur in summer as well as during the dormant season.

Peaches are pollinated by insects. However, most varieties are self-fruitful and do not require a pollinizer variety for cross-pollination. If fruit set is very heavy, they should be thinned by hand or fruit size will be reduced.

Peach trees respond well to fertilization, particularly with nitrogen. They also need regular watering or the plant will remove water from the developing fruit. However, watering and fertilization should be curtailed after harvest to allow trees to harden off for better winter protection.

Pest problems are moderately severe on peaches. The peach tree borer tunnels the lower trunk, and the green peach aphid can curl and distort new growth in the spring. Coryneum (shothole) blight is the most serious disease, and Cytospora canker is readily established in weakened or damaged trees.

***Common Disorders of Peaches***
Affecting leaves
Leaves curled
Green peach aphid (156)
White powdery covering on leaves
Powdery mildew (183)
Small holes in leaves
Coryneum (shothole) blight (186)
Trees decline in vigor, leaves yellow
X-disease (196)
Leaves yellow with veins darker green
Iron chlorosis (202)
Affecting twigs, branches
Clear, amber ooze from bark
Cytospora canker (186)
Branches die back
Cytospora canker (186)
Peach tree (crown) borer (123)
Shothole borer (99)
Small holes in bark of twigs, branches
Shothole borer (99)
Tips of twigs die back, twigs tunneled
Peach twig borer (122)
Oriental fruit moth
Affecting lower trunk
Clear, amber ooze from bark
Cytospora canker (156)

Amber ooze mixed with sawdust, often near or below soil line
Peach tree (crown) borer (123)
Affecting fruit
Fruit blemished, but not tunneled
Coryneum blight (186)
Powdery mildew (rusty spot) (183)
San Jose scale (160)
Plant bugs (147)
Flower thrips (136)
Hail injury
Bird damage (231)
Fruit swollen at flower end
X-disease (196)
Amber ooze from fruit (gummosis)
Cytospora canker (186)
Coryneum blight (186)
Mealy fruit
Overmaturity
Drought during fruit development
Fruit tunneled
Oriental fruit moth
Peach twig borer (122)
Walnut husk fly (134)
European earwig (141)

## Pears

Pears are less widely adapted to western conditions than apples, primarily because of their sensitivity to spring frost damage and fire blight. Careful selection of adapted varieties is essential for success with pears in the region. Rootstocks are available that can provide some dwarfing of pears, producing trees that will grow to 15 or 20 feet tall.

Some training of plant growth is essential for pears. Shoots tend to grow upward, which will produce weak crotches susceptible to breakage if not corrected. Pears also send up numerous nonfruit-bearing vegetative shoots, particularly in response to pruning. However, fruit is produced on slow-growing fruiting spurs, and their development is inhibited by the vegetative shoots. Larger fruit are produced on younger spurs. Spur development is encouraged by selectively pruning actively growing vegetative branches. All pruning should be done during the dormant season to limit spread of fire blight.

Pears require pollination by insects to set fruit. They are not self-fruitful, and the pollen must be from a compatible variety to allow for successful pollination. Pollination is sometimes poor because the flowering period is short and pear flowers contain relatively low quantities of attractive sugars in their nectar. If the fruit has set well, it is useful to thin it to one per cluster to increase fruit size and prevent alternate year, or biennial, bearing.

Fire blight is the most serious pest problem of pears. Developing multiple leaders early in its growth can better protect the tree, allowing infected branches to be cut away. The codling moth is a common concern throughout the region, and pear psylla occur widely west of the Rockies.

*Pears*

*Common Disorders of Pears*

Affecting leaves

- Leaves bronze-colored, may drop prematurely
  - Twospotted spider mite, McDaniel spider mite (163)
- Leaves with brown, scabby patches
  - Pearleaf blister mite (eriophyid mites) (164)
- Upper surface of leaves chewed, leaving veins
  - Pearslug (129)
- Leaves curled and chewed
  - Leafrollers (121)
- Leaves yellow and drop prematurely
  - Psylla shock (pear psylla injury) (161)
  - Iron chlorosis (202)
- Sticky material (honeydew) on leaves, fruit
  - Pear psylla (161)

Affecting fruit

- Fruit tunneled
  - Codling moth (120)
- Fruit deformed, old wounds with scarring
  - Leafrollers (121)
  - Oriental fruit moth
  - Climbing cutworms (116)
  - Hail injury
- Fruit deformed, old wounds form dimples
  - Plant bugs (147)
  - Stink bugs
- Spotting of fruit
  - San Jose scale (160)
  - Hail injury
- Fruit surface with brown coloring
  - Russet mite (eriophyid mites) (164)
  - Sooty mold growing on pear psylla honeydew (161)
  - Damage from oil-based sprays
- Fruit crop heavy one year, light the next
  - Alternate year (biennial) bearing

## Plums

Although a number of native plum species grow in the region, plums cultivated for eating are either European plum (*Prunus domestica*) or Japanese plum (*Prunus salicina*). European plums tend to be more winter-hardy than Japanese plums, approximating apples in this regard. European plums also bloom later, avoiding damage by spring frosts that can damage the Japanese plum crop in the West.

The European plum is fairly easy to grow and tolerates heavy soils. However, the rootstock on which the plum variety is grafted can affect hardiness. Dwarfing rootstocks are available.

Plums need only moderate fertilization for best production. However, regular watering is critical or fruit will be aborted. Plums are pollinated by insects, primarily bees. Most varieties are not self-fruitful and require a second pollinizer variety for cross-pollination.

Plums need little pruning. However, fruit is produced on one-year-old wood, and periodic pruning may be needed to stimulate production. Pruning is best done in early spring.

Plums have few problems with pests in the West. Cytospora canker moves readily into weakened and damaged limbs, and peach tree borer is commonly damaging in much of the region.

*Common Disorders of Plums*

Affecting leaves

- Leaves curled, undersides often with aphids
  - Green peach aphid (156)
  - Leafcurl plum aphid (155)
- Upper surface of leaves chewed, often leaving main veins
  - Pearslug (129)
- Leaves chewed, silken tent produced
  - Eastern tent caterpillar (119)
  - Fall webworm (119)
- White powdery covering on leaves
  - Powdery mildew (183)
- Small finger galls on leaves

Eriophyid mites (164)
Affecting trunk or branches
Clear, amber ooze from bark
Cytospora canker (186)
Amber ooze mixed with sawdust, often near soil line
Peach tree (crown) borer (123)
Branches die back, trees decline in vigor
Cytospora canker (186)
Peach tree (crown) borer (123)
Branches with circular scales
San Jose scale (160)
Small holes in bark of twigs, branches
Shothole borer (99)
Tips of twigs die back, twigs tunneled
Peach twig borer (122)
Affecting fruit
Fruit drops prematurely
Insufficient water during fruit set
Fruit pit infested by insect
Plum gouger (98)
European earwig (141)
Apple maggot (134)
Western cherry fruit fly (134)
Gum oozing from fruit
Plum gouger (98)
Cytospora canker (gummosis) (186)

## Raspberries

Several types of raspberries can be successfully grown in the region, although weather conditions often cause extensive winter injury. More marginally adapted types of brambles (blackberries, black or purple raspberries and loganberries) are very difficult to establish and require special winter protection.

Bearing of raspberries occurs either in the summer or fall. Fall-bearing varieties, which produce the fruiting cane during the growing season, are most easily managed. (A small, second crop is produced if canes of fall-bearing varieties are overwintered.) Summer-bearing varieties, which produce a biennial cane that does not fruit until the second season, are more susceptible to winter injury. Mulching for winter protection is required in most areas for the summer-bearing types.

Regular pruning is necessary with raspberries, both to remove old canes and to remove cane-boring insect pests. Fall-bearing types can be mowed at the end of the season, since they produce new suckers in the spring.

Raspberries can be damaged by several insect pests. Most important is the raspberry crown borer, which tunnels the base of the plants. Cane borers, such as the stemboring sawfly, are also common.

***Common Disorders of Raspberries***

Affecting canes
Canes wilt progressively from the bottom
Verticillium wilt (177)
Raspberry crown borer (125)
Entire canes wilt and die
Phytophthora root and crown rot (176)
Raspberry crown borer (177)
Rose stem girdler/bronze cane borer (100)
Wilting of cane restricted to top of plant
Spur blight/cane blight
Stemboring sawfly (127)
Affecting leaves
Leaves puckered and/or mottled yellow
Viral disease (171)
Iron chlorosis (202)
New leaves form tight cluster (rosette)
Viral disease (171)
Leaves flecked with small light spots, may drop
Twospotted spider mites (163)
Leaves chewed
Sawflies (129)
Leafrollers (121)
Affecting fruit
White spotting of fruit
Sunscald (198)
Fruit gouged by chewing
Grasshoppers (139)

## Strawberries

Strawberries are the most widely planted garden fruit and are easily grown. They are perennial, with plants usually persisting for about three years before they stop bearing fruit. The time of bearing tends to follow one of three patterns. June-bearing types produce a single large crop in late spring; ever-bearing varieties produce two main crops in spring and early fall, with little production in between; day-neutral varieties bear more consistently throughout the season. The performance of all strawberry varieties can differ in response to altitude and light intensity.

Strawberries are almost always started from rooted runners, although seed of some types is available. Strawberries require at least 8 hours of full sun, so plantings should be made in an open area, preferably with a south-facing exposure. They are fairly tolerant of various soils, but heavy, poorly drained soils can cause problems with root decay.

Although fairly cold-tolerant, strawberries should be protected from winter injury due to drying and freeze/thaw cycles. After soils have cooled in late fall, a mulch of straw or other nonmatting material should cover the plants. Early blossoms, particularly of June-bearing varieties, are susceptible to freezing injury. Delaying removal of the mulch until new growth has started can give maximum protection.

*Strawberries*

Strawberries have few diseases and insect pests. However, the ripe fruits are attractive to millipedes, slugs—and the occasional two-legged visitor.

***Common Disorders of Strawberries***

Whole plant affected
- Plants die back, cottony masses on roots
  - Root aphid
- Plants die back, roots decayed
  - Red stele root rot
  - Black root rot
- Plants fail to survive winter
  - Winter drying
  - Frost heaving (freeze/thaw) injury to roots

Affecting leaves
- Small dark spots on leaves
  - Leaf spot fungi (179)
- Leaves with powdery white covering
  - Powdery mildew (181)
- Leaves with yellow and green mottling
  - Viral disease (171)
- New growth small, crown area with tunneled streaks
  - Strawberry crownminer (99)
- Leaves chewed and curled
  - Leafrollers (121)
- New leaves curled, not crinkled
  - Cyclamen mite

Affecting fruit
- Fruit with small areas of tightly packed seeds
  - Cold injury (199)
  - Plant bugs (147)
- Fruit rotten
  - Botrytis (180)
- Fruit tastes moldy
  - Botrytis (180)
- Fruit with large areas chewed
  - Slugs (166)
  - Rabbits (226)
- Black beetles in ripe fruit
  - Sap beetles (103)
- White worms in fruit
  - Millipedes (165)

# Flowers and Herbs

**Common Disorders of Flowers and Herbs**

Whole plant affected
- Wilting
  - Fusarium wilt (178)

Affecting leaves
- New leaves curled
  - Cold injury (199)
  - Onion thrips (136)
  - Herbicide (2,4-D, etc.) injury (200)
  - Aphids (153, 154, 155)
- Leaves chewed
  - Grasshoppers (139)
  - Slugs (166)
  - Cutworms (115, 116, 117)
  - Hollyhock weevil (hollyhock only) (98)
  - Parsleyworm (parsley, dill, fennel only) (107)
  - Variegated fritillary (109)
  - Thistle caterpillar (108)
  - Flea beetles (90)
- Leaves with dark spotting
  - Leafminers (133)
  - Rust (189, 190)
  - Early blight (Alternaria) (179)
- Orange-red patches on lower leaf surface
  - Rust (189, 190)
- Leaves with white coating
  - Powdery mildew (181)
- Leaves generally yellow
  - Nitrogen deficiency (201)
  - Iron deficiency (202)
  - Greenhouse whitefly (152)
  - Aphids (153, 154, 155)
- Leaves with white flecks
  - Leafhoppers (147)
  - Spider mites (163)
  - Gladiolus thrips (gladiolus only) (138)
  - Onion thrips (136)
- Leaves distorted, puckered
  - Aphids (153, 154, 155)
  - Aster yellows (196)

Affecting flowers
- Flower head distorted
  - Aphids (154, 155)
  - Aster yellows (196)
- Flower buds chewed
  - Tobacco (geranium) budworm (111)
  - Hollyhock weevil (hollyhock only) (98)
- Flower petals chewed
  - Tobacco (geranium) budworm (111)
  - Earwigs (141)
  - Grasshoppers (139)
  - Blister beetles (102)
  - Bumble flower beetle (104)
  - Hollyhock weevil (hollyhock only) (98)
- Flowers speckled, deformed
  - Flower thrips (136)
  - Gladiolus thrips (gladiolus only) (138)
  - Onion thrips (136)
- Flower buds, petals soft and rotten
  - Botrytis (180)

# Bulbs and Irises

**Common Disorders of Bulbs and Irises**

- Flower bud emerges, fails to flower
  - Cold injury (199)
- Plantings produce small plants, flowers
  - Narcissus bulb fly (135)
  - Overcrowding
- Bulbs decay, die out
  - Narcissus bulb fly (135)
  - Freezing damage (199)
- Iris corms decay, die out
  - Iris borer
- Bulbs dug up from soil
  - Squirrels 224)
  - Voles (223)

# Roses

Roses are the most popular garden flower, and a tremendous number of types are offered through the nursery trade. Hybrid tea, miniature, floribunda, grandiflora, climber, polyantha and shrub classes are among those most adapted to the West.

Roses are best planted in sunny locations in early spring. Plantings should have good air circulation to reduce disease problems but ideally have protection from winds, which can damage and dry out plants. Almost all roses are grafted onto a rootstock. The bud union should be placed at, or slightly below, the soil surface.

For best flower production, roses need regular pruning. Spent blossoms should be removed to promote production on new flowers. Pruning is also useful to remove dead or diseased wood and to help shape the plant so that it gets good light penetration into the interior. Typically, the canes of hybrid teas, grandifloras and floribundas suffer a lot of winter injury and should be pruned back to living wood in early spring. (Living wood has green inner bark and a white center pith.)

Roses need fertilization to maintain good flower production. Fertilizer blends should contain fairly balanced amounts of phosphorus in relation to nitrogen. Early and mid-season fertilization is a common practice. Late-season fertilization should be avoided to prevent succulent new growth, which is not highly winter-hardy, from forming.

Winter protection and care are needed, particularly for the less hardy hybrid teas and grandifloras. Most critical is protection of the bud union, since plants that die back to this point will produce canes from the rootstock rather than from the desired variety. During winter, the bud union should be covered with insulating materials, such as loose soil or straw. Periodic watering during winter is also useful to prevent drying.

Roses are typically pest-intensive, with spider mites and powdery mildew the most common problems.

***Common Disorders of Roses***

Affecting leaves
- Powdery covering on leaves
  - Powdery mildew (183)
- Leaves with white spotting
  - Rose leafhopper (148)
  - Twospotted spider mite (163)
  - Rust (189, 190)
- Leaves with yellow or mottled colors
  - Viral diseases (171)
- Orange-red patches on lower leaf surface
  - Rust (189, 190)
- Top of leaf chewed
  - Roseslug (129)
- Even, semicircular cuts in leaf edge
  - Leafcutter bees (126)

Affecting canes
- Rose canes tunneled
  - Small carpenter bee (126)
  - Stemboring sawfly (127)
  - Bronze cane borer/rose stem girdler (100)
- Mossy or ball-like growths on stem
  - Gall wasps (127)
- Woody, tumorlike growth, usually on lower stem
  - Crown gall (192)

Affecting flowers
- Flower buds killed, fail to emerge ("blind" shoots)
  - Rose midge (135)
  - Fluctuating, cool temperatures
  - Rose curculio (97)
- Flower petals scarred
  - Flower thrips (136)
- Flower petals chewed
  - Earwigs (141)
  - Rose curculio (97)
  - Speckled green fruitworm (118)
- Flowers produced are "off-type"
  - Die back of canes to wild-type rootstock

# *What's Been Chewing in My Garden?*

After returning from a short vacation or an overnight trip, you may find that your garden has taken a change for the worse. Fruit may be harvested and plants damaged or even destroyed in a very short time. Although it may be of little solace, you can often determine the culprit.

**Cutworms.** Plants are damaged at night. Leaves or developing fruit may be chewed. Whole plants may disappear or be chewed near the soil line. The caterpillars are difficult to detect, but spend the day under cover near the damaged plants. Damage is limited almost entirely to spring.

**Caterpillars (cabbageworms, hornworms, etc.).** Irregular areas are chewed out of leaves and developing fruit. Often, feeding injuries are concentrated along leaf margins. Severe damage is preceded by "windowpane" feeding where upper leaf surface remains after lower leaf surface is chewed. Damage can occur on many types of plants during the growing season and in bursts as caterpillars become large and feed more. The caterpillars often feed during the day and the insects, or their droppings, may be readily detected.

**Colorado potato beetle.** Leaves of potatoes, eggplant, nightshade and related plants are irregular along the edge. New growth is favored by immature stages, which leave dark, smeary droppings on leaves. Adult beetles chew notches along the leaf edge. Feeding occurs during the day and the insect may be detected upon plant inspection.

**Grasshoppers.** Irregular chewing occurs along leaf edges. Bark of twigs may also be chewed, causing dieback. Damage is most severe in late summer. Often, a wide variety of plants are affected and the grasshoppers are easily detected.

**Flea beetles.** Small pits or shothole wounds are chewed in the leaves. Small seedlings can be killed when severely damaged. Damage is usually limited to a few types of closely related plants. The beetles are active and present on the plants during the day, but jump readily when disturbed. Damage is most severe during May and June, but can occur later.

**Mexican bean beetle.** Bean leaves are skeletonized, leaving the main veins. Feeding damage is concentrated on the underside of the leaf. Small yellow specks are frequently left behind. Adult or larval stages of the insect are usually detectable on the leaves.

**Slug sawflies (pear slug, rose slug).** Leaves are skeletonized, leaving the main veins. Feeding damage is concentrated on the upper surface of the leaf. Damage occurs during the day, peaking in mid to late summer. The insects are often present.

**Slugs.** Leaves are irregularly skeletonized, fruit may be chewed or tunneled and small seedlings can be destroyed. The presence of mucous slime trails is distinctive of slug injury. Damage occurs at

*Gray Garden Slug*

*European Earwig*

night, and slugs are difficult to detect during the day. Peak injury occurs in spring and early fall.

**Earwigs.** Soft plant parts are eaten, with preference given to materials such as flower petals and corn silk, which are hiding places used by earwigs. Feeding is concentrated along the edge of leaves or flowers and is usually shallower than feeding by grasshoppers or caterpillars. Damage occurs at night, with earwigs hiding in tight, dark places during the day.

**Raccoons.** Ripening fruit is particularly favored, including sweet corn, melons, cherries and grapes. Plant injury due to ripping claws (shredded corn husks, scooped-out fruit) is characteristic. Digging injuries may also occur. However, feeding on small fruit may show little evidence of other plant injury. Damage occurs at night, sometimes rapidly.

**Rabbits.** Green plants are cut at an oblique (about 45°) angle and damage can occur rapidly. In winter, lower branches or trunks may be gnawed in the region at or just above snow cover. Damage usually occurs at night.

**Birds.** Damage is limited largely to ripening fruit. In addition to fruit being removed and dropped, pecking injuries on some of the remaining fruit are characteristic. Damage occurs during the day but may be sudden when migrating flocks move to the plant.

**Deer.** Chewing injuries are extensive and plants often show distinctive shredding. During the growing season tender tips of sweet corn, heads of broccoli and cauliflower and other garden plants may be preferred. In winter, twigs are browsed up to a height of about 5 feet. Feeding occurs mainly after dusk, but the presence of tracks is often characteristic even if the animals are not observed.

CHAPTER 6

# Management of Common Garden "Bugs"

Bugs in the garden, bugging the plants. Although the word *bug* shares its origins with the bogeyman, the scarecrow and the bugbear, the twentieth-century bug has come to have many different definitions. Technically the word best applies to a group of insects, the Hemiptera or "true bugs," such as box elder bugs, squash bugs or bedbugs. (The idea that "all bugs are insects but not all insects are bugs" is commonly taught, with little permanent effect, in an introductory entomology class.) However, "bug" is more popularly used to describe any "wee little creature," such as the arthropods.

In terms of numbers and diversity, the joint-footed animals known as arthropods (Arthropoda phylum) are overwhelmingly the dominant form of animal life on earth. No other group of animals or plants comes close in diversity and number of species. Over 80 percent of all animals are arthropods (close to two million species), and over 80 percent of arthropods are insects. Mites, millipedes, pillbugs, lobsters and crabs are some of the other non-insect arthropods. While we don't have much problem with lobsters or crabs, several arthropods are serious pests. Others are highly beneficial by pollinating plants, recycling nutrients, destroying pests and quietly serving other important roles in nature.

Arthropods are distinguished by a unique set of characteristics that separate them from all other animals. These include a segmented body and external skeleton (exoskeleton), periodic shedding and replacement of the exoskeleton during growth (molting), and jointed appendages (legs, antennae, feet). Arthropods also have characteristic internal structures, such as a tubelike heart that runs down the back and a nerve cord that runs along the lower body.

## Insects

The insects are the largest group of arthropods and can be identified by means of certain additional characteristics that separate them from other arthropods. These include three body regions (head, thorax, abdomen), three pairs of legs and, often, a winged adult stage.

The various groups of insects are then recognized by such features as types of wings, mouthparts and changes they undergo during development (metamorphosis). Over twenty-five groups (orders) of insects exist, including the following common garden visitors.

- Moths and butterflies (Lepidoptera, the scale-winged insects)
- Beetles (Coleoptera, the hardened-winged insects)
- Flies (Diptera, the two-winged insects)
- Lacewings (Neuroptera, the nerve-winged insects)
- Bees, wasps, sawflies and ants (Hymenoptera)
- Grasshoppers, crickets and their kin (Orthoptera)
- Earwigs (Dermaptera)
- Aphids, scales, whiteflies and related insects (Homoptera)

- "True" bugs (Hemiptera, the half-winged insects)
- Thrips (Thysanoptera, the fringe-winged insects)

Wing types vary greatly among the insects. For example, beetles have a hardened pair of front wings, while moths and butterflies have wings covered with scales. Flies and mosquitoes have only a

## *Metamorphosis—"Change in Form"*

All arthropods, including insects, undergo changes in form as they grow and develop. This process is called *metamorphosis*. The forms arthropods take may vary dramatically as they develop. Two patterns of metamorphosis predominate: *simple* (or gradual) metamorphosis and *complete* metamorphosis.

Aphids, leafhoppers, earwigs, grasshoppers and true bugs are among the insects that undergo *simple metamorphosis*. Emerging from eggs, the immature stages of these insects are called *nymphs*. As the nymphs feed and grow, they periodically shed their skins and transform to the next stage, a process known as *molting*. Typically, insects go through three to seven molts before they are full-grown. In the final molt they transform to the adult stage. Both nymphs and adults of insects with simple metamorphosis feed in the same manner, are often found together and resemble each other. However, adult insects are sexually mature and often winged.

Moths, butterflies, beetles, flies, ants and lacewings are among the insects that undergo *complete metamorphosis*. Following egg hatch, they enter the immature stage as *larvae*. Frequently, it is the larval stage of these insects that causes the most injury to plants or animals, since they must feed heavily in order to develop. As they grow, the larvae molt and shed their skin repeatedly. After reaching full development, larvae shed their last larval skin and transform into a unique developmental stage, the *pupa*. Although insects in the pupal stage do not feed and appear to be inactive, dramatic changes in form occur that result in the adult stage. Among the insects with complete metamorphosis, larvae and adults may look strikingly different and may have very different feeding and behavioral habits.

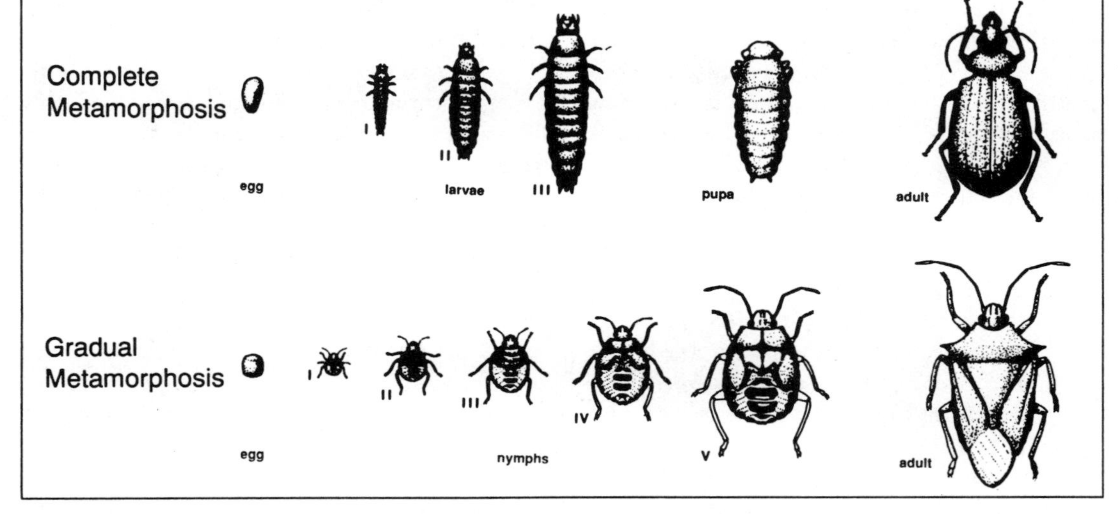

single pair of wings, in contrast to the two pairs of other insects, and some insects never produce wings. Mouthparts of insects are generally designed to either cut and chew or to suck fluids.

Grasshoppers, beetles, caterpillars, bees and wasps are typical of insects with chewing mouthparts. With these insects, a hardened pair of mandibles that work sideways is most important in grinding or cutting. A second set of mouthpart structures, the maxillae, also help to crush or handle food. In some insects, such as predatory green lacewings, the mouthparts form a sickle to pin and capture prey.

Insects with sucking mouthparts include aphids, leafhoppers, the true bugs, certain flies, moths and thrips. In most of these insects, the mouthparts (mandibles and maxillae) are greatly elongated and form a tube. These mouthparts allow them to suck nectar or remove fluids such as sap. The thrips, however, have a rasping-sucking mouth that allows them to puncture and drink the contents of the top layers of cells. Many of the insects with sucking mouthparts are important pests since they can transmit disease organisms to plants or inject toxic substances during feeding.

## Arachnids

Coming across mites, ticks, spiders and other "bugs," the gardener may consider them another type of insect. Actually, these are a separate group of arthropods, the *arachnids*, with very different habits and forms. The key feature of arachnids is their four pairs of legs. Most arachnids also have two distinctive body regions (cephalothorax and abdomen), although this is not very distinctive in the arachnids that are generally round in body shape, such as daddy longlegs.

The most damaging arachnids in a garden are the spider mites. Spider mites produce eggs. These hatch, often within a few days, and a very small larva emerges with only three pairs of legs. The larvae feed, later molt (to the nymph stage) and have the typical fourth pair of legs from that point on. The nymph feeds, grows and molts repeatedly until it reaches the mature adult stage.

## Other Common Arthropods

Other types of arthropods occasionally appear as pests in the yard and garden. *Sowbugs* and *pillbugs* (roly-polys) are land-adapted crustaceans, related to the shrimp and lobsters found in water. Sowbugs and pillbugs feed primarily on decaying plant materials, but can also damage seedlings and soft plant materials, such as strawberries, that lie on the ground. *Millipedes* (diplopods), the "1,000-leggers" of the garden, are another group of arthropods that feed on soft and decaying plant materials. *Centipedes* (chilopods), or "100-leggers," are also commonly found under rocks and debris. At night they actively hunt and feed on insects or other small arthropods.

## Slugs and Snails

Most gardeners sooner or later encounter a slug in the lettuce patch or gnawing on a precious tomato plant. Slugs, and the less common snails, are animals known as gastropods. As such, they are more similar to clams and mussels than to the other "bugs," such as insects and spider mites. Controls used for slugs and snails, including pesticides, are very different from those used for arthropods.

Slugs and most snails require moist conditions to survive and develop. Because of this limitation, problems with slugs and snails are relatively infrequent in the region, in contrast to more humid areas of the country. However, problems can be severe in some environments, such as greenhouses or lush garden areas. Snails and slugs can easily move to new locations on infested plant materials, the means by which all western garden slugs have spread to gardens.

Snails and slugs reproduce by eggs, which are round and clear, laid in small masses in soil cracks and around the base of plants. Eggs hatch in a few weeks and the young snails and slugs begin to eat.

Snails and slugs can live a year or more, but usually become semidormant when temperatures get too hot or cold.

# *Coleoptera (Beetles)*

## Mexican Bean Beetle (*Epilachna varivestis*)

**Damage:** The Mexican bean beetle is the most important insect pest of beans. Both the adult beetles and the spiny, yellow grub stage feed on the leaves. Typical damage is restricted to soft tissues on the leaf underside. Larger veins are avoided, producing a distinctive skeletonizing injury. Beans are quite resistant to defoliation injuries and often sustain little yield loss from damage. However, when bean beetles are numerous, loss of leaves during flowering and seed set can reduce bean yields and kill plants prematurely. Both adults and larvae will chew tender pods.

*Mexican bean beetle*

**Life History and Habits:** Mexican bean beetles overwinter in the adult stage in protected areas along the edges of plantings infested during the previous season. They straggle out from the overwinter sites over a period of several weeks in late spring. The beetles feed on the leaves, and the females begin to lay egg masses on the leaf underside. The orange-yellow larvae that emerge feed heavily for about a month before becoming full-grown. They then pupate on the underside of the leaf, often in small groups. One to two generations occur annually, depending on temperatures during development.

**Natural and Biological Controls:** Natural controls, such as predatory bugs, are important and usually reduce populations to nondamaging levels. These controls can be supplemented by annual releases of the parasitic wasp *Pediobius faveolatus* (see page 28). This wasp kills bean beetle larvae and has been an effective biological control in some areas.

**Cultural Controls:** Most damage occurs in July and August, so early planting can avoid most injury. Short-season varieties may prevent the buildup of a large population. Thus, most bush beans are less susceptible to damage than pole beans. After harvest, plants should be destroyed as soon as possible to kill remaining insect stages. Throughout the growing season, plants can tolerate some injury without yield losses. Beans are most sensitive to the effects of defoliation injuries when pods are being set.

**Mechanical Controls:** Handpick adults and crush eggs during plant inspections.

**Chemical Controls:** Adults and larvae can be controlled by some garden insecticides. However, many gardeners report that Mexican bean beetles appear to be quite resistant to insecticides. Younger slugs are most easily controlled.

**Miscellaneous Notes:** The Mexican bean beetle is a true lady beetle (Coccinellidae family), the only species of lady beetle in the region that feeds on plants. The adult can be separated from the beneficial lady beetles by its large size, yellow-orange color and numerous spots. Immature stages of the Mexican bean beetle are yellow and spiny, very different from the lady beetles, which feed on insects. The best way to separate "good" from "bad" in the lady beetle clan is to note which plant they have settled on. If it is a bean plant, be suspicious. If not, the lady beetles are probably beneficial insect predators.

# Flea Beetles

Flea beetles are common pests of almost all vegetable and flower crops. The adult beetles are usually small and shiny with very large rear legs that enable them to jump like a flea when disturbed.

About a dozen species of flea beetles in the West may damage garden plants. (Many other species also exist, but cause little injury to trees, weeds and other plants.) Although there is some sharing of tastes, each type of flea beetle has a decided preference for certain plants. For example, some flea beetles only feed on potatoes, eggplants and other members of the nightshade family (Solanaceae); others have a taste for broccoli, cabbage and other cole crops (Cruciferae).

Adult flea beetles are usually the most damaging stage to plants. The beetles chew small pits, or shothole wounds, in leaves. Small plants may be killed or stunted by these wounds. Wounds may also damage crops grown for greens.

The immature larval stage of flea beetles is rarely seen because it usually occurs underground. The larvae are small and wormlike, and most feed on plant roots. (A few, such as the grape flea beetle, chew and skeletonize leaves.) Larvae of flea beetles generally cause no significant plant injury. However, some species that feed on potatoes and other nightshade family plants can tunnel potato tubers. These produce fine, dark lines in the tubers, which may extend about 1/2 inch deep. They also feed on the tuber surface, leaving bumps and small scars. Tuber-rotting organisms sometimes invade through these wounds.

*Potato flea beetle*

| COMMON REGIONAL FLEA BEETLES AND THEIR HOST PLANTS | |
|---|---|
| **Common name and scientific name** | **Host plants** |
| Potato flea beetle (*Epitrix cucumeris*) | Tomato, potato and other nightshade family plants (Solanaceae family) |
| Tobacco flea beetle (*Epitrix hirtipennis*) | Tomato, potato and other nightshade family plants (Solanaceae family) |
| *Epitrix parvula* (no common name) | Tomato, potato and other nightshade family plants (Solanaceae family) |
| Tuber flea beetle (*Epitrix tuberis*) | Potatoes (damages tubers) |
| Eggplant flea beetle (*Epitrix fuscula*) | Eggplant and other nightshade family plants (Solanaceae family) |

| Common name and scientific name | Host plants |
|---|---|
| Western cabbage flea beetle (*Phyllotreta pusilla)* | Wide host range, primarily of cabbage family plants (Cruciferae family) |
| Palestriped flea beetle (*Systena blanda*) | Very wide host range including squash family (Cucurbitaceae family), beans, corn, sunflowers, lettuce, potatoes and many weeds |
| *Spinach flea beetle (*Disonycha xanthomelas*) | Pigweed, spinach and related plants |
| *Grape flea beetle (*Altica chalybea*) | Grape, Virginia creeper, elm and some fruit trees |
| *Apple flea beetle (*Haltica foliaceae)* | Apples |
| *Currant flea beetle (*Blepharida rhois)* | Currants, sumac |

*Larvae feed on leaves, rather than plant roots.

## Western Cabbage Flea Beetle (*Phyllotreta pusilla*)

**Damage:** Adult beetles feed on the foliage of many kinds of garden plants, causing small shothole feeding wounds. Mustard family plants are particularly favored, but damage can also occur to a variety of other plants, such as beets and lettuce. Seedling plants are particularly susceptible, and flea beetles can kill or retard plantings during establishment. Later in the season, shothole feeding by flea beetles can damage leafy vegetables, such as mustard and Chinese cabbage, prior to harvest. Sometimes red or brown areas develop around feeding wounds. Although flavor and nutritional value are not affected by this injury, the damaged leaves may be considered unattractive.

*Western cabbage flea beetle*

**Life History and Habits:** Western cabbage flea beetles overwinter in the adult stage in protected areas in and around garden plantings. They become active as temperatures warm in early spring and fly or jump to feed on host plants, often feeding first on wild mustards before becoming attracted to garden plants. Eggs are laid in soil cracks around the base of plants. The small, wormlike larvae feed on root hairs but do not appear to cause serious damage to garden plants. They then pupate in the root area, later emerging as beetles, which repeat the cycle. There are typically three generations per year at approximately 45-day intervals.

**Natural and Biological Controls:** Although the number of western cabbage flea beetles changes greatly from season to season, little is known about the specific natural controls that affect these insects. A native species of nematode is one naturally occurring biological control. Parasitic wasps attack the adults.

**Cultural Controls:** Since seedlings are at particular risk, using transplants or planting seeds in a well-prepared seedbed will hasten seedling growth and allow them to overcome insect injury. When possible, plantings in gardens may also use high seeding rates (with later thinning) or use susceptible trap crops, such as radish, daikon or Chinese cabbage, to divert insect attack and allow establishment of the main crop. Weeds in gardens that are also fed upon by flea beetles may be retained to further protect the crop. These weeds should be destroyed before they compete with garden plants or produce flowers. Because the adult beetles fly and feed on many different wild plants, crop rotations often have little effect on control of this insect.

**Mechanical and Physical Controls:** Most flea beetles overwinter under debris outside the garden. When this occurs, floating row covers or other screening can exclude the beetles during seedling establishment.

**Chemical Controls:** Control is usually needed only during plant establishment, although mature plantings of leafy crucifers can also be damaged. However, since beetles emerge from winter cover for several weeks, treatments need to be repeatedly applied. When flea beetle numbers are extremely high, insecticides cannot provide adequate protection of seedling plants.

## Palestriped Flea Beetle (*Systena blanda*)

**Damage:** Adult beetles chew small pits in the upper surface of leaves of many garden vegetables. Beans, beets, melons and sunflowers are among the plants most commonly damaged, but potatoes, squash, tomatoes, lettuce, and several common weeds (e.g., lambsquarters, redroot pigweed) also can be host plants. When numerous, palestriped flea beetles can kill seedlings and retard development of surviving plants. The larvae, which feed on plant roots, occasionally damage germinating seeds that are planted late. They have also been reported to tunnel sweet potato tubers.

*Palestriped flea beetle and inury*

**Life History and Habits:** The life history of the palestriped flea beetle has been little studied. Overwintering stages probably consist of a mixture of nearly full-grown larvae, pupae and perhaps some adult beetles. Development continues in spring, and beetles become active in late spring. Adult beetles can fly and feed for several weeks on leaves. Eggs are laid in soil cracks during early summer. Eggs hatch in about 10 to 15 days. Larvae feed on germinating seeds and plant roots, causing some injury to young plants. They pupate in the soil, and a second generation of adult beetles is active in late July and August.

**Cultural Controls:** Cultural conditions that promote vigorous seedling growth can allow plants to outgrow adult palestriped flea beetle injuries. Higher seeding rates, followed by thinning, can help dilute adult beetle attacks. Tillage can kill overwintering stages of the palestriped flea beetle in the soil. Keeping weeds under control also reduces egg-laying within a planting.

**Chemical Controls:** Palestriped flea beetles are difficult to control when abundant. They are susceptible to most garden insecticides but continually reinfest plantings since their development is staggered. Sulfur-containing fungicides, such as Bordeaux mixture, are repellent to flea beetles.

## Asparagus Beetle (*Crioceris asparagi*)

**Damage:** Adult beetles lay eggs on asparagus spears and chew the spears. Larvae feed on the ferns and can reduce plant vigor. Asparagus beetles are so abundant in some locations they can kill plants. Well-established plants are little damaged by this injury. However, young plants may not tolerate these injuries without yield reduction in the subsequent year.

**Life History and Habits:** Asparagus beetles overwinter in the adult stage around asparagus plantings. They become active in spring and fly to emerging spears and mate. The beetles chew on spears and females lay upright, dark eggs on the plant. Eggs hatch in about a week and the gray, grublike larvae feed on the ferns. The larvae become full-grown in about two to three weeks and drop to the soil to pupate. Adults emerge and repeat the cycle. Two or three generations are completed in a season.

**Natural and Biological Controls:** A tiny parasitic wasp (species of *Tetrastichus*) feeds on eggs and develops within asparagus beetle larvae. The wasp is common throughout much of the region and provides adequate control of asparagus beetles where it has become established.

**Cultural Controls:** Regularly pick spears in spring to eliminate early-laid eggs. Actively growing plants are rarely injured seriously by this insect, even when feeding injuries are extensive, so proper crop control can be important in managing asparagus beetles.

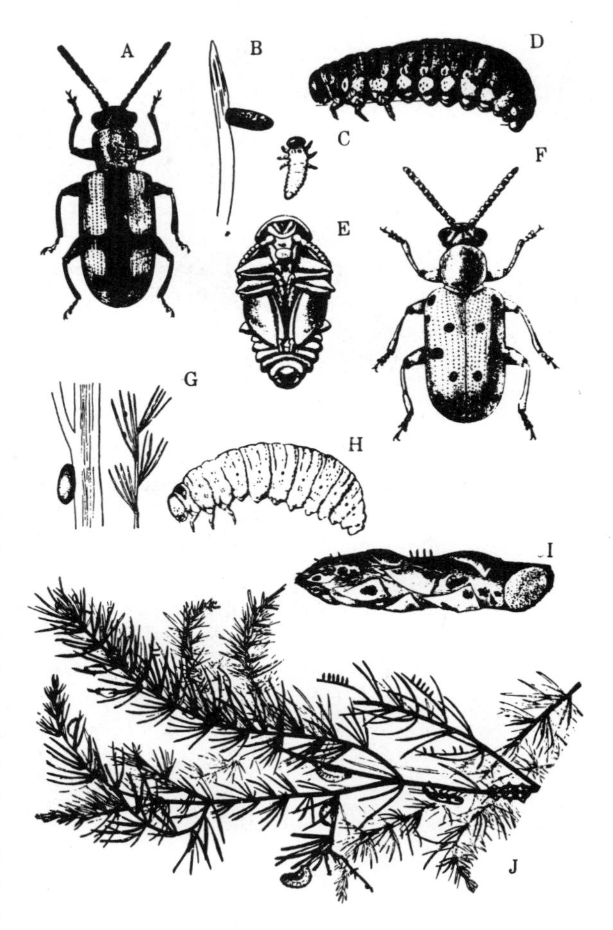


*Asparagus beetles. A-E: Asparagus beetle. A) Adult; B) Egg; C-D) Larvae; E) Pupa. F-H: Twelve-spotted asparagus beetle. F) Adult; G) Egg; H) Larva; I-J) Damage.*

**Chemical Controls:** Adult beetles and larvae are easily killed by several insecticides, although few are permitted for use on asparagus.

**Miscellaneous Notes:** The spotted asparagus beetle *Crioceris duodecimpunctata*, a bright-orange beetle with black spots, is also common on asparagus plantings. Since larvae of this beetle develop within asparagus berries and do not damage the fern, little injury is caused.

## Western Corn Rootworm (*Diabrotica virgifera*)

**Damage:** Adult beetles feed on and clip silks of corn. Although this damage is usually of little consequence, heavy infestations can occur in later plantings because beetles concentrate on these plants. Where silks are pruned severely during pollination, seed set can be poor. Beetles are also attracted to squash blossoms and can scar developing fruits. Larvae of the western corn rootworm feed on corn roots and can weaken plants, causing them to fall over.

**Life History and Habits:** Western corn rootworm beetles are present from late July through mid-September. Female beetles lay eggs during this time near the base of corn plants. Adults of both sexes may be found feeding on corn silks, tender corn leaves and squash blossoms. Eggs of the western corn rootworm hatch in June of the following year. The young, wormlike larvae require corn roots to develop and quickly starve if corn is not replanted within a few feet of where eggs are laid. If corn roots are found, the larvae feed on them, sometimes causing severe root pruning. Pupation occurs in the soil during July and adult beetles later emerge. There is one generation per year.

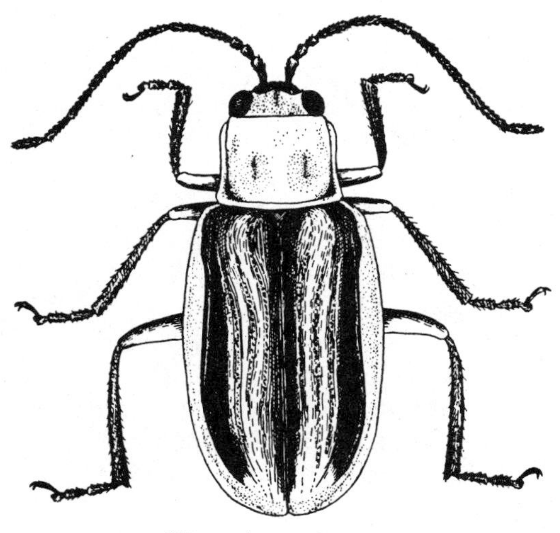
*Western corn rootworm*

**Natural and Biological Controls:** Eggs of the western corn rootworm are preyed upon by a number of soil organisms, such as predatory soil mites. Addition of organic matter, which increases populations of organisms on which predatory mites feed, has been shown to result in reduced survival of the western corn rootworm. Several trials conducted with insect-parasitic nematodes and fungi for control of corn rootworm larvae have had mixed success.

**Cultural Controls:** Crop rotation is extremely effective for control of the western corn rootworm since practically all eggs are laid around the base of corn plants in late summer. The larvae that hatch the following June starve if they are not near corn.

**Chemical Controls:** Under most conditions, corn silks readily outgrow corn rootworm feeding and can be pollinated even when some beetles are present. However, when high numbers of beetles are present (around five or more per ear tip), pollination can be disrupted, resulting in poor kernel set. Some insecticides (Sevin, rotenone) are registered for control of adult corn rootworm beetles.

**Miscellaneous Notes:** The western corn rootworm is very similar in appearance to the striped cucumber beetle, so the two species are commonly confused. However, habits differ greatly. Striped cucumber beetles are minor pests of cucumbers and melons grown in the region.

## Spotted Cucumber Beetle/ Southern Corn Rootworm (*Diabrotica undecimpunctata*)

**Damage:** The larval stages feed and tunnel the roots of many garden plants, particularly corn, bean and squash family plants. In addition, grasses and several weeds are hosts of this insect. Heavily infested plants grow poorly. However, the insect appears unable to overwinter in large numbers within the region and is not found west of the Rockies. Serious infestations are dependent on large migrations from the south, which are rare.

**Life History and Habits:** These insects overwinter in the adult beetle stage around gardens. Others migrate from southern states during the growing season. The beetles feed on developing seedlings and lay eggs in soil cracks. The emerging larvae feed on roots for one to two months. The adult beetles that emerge feed on pollen, corn silks and other plant parts, causing little injury. There is one generation per season.

**Cultural Controls:** Heaviest infestations occur in rich, moist soils with lush vegetation. Reduced watering systems, such as trickle irrigation, and good weed control should reduce infestations.

**Chemical Controls:** Treatments of soil insecticides at planting (e.g., diazinon) can control larvae if egg-laying occurs within a few weeks of planting. Insecticides applied to the leaves can control adults, although these treatments must be repeated because beetles migrate into gardens for several weeks.

## Striped Cucumber Beetle (*Acalymma vittata*)

**Damage:** The main damage is that adult beetles can transmit bacteria (*Erwinia tracheiphila*), which produce bacterial wilt in cucurbits (see page 194). Beetles can also transmit cucumber mosaic virus (see page 172). Infrequently, beetles may become abundant enough to cause serious chewing injury to seedlings or ripening fruit of cucumbers, melons or squash. Larvae develop on the roots of squash family plants. Root damage does not appear to be serious, although larvae may move into ripening fruit on the soil surface.

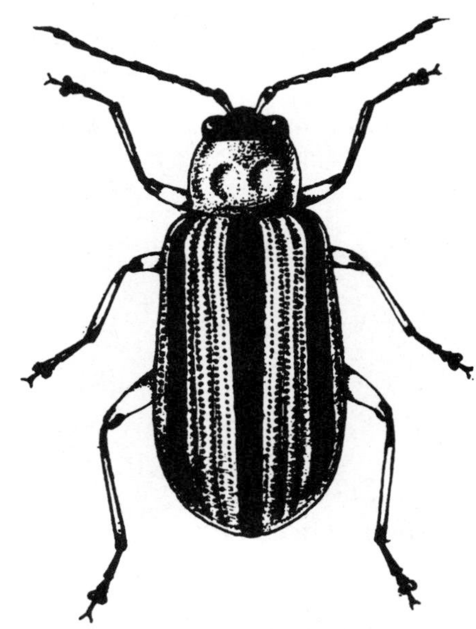

*Striped cucumber beetle*

**Life History and Habits:** Adult beetles overwinter around field edges. (Beetles that have previously fed on plants infected with bacterial wilt also harbor the bacteria in their digestive system.) As temperatures warm in spring, they become active and feed on leaves and flowers of several different trees and shrubs. As squash plants emerge, the beetles move to the crop and chew seedlings. Seedling damage can occur at this time, and plants may become infected with the bacteria. Eggs are laid in cracks around the base of the plants, and the hatching larvae feed on the roots for about a month. They then pupate in the soil, later emerging as adult beetles. Typically, there are two generations per season.

**Cultural and Mechanical Controls:** Where problems with the cucumber beetle and bacterial wilt are severe, make a thick original planting, later thinning surviving noninfected plants to the desired population. Row covers or mesh screening can exclude beetles and protect seedlings. Injury to fruits by tunneling of larvae occurs in very moist soil as fruits ripen. Limit water during this period to prevent this damage. Mulches or some other barrier slipped under the fruit will also prevent feeding.

**Miscellaneous Notes:** The striped cucumber beetle is commonly confused with the western corn rootworm. In fact, they may be found feeding together in late summer on blossoms of squash, melons or cucumbers. The striped cucumber beetle has straight (rather than wavy) dark stripes, is brighter yellow and has rows of small depressions on its wings. Another insect that is similar in appearance is the threelined potato beetle; the larvae of this insect feed on the leaves of tomatillo.

## Colorado Potato Beetle (*Leptinotarsa decemlineata*)

**Damage:** Adult beetles and the larvae chew on the foliage of potatoes, eggplant and tomatoes, reducing yield.

**Life History and Habits:** Colorado potato beetles overwinter in the adult stage under cover near plantings infested the previous season. They are capable of flying, and move back to fields in late spring as potatoes and other susceptible plants emerge. Eggs are laid during June, and the orange-red larvae feed on leaves for several weeks. After becoming full-grown, they drop from the plant and pupate in the soil. In about two weeks, adult beetles emerge and feed on plants. Beetles present early in the season may lay several egg masses, repeating the cycle. Later-emerging beetles feed for a few weeks and disperse to overwintering shelters without producing during the season.

**Natural and Biological Controls:** During mid-season, potato beetle eggs are often destroyed by ground beetles, and surviving larvae are prey for the twospotted stink bug *Perillus bioculatus*. The tachinid fly is also a common parasite of Colorado potato beetle larvae. New strains of *Bacillus thuringiensis* (*san diego* and *tenebrionis* strains) can control young larvae. Damage by the second generation of the beetle is dependent upon warm early-season temperatures, since in cooler years most beetles that develop during the first generation will not lay many eggs. Increasing numbers of beetles will lay eggs during the second generation if warm temperatures prevail early in the growing season.

**Cultural Controls:** Some hairy-leaved varieties of eggplant (e.g., Dusky Hybrid) are less preferred by the Colorado potato beetle. Early-maturing potato plantings can avoid much of the damage. It is sometimes possible to use existing infestations of hairy nightshade, a common weed, as a trap crop because this plant is highly preferred by the beetles. Collect and destroy beetles that feed on these trap crops.

**Mechanical Controls:** In small plantings, hand-pick adults and crush eggs.

**Chemical Controls:** Plants can tolerate low levels of leaf loss (at least 25 percent) with little or no yield reduction. Protection of plants is most criti-

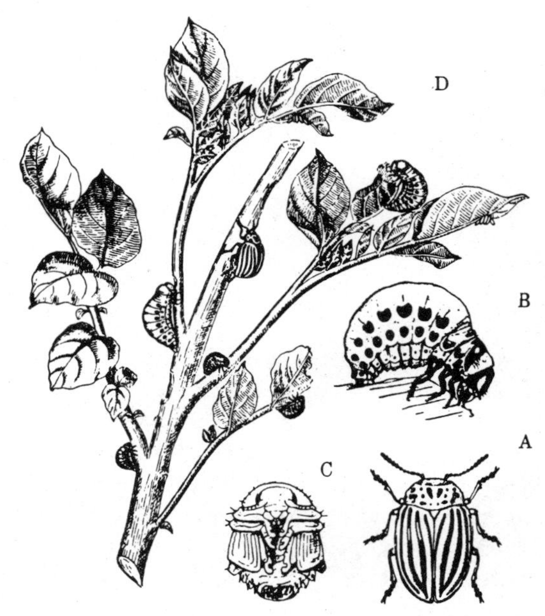


*Colorado potato beetle. A) Adult; B) Larva; C) Pupa; D) Damage.*

cal as new tubers are forming. Most insecticides registered for use on potatoes are effective for control of the Colorado potato beetle in the West. However, this insect has become extremely resistant to insecticides in the eastern United States.

**Miscellaneous Notes:** The Colorado potato beetle is a very famous insect that has spread widely and devastates crops through eastern North America and most of Europe. The name of a western state was attached to this species since it was native to the region and thought to have been first described in Colorado. However, a preexisting description was recently discovered in Iowa, suggesting that it should be called the "Iowa potato beetle."

**Related Insects:** The threelined potato beetle (*Lema trilineata*) feeds on the leaves of tomatillo. Adults are striped and resemble the striped cucumber beetle. Larvae are gray and cover themselves with excrement.

## Rose Curculio (*Merhynchites bicolor*)

**Damage:** The rose curculio damages roses by making feeding punctures into flower buds, resulting in ragged flowers. During periods when the buds are not common, feeding occurs on the tips of shoots, killing or distorting the shoot. The bent neck condition of roses can be caused by this insect.

*Rose curculio*

**Life History and Habits:** The adult stage of the rose curculio is a red snout beetle (weevil) with a black beak. They become active in late spring and lay eggs in developing flowers. The larval (grub) stage feeds on reproductive parts of the flower. Blossoms on the plant, including those clipped off by the gardener, are suitable places for the insect to develop. When full-grown, the grubs fall to the soil and pupate. There is one generation per year.

**Mechanical Controls:** Regular handpicking and removal of spent blossoms will prevent populations from developing. However, other brambles are hosts and can serve as additional sources of the insect. Adult weevils drop readily from plants and feign death when disturbed. Where plantings allow, shake plants over a collecting container to speed hand collection of the rose curculio.

**Chemical Controls:** The rose curculio can be controlled with most insecticides, applied during late May and June.

## Cherry Curculio (*Tachypterellus consors ssp. cerasi*)

**Damage:** Early in the season the adults chew small holes in the base of flowers and cause abortion of developing fruit. The skin of older fruit is pitted by this injury. Eggs are inserted into fruit and larvae tunnel the fruit, ultimately feeding on the pit. Damage is limited to sour cherries and chokecherries.

**Life History and Habits:** The cherry curculio spends the winter as an adult snout beetle (weevil) around trees infested the previous

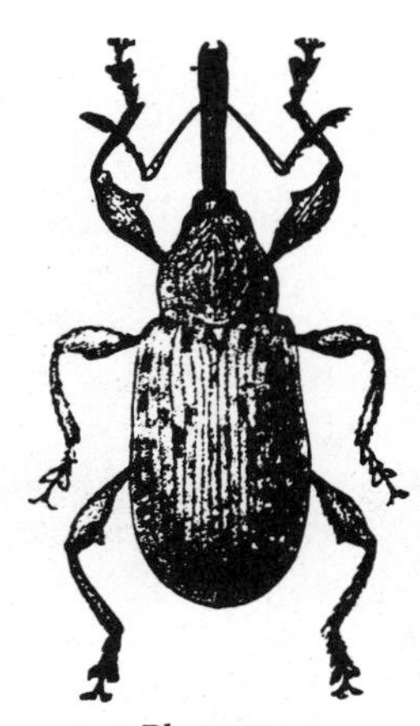

*Plum gouger*

season. They fly to trees in spring as flower buds form and feed on the buds and flowers. After several weeks, the developing fruits appear and females insert eggs into them. The young larvae feed and hollow the pit, then pupate within it, and leave as the cherries ripen. There is one generation per year.

**Natural Controls:** Several types of parasitic wasps commonly attack the cherry gouger.

**Mechanical Controls:** Adult weevils can be shaken from trees and collected on sheets. However, they are difficult to see and play dead when disturbed.

**Cultural Controls:** To prevent the insect from developing, damaged fruit should be picked and destroyed as it is observed.

**Chemical Controls:** Little work has been done on the chemical control of this insect. Effective insecticides, applied before flowering, should provide control.

**Related Species:** Related species of small weevils attack plum (plum gouger) and apple (apple curculio) but rarely cause serious injury to these trees. The plum curculio, a fruit-infesting weevil that seriously damages plums, apples and apricots in eastern states, does not occur in the West.

The apple curculio, *Tachypterellus quadrigibbus*, feeds on *Amelanchier* (serviceberry, shadbush) and apple trees. It is difficult to detect since it remains motionless, with beak raised up, when disturbed and closely resembles bud scales or dried petals. The apple curculio makes small holes in leaves and feeding punctures in fruit. Fruit injuries made in late spring heal over, resulting in dry, sunken areas on the fruit. Similar to the cherry gouger, the females excavate holes in which eggs are laid. Larvae and pupae develop only within the shed apples (growing apples kill them). Only some varieties of apples are susceptible, including Delicious. Cleaning up June drop apples can control the insect but will not eliminate spring feeding damage during the first year.

The plum gouger, *Anthonomus scutellaris*, feeds on hard types of plums. Adult weevils make numerous feeding punctures in the developing fruit, many of which result in a flow of clear ooze from the wound. Eggs are laid in some of the punctures, and the larvae develop within the pit. This species apparently spends the winter as an adult within the pit. Removal of damaged fruit should help control this species.

## Hollyhock Weevil (*Apion longirostre*)

**Damage:** The hollyhock weevil feeds on the seeds, buds and leaves of hollyhock. Small feeding punctures in developing leaves and petals give the plant a ragged appearance. The larvae consume the seeds of the plant.

**Life History and Habits:** Hollyhock weevils spend the winter either in the adult stage, in protected areas around hollyhock plantings or within seeds. They move to hollyhock plants in late spring and chew small holes in the buds of the plants. During this time they are commonly observed mating, the female being identifiable by an extremely long beak. As flower buds form in June, the females chew deep pits in the buds and lay eggs. The grub stage of the insect feeds on the developing embyro of the seed. After feeding is completed, it pupates within the seed. Adults usually emerge in August and September, but some remain within the seeds, emerging the following spring. There is one generation per year.

**Cultural Controls:** Mature seed pods should be removed as they are produced to prevent the insects from completing development.

**Mechanical Controls:** Hollyhock weevils drop readily when disturbed and can be easily collected by shaking plants over a tray. Collections should be made in spring when the weevils first move to the plant and before eggs are laid.

**Chemical Controls:** Recent information on control of this insect is lacking. Sprays are best timed for late spring before flower buds become well developed and eggs are laid.

## Strawberry Root Weevil (*Otiorhynchus ovatus*)

**Damage:** Larvae feed on larger roots of strawberries and may damage the crown. Injured plants are stunted with leaves bunched and deep-colored. Most damage occurs in spring as larvae become full-grown. Some damage can also occur in fall, particularly during unusually warm years. Adults feed on leaves and fruit. Adult feeding is much less injurious than larval feeding. However, adult strawberry root weevils are a common nuisance invader of homes during summer, particularly at higher elevations. This insect feeds on the roots of many other herbaceous plants but is rarely damaging.

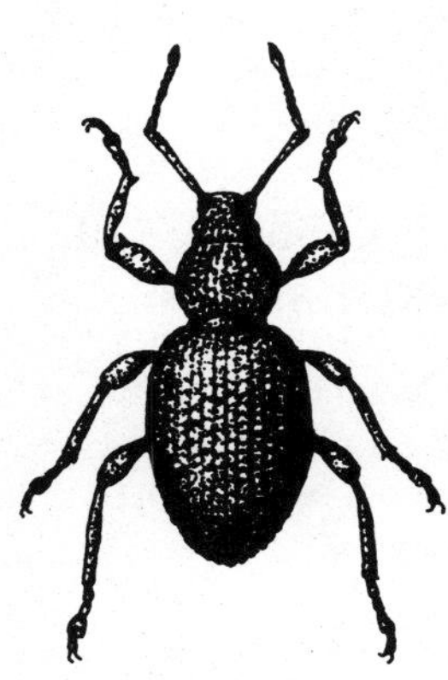

*Strawberry root weevil*

**Life History and Habits:** The strawberry root weevil survives the winter in the larval stage around the roots of plants. Some adults may also survive in protected locations. Most adults emerge during June, typically during berry-picking season. They feed about 10 to 14 days before beginning to lay eggs. Egg-laying continues throughout summer. Only females are known from this species and they reproduce asexually. The wing covers are fused and they are unable to fly.

**Cultural Controls:** Destruction of old beds can kill developing larvae. This is best done after eggs have been laid. New beds should be located as far as possible from previously infested plantings.

**Biological Controls:** Insect-parasitic nematodes have been effective for control of larvae of related root weevils. These should be applied as soil drenches and watered into the root zone.

**Chemical Controls:** Relatively little control work has been done with this insect, although information suggests it is difficult to control. Sprays of insecticides applied to kill adults should be more effective than larval treatments.

**Related Insects:** The rough strawberry root weevil, O. *rugostriatus*, and the black vine weevil, O. *sulcatus*, are also root weevils that occasionally feed on strawberries. In many areas, the latter species is also a serious pest of various shrubs such as euonymus, yew and rhododendron.

The strawberry crownminer, the larva of a small moth, tunnels strawberry crowns. It can be differentiated from the strawberry root weevil by its habit of making tunnels packed with excrement in the crown area, in contrast to the root damage of the strawberry root weevil.

## Shothole Borer (*Scolytus rugulosus*)

**Damage:** The grublike larval stage tunnels under the bark of twigs and branches of peach, plum and cherry trees. (Apple and pear are less common hosts.) This produces a girdling wound that can weaken, and sometimes kill, the plant beyond the damaged area. Trees in poor health are much more susceptible to shothole borer damage. When the adult beetles emerge through the bark, they chew small exit holes—the shotholes that are commonly observed. The branches may ooze gummy sap where the beetles make entry tunnels.

**Life History and Habits:** Shothole borers spend the winter as grublike larvae under the bark of trees. They continue to develop, pupate and change to the adult stage in late spring. The small (1/10 inch), brown adult beetles emerge in June and July. After mating, the females seek out tree branches that are in poor health and chew out a one to two-inch-long egg gallery under the bark. Eggs are laid along the gallery, and the newly hatched larvae later feed under the bark, making new galleries away from the central one. (The

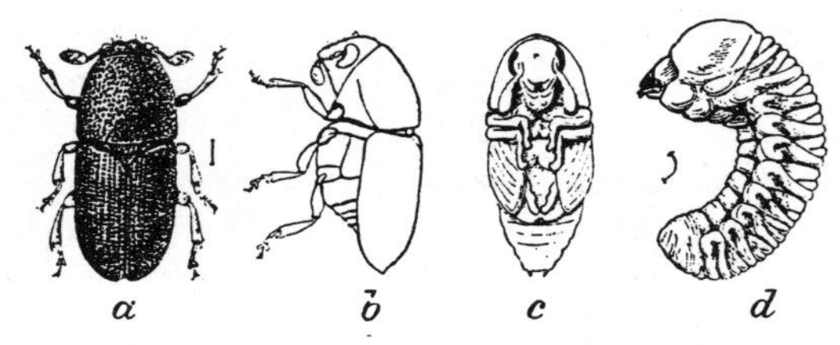


*The shothole borer. A) Adult or beetle; B) Side profile; C) Pupa; D) Larva.*

pattern made by the egg and larval galleries is useful for diagnosing shothole borer infestations.) There are usually two generations per year. The larvae become full-grown by midsummer, and a second generation is produced at that time.

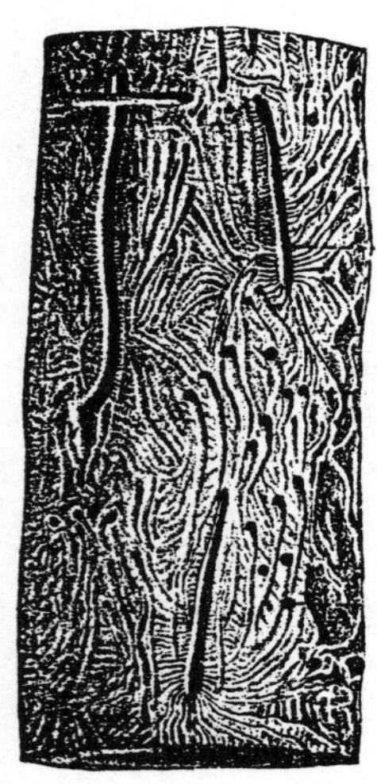

*Galleries of the shothole borer on twig under the bark*

**Cultural Controls:** Shothole borers rarely attack, and survive poorly, in trees that are actively growing. Serious damage by shothole borers can be almost entirely avoided by growing trees under favorable conditions. Trees stressed by drought, winter injury, poor site conditions, wounding injuries or other problems are at greater risk of shothole borer damage. Regularly prune dead or dying branches and limbs in which shothole borers breed. Since the insects can continue to develop in pruned wood, remove or destroy it before adult beetles emerge.

**Chemical Controls:** On nonbearing trees, some insecticides are registered that control shothole borers (and other bark beetles). These need to be applied to the branches before adult beetles have tunneled into the tree and laid eggs (early to mid-June).

## Rose Stem Girdler/Bronze Cane Borer (*Agrilus rubricola*)

**Damage:** The flathead borer stage (larva) makes meandering tunnels under the bark of rose and caneberry (raspberry, blackberry, etc.) plants. It has also been recovered from currants. Characteristically, a swollen area develops around the wounded area of the stem. Canes often die back or break at these wounded sites, sometimes several weeks after the tunneling occurs.

**Life History and Habits:** The rose stem girdler overwinters as a partially developed borer under the bark of the canes. In spring it resumes feeding and pupates within the plant. The adult stage is a bronze-colored beetle that emerges in late May and June. Eggs are laid during this period, usually near the base of leaves. The larvae that hatch tunnel under the bark and move upward in the plant. There is one generation per year.

**Cultural Controls:** Canes that show evidence of injury should be removed before mid-spring to destroy the insects before they emerge.

**Chemical Controls:** Effective chemical controls have not been demonstrated. If attempted, they are best applied in late spring to kill adult beetles before they have laid eggs. Larvae within stems cannot be controlled with insecticides.

**Related Species:** Another cane borer, *Agrilus polistas*, is common in caneberries, currants and other woody plants grown east of the Rockies. The rose stem girdler is also related to the rednecked cane borer, *Agrilus ruficollis*, an eastern species that causes similar injury to caneberries. Other common borers that attack small fruits and roses include the stemboring sawfly, raspberry crown borer and currant borer.

*Tunneling by the rose stem girdler in a raspberry cane*

## Flatheaded Apple Tree Borer (*Chrysobothris femorata*)

## Pacific Flathead Borer (*Chrysobothris mali*)

Two closely related species occur within the region. The Pacific flathead borer predominates in the West; the flatheaded apple tree borer in the East. However, their ranges overlap.

**Damage:** The immature stage (flatheaded borer) tunnels under the bark of trunks and larger branches. Large areas of bark die beyond the wounds, weakening and sometimes girdling trees. Apple is the most common fruit tree damaged by flathead borers. However, a wide variety of trees can be attacked, including pear, maple, rose, willow and sycamore. Newly transplanted trees and trees in poor health are at greatest risk of attack. Damage is generally first concentrated on the sun-exposed sides of the tree but can later expand. The meandering mines are packed with a fine sawdust, which is diagnostic for these and other flatheaded borers. A slight darkening of the bark and depressions can be external evidences of infestation.

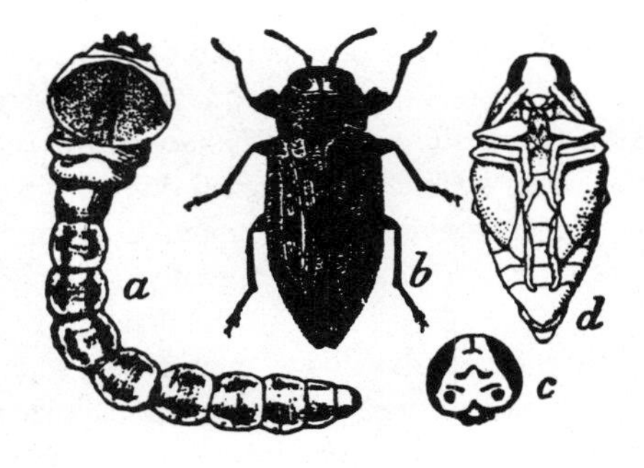


*Flatheaded apple tree borer. A) Flatheaded larvae; B) Mature beetle; C) Close-up of head; D) Pupa.*

**Life History and Habits:** The flatheaded apple tree borer and the Pacific flathead borer have similar life histories. Winter is spent in the larval (flathead borer) stage, under the bark. They continue to grow and feed during late winter and early spring, then bore into the heartwood of the trunk and pupate. The adult beetles (known as metallic wood borers) emerge between mid-May (shortly after apple bloom) and August. During this period, the individual beetles remain active for several weeks and feed on leaves. They may also be observed sunning themselves on the bark during the day. Mated females search the trees and lay eggs in cracks and crevices of large branches or trunks. Trees in poor health are preferred, and most eggs are laid near areas damaged by bruising, sunscald or other injuries. The larvae later emerge by chewing through the bottom of the egg and tunneling directly into the tree. They continue to feed for several months, becoming dormant during the cold season. There is one generation per year.

**Cultural Controls:** Attacks by flatheaded apple tree borers and Pacific flathead borers are concentrated around wounds and on trees in poor health.

Healthy trees are less attractive to egg-laying females, and the tunneling larvae often are killed by tree defenses, such as oozing sap. By maintaining trees in a vigorous condition and preventing injuries, problems with these insects can be avoided. Where sunscald injuries are likely, shade the lower trunk of young trees or use tree wrap in winter to prevent this injury. (This wrap must be removed in spring since it can attract egg-laying during the season.) Whitewashing the trunks can also reduce injury and attacks. Dying trees should be removed and newly cut wood should not be kept near susceptible trees, since large numbers of borers can develop in these.

**Mechanical Controls:** Once the borers are in the trunk, digging them out in late summer or early fall is the only control. This is difficult to do without causing additional tree injury.

**Chemical Controls:** Preventive use of insecticides, applied to trunks and branches before egg-laying, has sometimes provided good control, although few insecticides are permitted for use on bearing trees. Use of fumigants (e.g., paradichlorbenzene) has not been effective against this borer.

**Related Insects:** The roundheaded apple tree borer (*Saperda candida*) occasionally damages apple and pear trees. The larvae of this borer are usually found near the base of the tree and produce a coarse, stringy sawdust that is sometimes pushed out of the trunk. Controls are generally similar, although hand worming and use of fumigants are more successful because the larvae make an opening to the outside while tunneling. Preventive insecticide sprays, if used, should be applied somewhat later, in June through August.

## Blister Beetles (species of *Epicauta* and *Meloe*)

**Damage:** Blister beetles appear during June and early July and feed on leaves and flowers of plants, sometimes causing serious damage. Most often, injury occurs to potatoes or legumes (beans, peas, etc.), but many different garden plants can be attacked. Late in the season, flowers may be particular favorites of blister beetles, and petals are shredded by feeding. Blister beetles usually feed in groups, massing in a small area of the garden. They appear—and disappear—quite abruptly. An abundance of blister beetles often follows high populations of grasshoppers, whose eggs are a primary food source for blister beetle larvae.

**Life History and Habits:** Most blister beetles develop by feeding on grasshopper egg pods within the soil. The adult beetles lay groups of eggs in the soil during midsummer. The newly hatched larvae are highly active and burrow through the soil, seeking

*Blister beetle (Epicauta species)*

grasshopper eggs. When eggs are found, the young blister beetles feed and develop rapidly. They then form a cell in the soil and spend the winter as a fully developed larva or pupa. Adults emerge in late spring the following season. Some blister beetles feed on different insects. For example, one group of large, fat, black blister beetles known as oil beetles (species of *Meloe*) feed on ground nesting bees rather than grasshoppers.

**Mechanical Controls:** Blister beetles can be excluded from susceptible plants by shade cloth or floating row covers. The beetles can also be handpicked, but they produce a chemical that can injure tender skin.

**Chemical Controls:** Blister beetles are easily controlled with several garden insecticides (e.g., Sevin, rotenone).

**Miscellaneous Notes:** Blister beetles produce a protective toxic chemical, cantharidin, which is released when the insects are crushed. Cantharidin can cause blistering of tender human skin. Blister beetles are highly toxic to some livestock, notably horses, which sometimes eat them in alfalfa hay. Many different types of blister beetles occur in the region, and the amount of cantharidin produced varies widely among these species.

## Dusky Sap Beetle (*Carpophilus lugubris*)

**Damage:** The adult gray or brown beetles are attracted to sweet corn and invade ears as they begin to form. The beetles chew on kernels in the ear tip area and begin to lay eggs as silks start to wilt. The wormlike larvae further tunnel the corn ear. The dusky sap beetle is most commonly attracted to ears that have already been damaged by corn earworm or some other injury. However, they can infest and damage healthy ears. Problems are most severe on the earliest-planted corn. Sap beetles can also be found in damaged fruit and vegetables. However, they are not primary pests of these crops and are only found after some decay has occurred.

**Life History and Habits:** Adult beetles are active during much of the year. They feed primarily on decaying plant materials, flowers of yucca and other plants, oozing wounds of trees and damaged fruit. Larvae are scavengers on these types of foods and develop in about four weeks, after which they drop to the ground and pupate. There are probably two to three generations per year.

**Cultural Controls:** Sap beetles may develop high local populations early in the season where abundant foods are available. Decaying plant materials should be eliminated by composting or tillage. However, the beetles are quite migratory and will readily move short distances throughout a neighborhood. Sweet corn varieties with tight, long husks will help exclude sap beetles as well as corn earworms. These aren't readily available but include Victory Golden, Golden Security, Tender Joy, Stowell's Evergreen, Country Gentleman, Early Sunglo and Trucker's Favorite. Planting later-maturing sweet corn will also avoid some problems.

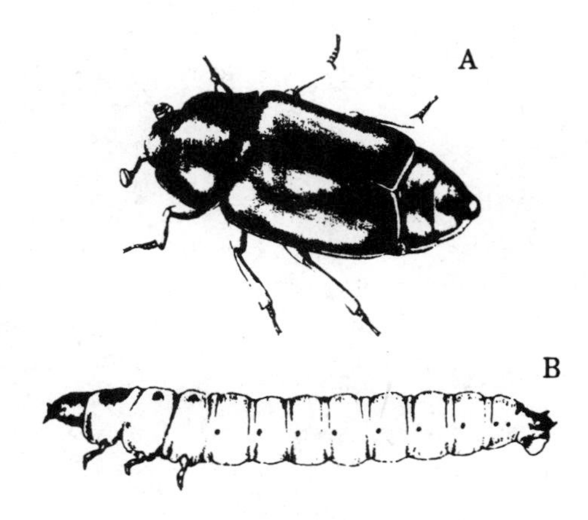

*Dusky sap beetle. A) Adult; B) Larva.*

**Chemical Controls:** Some insecticides (e.g., Sevin/carbaryl) can control sap beetles before they enter ear tips. Treatments should concentrate on the period of early silking, when most beetles move to the ears. Control of corn earworms (see page ) will prevent many attacks on sweet corn by sap beetles.

**Related Insects:** Several other sap beetles are commonly found in the region, primarily the shiny, black spotted "picnic beetles" in the genus *Glischrochilus*. These scavengers are often seen around yards feeding on overripe strawberries, tree sap or other fermented materials.

## Bumble Flower Beetle (*Euphoria inda*)

**Damage:** Bumble flower beetles rarely cause serious injury in gardens, but they do attract atten-

## *Gather Round the Cider Barrel— Insects That Prefer Damaged Plants*

Several insects around the garden have a penchant for vegetables and fruits that have already been damaged by hail, insect wounding or disease. These wounded areas favor development of yeasts and bacteria, which ferment the plant sugars, producing a heady brew of attractive chemicals (and alcohol), which these insects favor.

Sap beetles are perhaps the most common insects that frequent these "microbreweries." However, the most conspicuous insect visitors are the bumble flower beetles, a type of hairy June beetle that is most common in August and September. (During June it spends its time as a white grub, feeding on horse manure and other organic matter.) Considering its fondness for fermented plant sap and wounded fruits, the bumble flower beetle has a particularly appropriate scientific name—*Euphoria*.

Other garden "bugs" also gravitate toward wounded plants, and an open melon or damaged ear of sweet corn attracts a large number of different insects. Slugs are well known for their attraction to fermented baits such as beer. Millipedes and sowbugs also move to wounded areas of plants.

These garden pests may be on the scene when plant injuries are discovered. However, the gardener should be careful to separate the damage caused by the pest from the underlying problem (e.g., wounds or disease) that may have "set the table" for them.

*Bumble flower beetle*

tion. Large masses of beetles can sometimes be found in late summer clustered on oozing sap or in damaged corn or melons. Earlier in the season, the grub-stage larvae are commonly found feeding on manure. The large, hairy beetles sometimes feed on flowers, including strawflower, sunflower and daylily. They may transmit bacterial diseases to some of these plants, which can produce wilting.

**Life History and Habits:** The overwintering state is the adult beetle. These are broadly oval, about 1/2 to 5/8 inches long and densely covered with yellowish, brown hairs. In spring the beetles usually lay eggs in fresh manure (particularly horse manure), rotten wood or compost, and the C-shaped grubs develop in the decaying organic matter. As they feed, they form small, packed chambers in which they later pupate. The grubs are commonly found in gardens fertilized with manure or compost but do not feed on plant roots. The adults emerge in mid- to late summer and feed on a wide variety of sweet or fermenting liquids. They are commonly attracted in late summer to the bacterial ooze produced by infection of many trees. They may also occasionally damage ripening corn, ripe apples, grapes, melons and peaches. The pollen and nectar of flowers such as sunflower, strawflower and daylily may be food plants. There is one generation per year.

**Chemical Controls:** No controls have been developed and would rarely be of benefit. Because the adults are so mobile, insecticides are likely to have little effect. Handpicking beetles as they are observed is probably the best control. Larvae are not damaging to plants.

# *Lepidoptera (Moths and Butterflies)*

Although the caterpillar stage of the butterfly can cause damage to fruits and vegetables, some gardeners actually make a conscious effort to attract and retain butterflies in and around the garden. Butterfly gardening is the purposeful culture of plants and other foods used by native butterflies in the region. It has long been practiced in Europe (particularly England) but only recently has begun to attract attention in North America. However, the tremendous number of highly attractive butterflies common to the West can make this a very satisfying gardening exercise.

REGIONAL BUTTERFLIES AND THEIR FOOD SOURCES

| Butterfly | Food of the caterpillar stage | Some adult foods |
|---|---|---|
| Two-tailed swallowtail *Papilo multicaudatus* | Ash, chokecherry | Geranium, thistle, milkweed, larskspur, cosmos |
| Western tiger swallowtail *Papilio rutulus* | Willow, cottonwood, chokecherry | Zinnia, lilac, butterfly bush, thistle, milkweed |
| Black swallowtail *Papilio polyxenes* | Dill, parsley, carrot | Zinnia, thistle, butterfly bush |
| Painted lady *Vanessa cardui* | Sunflower, thistle, hollyhock | Grape hyacinth, cosmos, zinnia, alfalfa, many flowers |
| Variegated fritillary *Euptoieta claudia* | Pansy, other flowers | Rabbitbrush, gaillardia, beeplant |
| Wood nymph *Cercyonis pegala* | Grasses | Rabbitbrush, clematis, thistle |
| Weidemeyers admiral *Limenitis weidemeyerii* | Willow, aspen, cottonwood | Sap flows, dung, snowberry |
| Hackberry butterfly *Asterocampa celtis* | Hackberry | Rotting fruit, sap flows |
| Mourning cloak *Nymphalis antiopa* | Elm, aspen, willow, cottonwood | Rabbitbrush, milkweed, sap flows |
| Checkered skipper *Pyrgus communis* | Mallow, hollyhock | Verbena, dandelion thistle, aster |
| Melissa blue *Lycaeides melissa* | Wild licorice, alfalfa, etc. | Beeplant, sweetclover |

## Imported Cabbageworm (*Pieris rapae*)

**Damage:** Caterpillar stages chew on leaves of broccoli, cabbage and related plants. Feeding injuries and contamination of edible heads and leaves by insects or their fecal droppings can destroy plants for most uses. This is the most common and destructive cabbageworm in the garden.

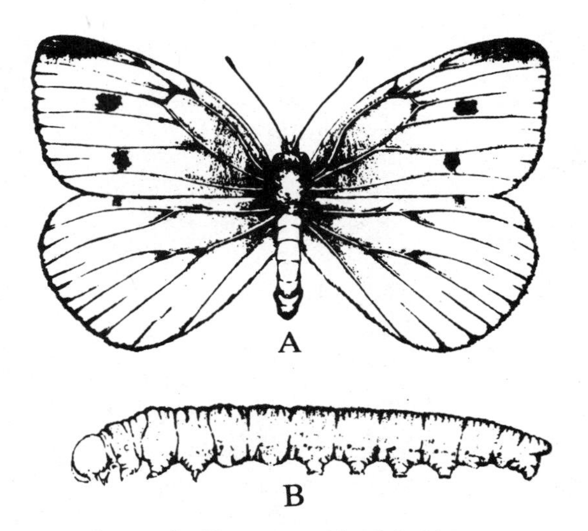


*Imported cabbageworm. A) Adult; B) Larva.*

**Life History and Habits:** The imported cabbageworm is extremely common throughout the region and feeds on many cabbage family plants (crucifers). The overwintering stage is the pupa, which often occurs in protected sites several yards away from the host plants. The adult insects (the familiar cabbage butterflies) emerge in mid-spring and females lay yellow, bullet-shaped eggs on leaves. The larvae hatch in 3 to 5 days and begin to feed on the plants, often first remaining on the outer leaves. Later, they tunnel into the head. Larvae are full-grown in two to three weeks. Pupation occurs in the vicinity of the plant. Three generations typically occur per growing season.

**Natural and Biological Controls:** Several naturally occurring predators and parasites can greatly reduce infestations, including ground beetles, paper wasps, spiders and parasitic wasps. One of the most common parasitic wasps is *Apanteles glomeratus*, which produces small masses of yellow cocoons after emerging from the caterpillar. Young caterpillars are also very susceptible to drowning following heavy rains or overhead irrigation. *Bacillus thuringiensis* is very effective against larvae. Periodic releases of trichogramma wasps can allow this insect to parasitize and destroy some cabbageworm eggs.

**Cultural Controls:** Caterpillar populations tend to increase later in the season, so early plantings can escape more serious injury. Although no cabbages are highly resistant to imported cabbageworms, red cabbage varieties are less preferred than are green varieties. Caterpillars are usually more damaging to cabbage and broccoli than other cabbage family plants. Ground covers and straw mulches provide daytime cover, which can improve the garden habitat for cabbageworm predators such as ground beetles. Interplantings of various herbs or flowers have been recommended for control in gardening publications for many years, but numerous experimental trials conducted throughout North America have repeatedly shown this recommendation to be without value.

**Mechanical and Physical Controls:** Overhead watering or periodic hosing of plants with a forceful jet of water can kill many small caterpillars. Handpicking is possible in small plantings. Insects may be excluded by use of plant covers, such as floating row covers, which prevent the butterflies from laying eggs on plants.

**Chemical Controls:** After plants have become established, they can tolerate considerable defoliation (up to 50 percent) with little or no loss in yield. However, as plants begin to produce heads, injury is more serious and damage to the edible heads can occur. Several insecticides (including *Bacillus thuringiensis* products) are registered that can usually provide control on crucifers. Control is more difficult when older caterpillars are present since they tend to tunnel more deeply into the plant. Adequate coverage of plants can become difficult on varieties with large leaves. Because plants have waxy leaves, addition of a small amount of soap or detergent can usually improve coverage.

**Related Species:** A closely related cabbageworm sometimes found in gardens is the yellow and black-spotted southern cabbageworm (*Pieris protodice*). The adult stage of this is known as the checkered white butterfly. Other common caterpillars that attack plants of the cabbage family are the cabbage looper, diamondback moth and zebra caterpillars.

## *Parsleyworms*

One of the more entertaining insects found within the yard and garden is the parsleyworm. As its name suggests, the caterpillar commonly feeds on parsley but can also be found on related plants such as dill, fennel and carrots.

Parsleyworms start out as small black caterpillars that resemble bird droppings. As they get older and larger, they undergo a dramatic color change, becoming conspicuously striped with white, yellow and black. Perhaps most interesting, parsleyworm caterpillars can display a Y-shaped horn from behind their head when disturbed. This is assumed to be a defense against predators, such as birds, since it has a powerful smell—compared by some to rancid butter. However, this very behavior also makes them favorites for children to collect and observe.

Parsleyworm caterpillars save their best displays for the adult stage. After pupating in a camouflaged corner of the garden, the black swallowtail emerges. This butterfly is one of the largest and most attractive species that graces our yards and gardens.

### Parsleyworm/Black Swallowtail Butterfly (*Papilio polyxenes asterius*)

**Damage:** The parsleyworm is a brightly colored (black, white and yellow) caterpillar that feeds on leaves and flowers of parsley, dill, fennel and occasionally celery, carrots and parsnips. They also may clip the heads and feed on developing dill seeds.

**Life History and Habits:** The parsleyworm spends the winter as a pupa attached to tree bark, sides of buildings or other protected locations. The adult, known as the black swallowtail butterfly, emerges during late May and early June and lays eggs on plants in the carrot family (Umbelliferae). After hatching, the young parsleyworms start out as small black caterpillars that resemble bird droppings. As they get older and larger, they undergo a dramatic color change, becoming conspicuously striped with white, yellow and black. After they become full-grown, they wander away from the plant to find a place to pupate. They then form the pupa within a grayish chrysalid that blends in color with the background. Adult butterflies emerge after about two weeks and feed on nectar. They then mate, lay eggs and produce a second generation in midsummer. There are two generations per year.

**Biological Controls:** Several general predators, such as spiders, kill many of the larvae. Parasitic

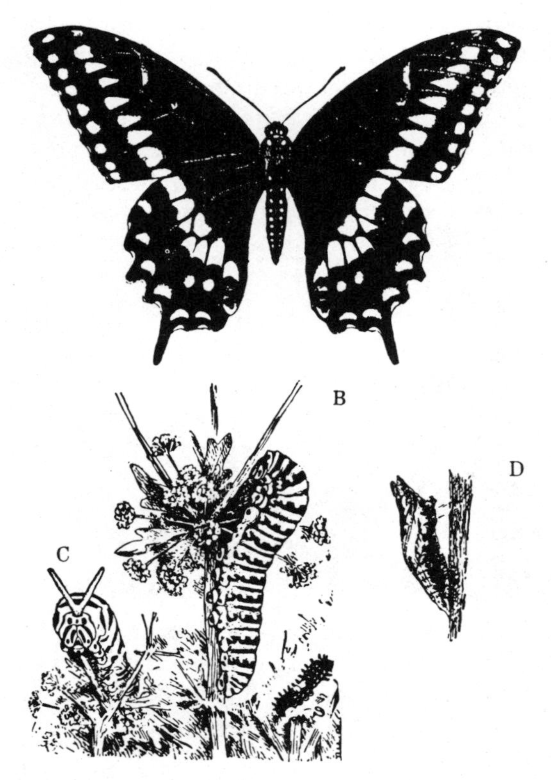


*Parsleyworm. A) Adult (black swallowtail butterfly); B-C) Larva; D) Pupa.*

wasps kill some caterpillars, and tachinid flies attack the pupae. The parsleyworm is very susceptible to *Bacillus thuringiensis*.

**Mechanical Controls:** The caterpillars are brightly colored and can easily be picked off plants if damage becomes objectionable. However, they do defend themselves by everting a pair of fleshy horns that emit a disagreeable odor.

## Painted Lady/Thistle Caterpillar (*Vanessa cardui*)

**Damage:** The thistle caterpillar will feed on a fairly wide range of plants, with over 100 species from several plant families reported as hosts. By far the most common host is the Canada thistle, and it can be very beneficial in controlling this serious weed. However, some other garden plants may be fed on, including sunflower, globe thistle, Jerusalem artichoke, globe artichoke, hollyhock, common mallow, lupins, soybean, basil, artemisia and geranium.

**Life History and Habits:** The adult stage of the insect is known as the painted lady, *Vanessa cardui*. It is generally orange with irregular black and white spotting on the wings. This species is one of the most common butterflies found in the Rocky Mountain region. However, it is an annual migrant, spending the winter in more southerly areas, including Mexico. In spring, migrations are northward; in fall, to the south. The adult butterflies feed only on nectar and other fluids, sometimes drinking from damp spots on the ground. Females lay eggs on host plants. The Canada thistle is by far the most preferred host, giving rise to the alternative name—the thistle caterpillar. The caterpillar stage is generally black with some lighter flecking and rather spiny. The caterpillars feed on leaves and often produce a little webbing. When abundant they can extensively defoliate plants, and once the food plant is destroyed, the caterpillars will migrate. Pupation occurs in a silvery chrysalid, usually some distance from the food plant. Adults emerge about one week after pupation. Several overlapping generations are produced annually.

**Mechanical Controls:** Individual larvae can be handpicked. Although covered with fleshy spines, they are harmless.

**Natural and Biological Controls:** Thistle caterpillars are usually heavily parasitized by wasps and tachinid flies, so outbreaks tend to be short-lived. In addition, since the butterflies are so dispersive, they typically move out of the area where they earlier developed. Thistle caterpillars are susceptible to *Bacillus thuringiensis*.

**Related Insects:** Another annual migrant is the variegated fritillary, *Euptoieta claudia*. The brightly colored larvae of this butterfly most commonly feed on pansies but also can be found on lobelia and petunia.

## Cabbage Looper (*Trichoplusia ni*)

**Damage:** The cabbage looper commonly damages cabbage family plants in a manner similar to the imported cabbageworm. In addition, it may feed on a wide range of other garden plants, including potatoes, peas, lettuce, spinach, nasturtiums, and carnations. Damage to these latter plants is usually minor. The highest numbers of insects usually occur late in the growing season.

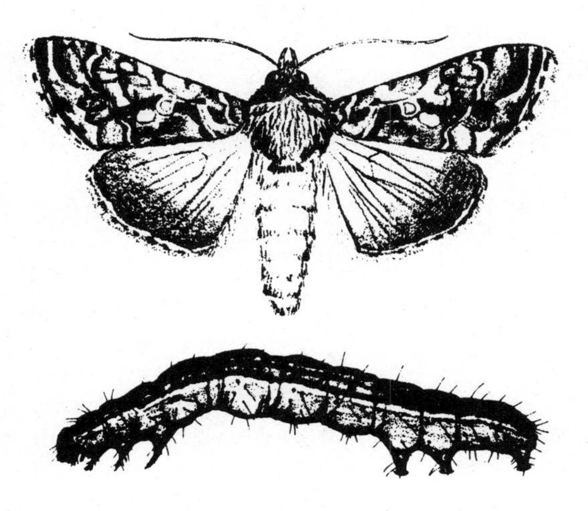

*Cabbage looper. Adult (top), larva (bottom).*

**Life History and Habits:** The cabbage looper overwinters as a pupa within a loose cocoon, located on plant debris around plantings of the previous year. The adult moths emerge in midspring. At night, females glue eggs singly to the leaves of plants. The caterpillars feed on the plants, becoming full-grown in about three to four weeks. Pupation occurs near the host plant. There are typically two to three generations per year.

**Natural and Biological Controls:** Cabbage loopers are preyed upon by several garden predators, such as damsel bugs, paper wasps and spiders. Tachinid flies and parasitic wasps commonly kill the larvae and pupae, although not until the insect has fed extensively. A wilt viral disease can be very destructive to the insect, tending to occur most frequently late in the season. Parasitism of eggs by trichogramma wasps occurs at low levels.

Trichogramma wasps can be introduced into a planting as a supplemental control. Cabbage loopers are also susceptible to *Bacillus thuringiensis*. Caterpillars that are dead and dying from viral disease may be collected and sprayed on plants to spread the disease.

**Mechanical Controls:** Row covers or other materials can exclude the moths from the plants and prevent egg-laying. Tilling debris may kill some of the overwintering pupae, although the moths are very migratory and can infest plantings from long distances.

**Chemical Controls:** Several insecticides, including *Bacillus thuringiensis*, can control cabbage loopers. However, treatments are most effective against small caterpillars. Older caterpillars are quite resistant to many insecticides and tunnel into plants, making control difficult.

**Miscellaneous Notes:** The cabbage looper gets its name from its distinctive looping manner of walking, since it lacks a few pairs of the abdominal prolegs found on other cabbageworms. Although other related species of loopers are found on vegetables (e.g., celery looper, alfalfa looper), the cabbage looper is the only species that attacks crucifers (cabbage family plants).

## Corn Earworm/Tomato Fruitworm (*Heliothis zea*)

**Damage:** The larvae feed on the ear tips of sweet corn, causing extensive damage. Pepper and tomato fruits may also be tunneled and destroyed. The corn earworm is considered one of the most destructive insects in the United States, attacking many of the most important field and vegetable crops.

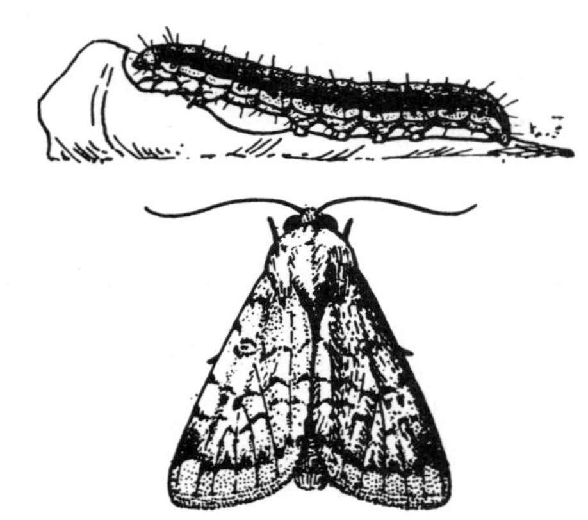

*Corn earworms. Larva and adult.*

**Life History and Habits:** The corn earworm is not able to survive western winters, except in some of the warmer southern areas of the region. Almost all infestations arise from annual migrations of the adult moths from the southern U.S. and Mexico. The migration flights typically occur sometime during June. The female moths fly at night, laying eggs singly on suitable host plants. On sweet corn, eggs are laid on green silks; on tomatoes and peppers, the eggs are usually laid on the new leaves near flowers of developing fruit. Eggs hatch in about 2 to 5 days, and the young caterpillars begin to tunnel into the plants. Corn earworm caterpillars usually feed for about four weeks before becoming full-grown. They vary in color from pale-green to pink or even black. Since they are highly cannibalistic, it is fairly uncommon to find more than one within a single corn ear. When full grown, they drop from the plant,

*Corn earworm on sweet corn ear tip*

construct a small cell in the soil, and pupate. Adults emerge in 10 to 14 days. In most of the West, two generations occur annually. The first generation tends to damage the earliest planted sweet corn. The second generation peaks during September harvasts.

**Natural and Biological Controls:** General predators, such as minute pirate bugs and damsel bugs, feed on the eggs. Corn earworm larvae are also highly cannibalistic and will readily eat each other. Caterpillars are susceptible to *Bacillus thuringiensis*, which can be an effective treatment on peppers and tomatoes. Its effectiveness on sweet corn is marginal, since the insect tunnels into untreated parts of the silk and ear tip soon after hatching. Insect-parasitic nematodes injected into the ear tips kill existing larvae, but their use as a preventive treatment has not been successful in areas where hot summer temperatures can inactivate the nematodes. Releases of trichogramma wasps can kill some of the eggs and help suppress infestations when used repeatedly during periods of egg-laying.

**Cultural Controls:** Varieties of sweet corn with husks that tightly cover the ear tip are more resistant to corn earworms. Since sweet corn is highly favored, it can be used as a trap crop to divert the egg-laying moths from less susceptible crops such as peppers and tomatoes.

**Mechanical and Physical Controls:** Injection of a small amount of mineral oil into the silks can control caterpillars. Because oil treatments can interfere with pollination, they should be made after silks begin to brown to avoid injury. Some vegetable oils, such as soybean, can provide some earworm control, but they are less effective than mineral oil.

**Chemical Controls:** Sweet corn becomes attractive to egg-laying moths from the time silks first emerge from the plant until they turn brown. Eggs are laid on the silk at that time. Insecticide applications should be timed for this critical period of silking and directed to the ear tip zone. Since the insect usually overwinters in southern states, its occurrence in much of the region is irregular. Pheromone traps or light traps can monitor flights and help determine if large numbers have moved into the area. Where pheromone traps are used, trap design is important and Heliothis models should be used. Corn earworm pheromone is also very chemically unstable and must be replaced frequently (at one to two week intervals) when removed from the refrigerator.

**Miscellaneous Notes:** The corn earworm is called the tomato fruitworm when it attacks tomatoes. On cotton, it is known as the cotton bollworm. Its wide range and destructive habits make it the most destructive agricultural insect pest in the United States.

## Tobacco Budworm/"Geranium Budworm" (*Heliothis virescens*)

**Damage:** Most damage is caused by tunneling of flower buds, which prevents or shortens flowering. Young larvae tunnel primarily small buds, while larger caterpillars feed on flower petals, chewing the reproductive ovaries of the flowers. Geranium and petunia are most commonly damaged, but a wide variety of other flowers (e.g., nicotiana, ageratum, dandelion, marigold) are also fed on. Although its accepted scientific name is the tobacco budworm, its fondness for geranium and other flowers has caused it to be called the geranium budworm.

**Life History and Habits:** The tobacco budworm overwinters as a pupa in an earthen cell buried a few inches beneath the soil surface. During late spring, adults emerge. The adult stage is a pale green moth, active in the early evening. Migrations of moths into more northern areas are sometimes responsible for summer infestations that

*Tobacco budworm/"geranium budworm"*

occur outside areas where the insect normally survives winter conditions. The moths glue eggs singly onto flower buds and leaves. The young caterpillars emerge and feed on buds, flowers and, rarely, leaves. They become full-grown in about three weeks, causing extensive injury. They then drop to the soil and pupate. There are probably two generations per season, with the highest caterpillar populations occurring in August and September. The damaging caterpillars are often observed on the plant. These are highly variable in color, ranging from pale brown or light green to red or black. The color of the flower on which they are feeding determines, in part, the caterpillar color. The tobacco budworm is a very serious pest of cotton and tobacco in more southern areas. It is generally considered to be a warm-weather, subtropical insect generally restricted to areas below northern Texas and New Mexico. Although the overwintering pupae cannot survive temperatures below 20°F, local "microclimates" in and around homes appear to allow the insect to survive farther north (e.g., Denver, Grand Junction). In these areas, problems tend to be most serious following mild winter conditions.

**Natural and Biological Controls:** Cold winter temperatures, which freeze soil where the pupae overwinter, are the most important natural control of tobacco budworm in the region. Local populations of budworms can be killed by a severe freeze. During the summer, several biological controls are active, including tachinid flies and general predators such as ground beetles, damsel bugs and minute pirate bugs. The tobacco budworm is susceptible to *Bacillus thuringiensis* (Bt). However, since the caterpillars tend to tunnel plant buds, there is often little chance for the insect to eat enough to be killed by this disease. Bt is more effective on plants where the caterpillars feed on petals (e.g., petunia) than when they restrict feeding to buds (e.g., geranium).

**Cultural Controls:** Because the insects survive winters in soil, potted geraniums should not be brought indoors. (However, cuttings can safely be taken, since these should not be infested with the insect.) Tilling the garden bed in fall will further expose pupae to freezing winter temperatures. Some plants appear to be resistant to tobacco budworm. For example, ivy geraniums are rarely injured, in contrast to standard geranium types.

**Mechanical Controls:** Handpicking the caterpillars is often the best control in small plantings.

**Chemical Controls:** The tobacco budworm is difficult to control with most insecticides because it has become insecticide-resistant. Younger caterpillars are much more susceptible than older caterpillars.

## European Corn Borer (*Ostrinia nubilalis*)

**Damage:** The caterpillars tunnel into the stems and fruit of many vegetables, including corn, peppers, beans and potatoes. Stem tunneling may cause plants to wilt and die back beyond the wounded area, although little damage results from

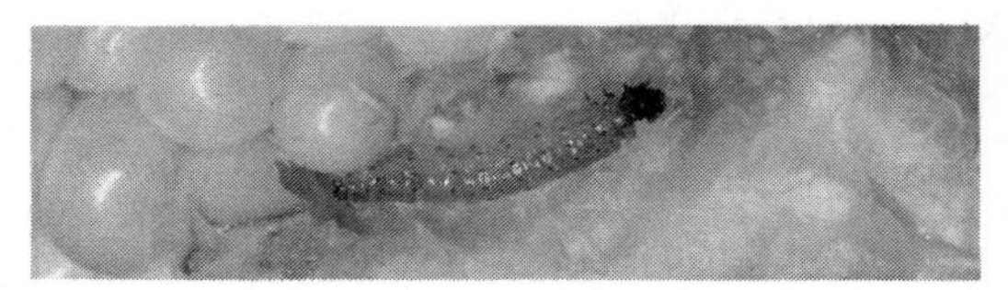

*Larva of European corn borer*

this injury. Corn may be tunneled from the tip, as with corn earworms, but the caterpillars also may bore from the stem or side of the ear. Beans tend to be tunneled from the side; peppers near the top of the fruit. Sap beetles often move into tunnels damaged by European corn borers. These caterpillars are most easily distinguished from the corn earworm by their lack of striping and smaller size. The manner in which they tunnel the plant is also characteristic. European corn borers are limited to areas east of the Rockies and are most common on the eastern edge of the region.

**Life History and Habits:** The European corn borer spends the winter as a full-grown caterpillar within cornstalks and other plant debris. In spring it pupates, later emerging as a light brown moth that mates and lays eggs. Eggs are laid in masses, overlapping in a manner that resembles fish scales. The young caterpillars emerge in about 10 days and wander about the plant. They feed on the leaf surface for several days and later bore into leaf veins, stems and fruit. Once inside the plant, they continue to feed for about two to three weeks before pupating. A second generation, with egg-laying during late July and August, is produced in most areas, except the most northern region. Corn less than 18 inches tall is rarely attacked by the first generation of European corn borers. The second generation tends to concentrate egg-laying on pollinating plants. Corn that matures in mid-season can largely escape injury.

**Natural and Biological Controls:** Young stages of the European corn borer are very susceptible to high temperatures and drowning during rains. Survival of the caterpillars is usually very poor until they have tunneled into the plant. Eggs and small larvae are commonly fed on by minute pirate bugs and other general predators, such as damsel bugs. Parasitic wasps and fungus diseases sometimes kill substantial numbers of older caterpillars. The European corn borer is susceptible to *Bacillus thuringiensis*. However, due to its habit of tunneling plants, it often does not have an opportunity to eat enough Bt to die before it enters the interior of the plant. The most effective use has been treating corn before tasseling (whorl treatment), which kills caterpillars that are feeding on the unraveling leaves.

**Cultural Controls:** Rototilling or plowing can kill caterpillars that overwinter in the garden. However, European corn borer moths are strong fliers that will spread to new areas the following season.

**Mechanical Controls:** Overhead irrigation, or hosing plants during periods when eggs are hatching, kills many of the young borers exposed on the plants.

## *Hornworms*

Running across a hornworm in the vegetable patch is one of the more startling garden encounters. Hornworms are the largest common garden insects and reach a length of 3 to 4 inches when full-grown. They are also marked with a fearsome-looking horn on the hind end. (Its only known function is to scare gardeners.)

Diagnosis of problems with the tomato hornworm is relatively easy: the rapid disappearance of leaves stripped to the stem, followed by the appearance of large dark green droppings around the base of the plant. However, the caterpillars are remarkably difficult to find, being extremely well camouflaged to blend with the foliage. Eggs and young caterpillars may be present in early summer, but most garden problems are observed during mid- to late summer as the caterpillars rapidly increase in size—and appetite. Since the caterpillars eat most of their food at the very end of the caterpillar stage, gardens can be devastated in short order (usually when you are away for a 3-day vacation).

The tomato hornworm is only one member of the hornworm family of caterpillars (Sphingidae). Indeed, throughout most of the region, the tobacco hornworm is at least as common on tomatoes. Other common hornworms feed on trees (elm, ash and poplar, particularly), grape and Virginia creeper (a "hornless" hornworm, the Achemon sphinx) and weeds such as purslane (the common whitelined sphinx). The full-grown hornworms tend to wander away from the plant on which they develop and are most often seen crossing the yard late in the season.

Adult stages of the hornworms are heavy-bodied, strong flying insects known as sphinx, hawk or hummingbird moths. The latter name comes from their resemblance to hummingbirds as they feed from deep-lobed flowers at dusk.

*Sam Cranshaw checks out a hornworm*

**Chemical Controls:** The European corn borer is difficult to control with insecticides, since the caterpillars tunnel into the plants. Treatments should be applied when eggs are hatching. Small shotholes in emerging corn leaves are a useful indication of early corn borer feeding.

## Limabean Pod Borer (*Etiella zinckenella*)

**Damage:** The caterpillars of the limabean pod borer tunnel and feed on the developing seed pods of many legumes, including beans, black locust and vetch. Bush lima beans are particularly susceptible to injury. Problems are limited to the southern and western areas of the region.

*"Hummingbird moth" visiting flower*

**Life History and Habits:** The limabean pod borer overwinters as a pupa in the old pods and plant debris of the previous crop. The adult moths emerge in early spring and seek susceptible plants. The pale green or reddish caterpillars feed on developing plants, infesting pods when they are produced. There is one generation per year.

**Cultural Controls:** Till and destroy old plant debris to eliminate overwintering stages of the insect. Where the growing season allows, delay planting to reduce infestation.

## Tomato Hornworm (*Manduca quinquemaculata*)

## Tobacco Hornworm (*Manduca sexta*)

**Damage:** Two similar species of hornworm are common in gardens: the tobacco hornworm and the tomato hornworm. Caterpillars chew leaves and plants can be defoliated rapidly. Fruits may also be chewed. Tomatoes are particularly susceptible to injury, but other related plants, such as peppers and potatoes, are occasionally infested.

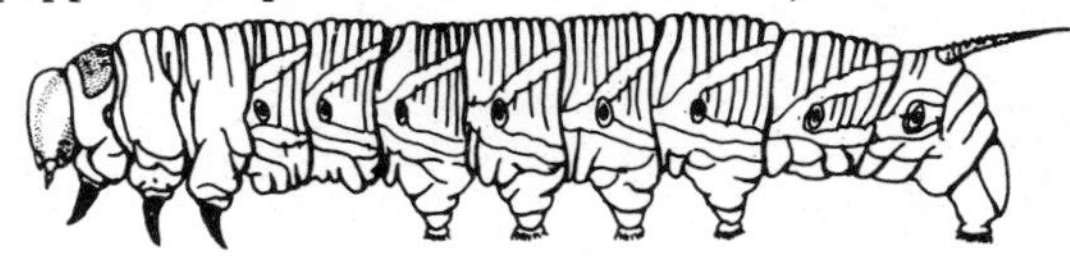

*Tomato hornworm larva*

**Life History and Habits:** Tomato and tobacco hornworms overwinter as pupae in the soil in the vicinity of gardens. The adult moths (a type of sphinx, hawk or hummingbird moth) emerge in late spring. They are strong fliers and appear after dusk to feed on nectar. Female moths lay large pearly eggs on the upper surface of leaves. The young caterpillars hatch and feed on the plant for a month or more. They have tremendous appetites and consume large amounts as they grow older and larger. After feeding, they wander away from the plant and pupate in the soil. There is usually one generation per season, although a few moths will emerge and produce a second generation.

**Natural and Biological Controls:** Hornworm eggs and small larvae are killed by many general predators common to the garden, such as spiders, predatory stink bugs and damsel bugs. The most conspicuous natural hornworm enemy is a parasitic wasp that develops within the caterpillars. When the developing wasps emerge from the hornworm caterpillar, they remain attached to their host and spin distinctive pale-colored cocoons. Despite their large size, hornworms are extremely susceptible to *Bacillus thuringiensis*.

**Mechanical and Physical Controls:** Hornworms can be handpicked off plants or cut with scissors during regular plant inspections. Larvae are most easily located in early morning, often on the exterior of the plant.

**Chemical Controls:** Hornworms are easily controlled with most insecticides registered for use on tomatoes.

**Miscellaneous Notes:** The tobacco hornworm is more common in regional vegetable gardens than the tomato hornworm, but both may be found together and have similar habits. The tomato hornworm has a green horn with black sides, while that of the tobacco hornworm is red. The caterpillars are also differentiated by the white striping along their sides. These form a series of Vs with tomato hornworms, while they are diagonal dashes on the tobacco hornworm. Occasionally, dark forms of the tomato hornworm occur. The caterpillars of these are much darker than the normal green, although the adult moths differ little in appearance. Several other species of hornworms common to the yard and garden do not feed on vegetable plants. The largest species is the giant poplar sphinx, which develops on cottonwoods and poplars. The whitelined sphinx is a particularly common species and feeds on a number of weeds such as purslane. A hornless hornworm, the larva of the Achemon sphinx moth, is also present in the West and feeds on grape and Virginia creeper.

## *Cutworms*

Caterpillars that cut and kill seedling plants around the soil line are usually described as cutworms. Not a very exact term, this habit is shared by many different caterpillar species within the cutworm family (Noctuidae), which also includes garden pests such as the cabbage looper and corn earworm. In addition, there are climbing cutworms that climb plants and chew leaves. Cutworms typically feed at night and hide in soil cracks, dirt clods and other cover during the day.

As might be expected, cutworms have a variety of habits. Most of the damaging cutworms in the region have one generation per season and lay eggs during late summer among weeds and other dense vegetation. However, some cutworm moths do not survive area winters and annually migrate into the region. This includes the black cutworm, beet armyworm and "true" armyworm. Problems with these insects are more sporadic than with cutworms that overwinter in the region.

Adult stages of cutworms are moderately large gray or brown moths. Some cutworm moths become quite abundant at times and are popularly known as millers. One species, the army cutworm, is the common nuisance miller moth of Colorado and Wyoming. However, the term *miller moth* may be used to describe any type of nuisance moth that occurs in large numbers.

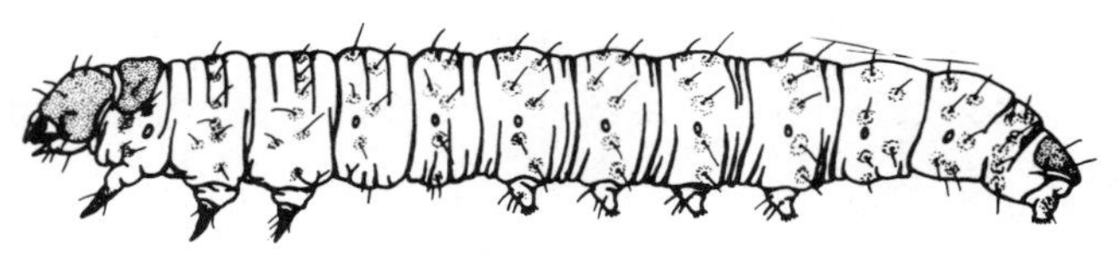

*Larva of pale western cutworm*

| SOME COMMON REGIONAL CUTWORMS | |
|---|---|
| **Common name** | **Comments, habits in West** |
| Army cutworm (*Euxoa auxiliaris)* | Winters as small larva with peak injury in early spring. Adult stage is the common "miller" of Colorado and Wyoming. One generation per year. |
| Armyworm (*Pseudaletia unipuncta)* | Winters in southern United States and Mexico; an annual migrant. Climbing cutworm with peak injury in mid- to late summer. Fairly uncommon in West. Probably one generation per year. |
| Beet armyworm (*Spodoptera exigua)* | Winters in southern United States and Mexico; an annual migrant. Climbing cutworm with peak injury in mid- to late summer. Probably two generations per year in West. |
| Black cutworm (*Agrotis ipsilon)* | Winters in southern United States and Mexico; an annual migrant. Feeds at or just below the soil surface and is the most aggressive species for cutting plants. Peak injury typically occurs in late spring. Probably two generations per year in West. |
| Clover cutworm (*Scotogramma trifolii)* | Winters as a pupa, emerging to lay eggs in late spring. Feeds as a climbing cutworm, primarily on legumes. Two generations per year. |
| Darksided cutworm (*Euxoa messoria)* | Winters as eggs which hatch in spring, the caterpillars causing peak injury in late spring. Typically feeds around the soil surface. One generation per year. |
| Speckled green fruitworm (*Orthosia hibisci)* | Winters as a pupa which lays eggs in early spring. A climbing curtworm causing peak injury in June. One generation per year. |
| Spotted cutworm (*Amanthes c-nigrum)* | Winters as a pupa, emerging and laying eggs in late spring. Mostly a climbing cutworm-type with very wide host range. Probably two generations per year. |

| Common name | Comments, habits in West |
|---|---|
| Variegated cutworm (*Peridroma saucia*) | Winters as a pupa, emerging and laying eggs in late spring. Mostly a climbing cutworm-type with very wide host range. Probably three generations per year. |
| Western bean cutworm (*Loxagrotis albicosta*) | Overwinter as full-grown larvae but do not transform to adult until July. Peak injury in August. One generation per year. |

## Army Cutworm (*Euxoa auxiliaris*)

**Damage:** Army cutworms cut and feed on seedlings of a wide variety of garden plants, including tomatoes, beans and corn. They are one of the more common of the garden cutworms in the Rocky Mountain and high plains region, as well as an occasional pest of winter wheat and alfalfa. Their name is derived from their occasional habit of migrating in large bands during outbreaks. The adult stage of the army cutworm is the common miller moth of the high plains region and eastern Rockies.

*Army cutworm, larva of the "miller moth"*

**Life History and Habits:** Eggs of the army cutworm are laid in late summer and early fall, particularly in areas of dense growth. The eggs hatch shortly after being laid, and the young caterpillars then feed on weeds and garden plants. Because of their small size, injury does not occur during fall. However, in spring the developing caterpillars are still present and can seriously damage young plants that emerge or are transplanted in the garden. After feeding for several weeks, the caterpillars dig a small chamber underground and pupate. They emerge as the miller moth in late May and June. The adult moths then migrate to higher elevations and feed on the nectar of flowering plants. They do not lay eggs or reproduce during these flights, but can be a serious nuisance pest in homes and cars. Ultimately the moths reach the mountains, where they spend the summer. By midsummer they begin to make a return migration to the plains, laying eggs to repeat the cycle. There is one generation per year.

**Natural and Biological Controls:** Cutworms are preyed upon by several garden insects, such as ground beetles, rove beetles and spiders. Mulches and other covers that favor these beneficial species can help with cutworm control. Toads and snakes can also prey on cutworms. Parasites of cutworms include tachinid flies and various parasitic wasps. Although they are a type of caterpillar, most cutworms are relatively immune to *Bacillus thuringiensis*.

**Cultural Controls:** Weed control is the most important means of avoiding cutworm problems. Weedy gardens attract egg-laying moths during late summer. Tillage in the garden during fall and spring can destroy many cutworms. Cultural practices that promote rapid seedling growth and use of transplants can help plantings outgrow cutworm damage. Well-established plants are generally little damaged by cutworms. Some weeds in spring can be retained temporarily as alternate food for young cutworms. This will divert some of the feeding from garden plants. The weeds should be removed and destroyed before they compete with garden plants or produce seeds.

**Mechanical and Physical Controls:** Young transplants can be protected by various barriers that prevent the cutworm from reaching the plant. These can be constructed of various materials, such as cut milk jugs or paper cups. Outdoor lighting should be minimized during the egg-laying period since lighting can attract the adult moths.

**Chemical Controls:** Several garden insecticides sprayed on susceptible plants can provide cutworm protection. Various baits are also available which can be spread around the base of plants. Baits generally provide a more selective type of treatment than sprays or dusts, since fewer beneficial organisms are likely to contact the insecticide.

## Speckled Green Fruitworm (*Orthosia hibisci*)

**Damage:** The caterpillars cause most injury when chewing on developing apple or pear fruit in late spring. Damaged fruit often drops from the tree. However, fruit with small injuries often continues to grow but becomes misshapen and develops a corky area around the old chewing wound. This fruit damage is similar to that produced by some leafroller caterpillars. (See fruittree leafroller, page 121.) Rosebuds, cherry, plum, crabapple, apricot and hawthorn are less commonly damaged. The speckled green fruitworm also develops on the leaves of many other trees and shrubs but does not cause significant injury to these plants.

**Life History and Habits:** The speckled green fruitworm is a climbing cutworm. It spends the winter underground in the pupal stage. Adult moths emerge very early, in March or April, and females lay eggs near plant buds. The caterpillars feed on the new leaves and remain on the tree throughout their development. If fruit is present, they may also chew these. By early June, they become full-grown and drop from the plant. They often wander a short distance, then dig into the soil and pupate. There is one generation per year.

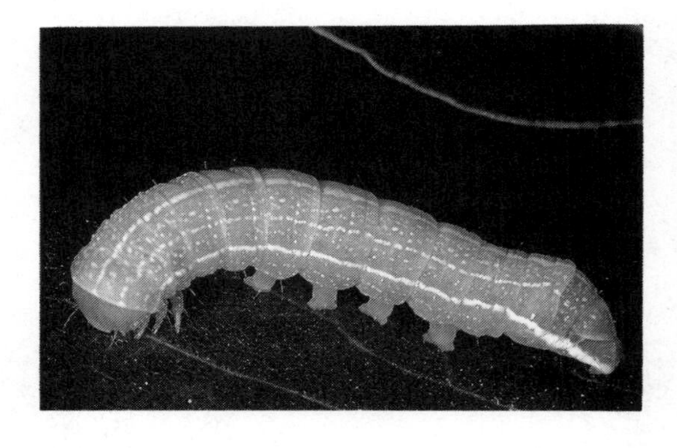

*Speckled green fruitworm*

**Natural and Biological Controls:** Several species of parasitic wasps and tachinid flies attack the speckled green fruitworm and usually prevent the insect from causing serious damage in unsprayed garden plantings. Speckled green fruitworm is also susceptible to *Bacillus thuringiensis*.

**Chemical Controls:** Sprays are rarely needed for control of speckled green fruitworms in backyard plantings. However, it is useful to check the emerging leaves and blossoms for evidence of chewing injury. Several fruit tree insecticides can control the speckled green fruitworm. These should be applied after petal fall or prior to blossoming (pink buds).

**Miscellaneous Notes:** Several other fruitworms and climbing cutworms may cause occasional damage to fruit trees. Most species remain hidden at the base of the tree during the day, climbing up to feed each night. These can be controlled by spot sprays of insecticides around the base of the tree trunk or can be concentrated under boards or other covers for collection and disposal. Barriers on the trunk can also prevent the cutworms (and earwigs) from climbing trees.

## Eastern Tent Caterpillar (*Malacosoma americanum*)

**Damage:** The caterpillars produce a tight tent in the crotch of various trees and shrubs, particularly fruit trees such as cherry, plum, apricot and apple. They rest around the tent during the day and crawl away during the night to chew on leaves. Developing fruit may also be chewed and damaged. When very abundant, tent caterpillars can destroy many of the leaves and weaken the tree.

**Life History and Habits:** Tent caterpillars spend the winter in shiny brown egg masses glued to twigs. The eggs typically hatch in late April or early May, about the time apple leaves begin to emerge. The caterpillars stay together as a group and spin a small tent in the fork of branches. As they feed and grow, the tent enlarges, becoming quite conspicuous after about a month. After becoming full-grown, the caterpillars wander to find places to pupate. They spin a white cocoon on trunks or other sheltering objects around the infested tree. The adult moths emerge about two weeks later, mate, and the females lay egg masses. There is one generation per season.

**Natural and Biological Controls:** Tent caterpillars are attacked by many insects, such as tachinid flies and parasitic wasps. Birds fledging young also seek caterpillars as food. These natural controls typically prevent tent caterpillars from causing serious damage. Tent caterpillars are very susceptible to sprays of *Bacillus thuringiensis*.

*Colony of tent caterpillars*

**Mechanical Controls:** The caterpillars and tents can easily be pulled out of trees and destroyed.

**Chemical Controls:** Tent caterpillars are susceptible to most insecticides used for control of codling moths and other fruit pests.

**Miscellaneous Notes:** The eastern tent caterpillar is limited to areas east of the Rockies. However, other tent-making caterpillars occur in the region. The western tent caterpillar commonly damages aspen in parts of the region and has habits similar to those of the eastern tent caterpillar. Perhaps the most common tent-making caterpillar is the fall webworm, which makes large loose tents on the outer branches of cottonwoods, poplars, chokecherry and other plants in midsummer.

## Codling Moth (*Cydia pomonella*)

**Damage:** Larvae tunnel into the fruit of apple, pear, and crabapple trees. (It is almost always the worm in a wormy apple.) Less commonly, it may damage other fruits, including apricot and peach. It is the single most important insect pest of tree fruits in the western United States.

**Life History and Habits:** Codling moth larvae spend the winter inside a silken cocoon attached to rough bark or other protected locations around the tree. With warm spring weather, they pupate and later begin to emerge around blossom time as small (about 1/2 inch), gray moths. The spring appearance of this adult stage may occur primarily over the course of one or two weeks, but can be much more prolonged if weather is cool. During periods when early evening temperatures are warm (above 60°F) and it is not windy, the moths lay small white eggs on the leaves. The hatching larvae may first feed on the leaves, but then migrate to the fruit, usually entering the calyx (flower) end. They tunnel the fruit, feeding primarily on the developing seeds. After about three to four weeks, the larvae become full-grown, leave the fruit and crawl or drop down the tree to spin a cocoon and prepare to pupate. After about two weeks most, but not all, of the pupae develop to produce a second generation of moths. The remaining moths stay dormant, emerging the following season. (For example, in western Colorado, only about two-thirds go on to produce a second generation, and fewer than 50 percent of their progeny go on to produce a third generation.) These moths lay eggs directly on the fruit, and damage by the larvae to fruit is greatest at this time. Becoming full-grown, the larvae emerge from the fruit and seek protected areas to pupate. In the warmer, southern parts of the region, a small third generation is produced in late summer. This one generally causes much less damage to apples and pears than the earlier generations.

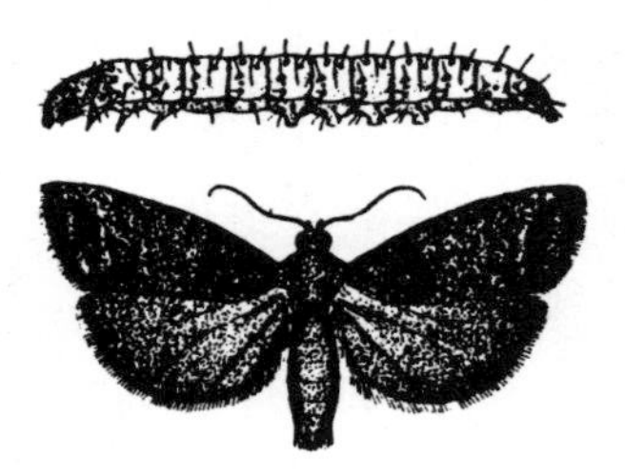

*Codling moth. Larva and adult.*

**Natural and Biological Controls:** The codling moth has many natural enemies, although these biological controls often are not adequate to provide satisfactory control. Birds, notably downy and hairy woodpeckers, will feed on the larvae and pupae in cocoons. Perhaps most important, codling moth larvae and pupae are often killed by several types of parasitic wasps. The activity of these wasps has been improved in some areas by the presence of nearby flowering plants that provide alternative foods. Codling moth larvae also are attacked by several general predators, such as ground beetles and earwigs. Some additional control is possible with repeated applications of *Bacillus thuringiensis*, timed for periods when moths are laying eggs. However, since it acts as a stomach poison, it must be applied shortly before the larvae enter the fruit. Trichogramma wasps, available for mass release, can parasitize and kill codling moth eggs.

**Cultural Controls:** Keeping loose bark scraped from trees and removing debris from around the tree can eliminate shelter used by the insects when they pupate. This will cause them to be more exposed to birds and other natural controls. Since young caterpillars often need some leverage to help them cut into the fruit, many of them enter where two fruits are touching. Thinning apples to prevent this can reduce the survival of the delicate young larvae and control damage to remaining fruit.

**Mechanical Controls:** Adult moths can be attracted to baits and trapped. A typical design is a gallon jar baited with a pool of fermenting molasses and water (1:10 to 1:15 dilution is suggested). This attracts both male and female moths. (Pheromone traps capture only male moths.) Coarse screening can be useful to exclude visiting honeybees. However, raccoons readily feed on this mixture and will often destroy traps.

Pupal stages of the moth can be concentrated by placing bands of corrugated cardboard or burlap around the trunk. They can then be easily collected and destroyed.

**Chemical Controls:** Insecticides should be applied during periods of peak egg-laying. Two damaging generations of codling moths are common, one in late spring (after petal fall) and the other in midsummer. Feeding by the larvae of the latter causes most fruit damage. Use of pheromone traps containing the sex attractant of the female moth can improve treatment timing. (See discussion of pheromone traps in Chapter 3.) Using this technique, insecticides are most effectively used 10 to 14 days after peak flights are detected in pheromone traps. Experimental work indicates that pheromones may be effectively used in the male confusion method for control of codling moths. This involves permeating the air with the sex attractant used by the female to attract males, reducing successful mating. In backyard settings, this would be most effective if apple and pear trees were isolated from other sources of codling moths (including crabapple trees), which could otherwise migrate into the yard. These are not yet available to gardeners, as they are only beginning to be marketed to commercial growers on an experimental basis.

## Fruittree Leafroller (*Archips argyrospilus*)

**Damage:** Caterpillars of the fruittree leafroller feed on leaves of various fruit trees, shade trees and shrubs. The caterpillars first tie together the leaves and feed by skeletonizing the leaf, avoiding the larger veins. Leafroller outbreaks are rare but can result temporarily in large areas of leaf loss. On fruit trees, the developing fruit may also be included within the tied leaves and chewed, later resulting in deep fruit scarring.

**Life History and Habits:** The fruittree leafroller overwinters as eggs laid in masses on twigs and small branches. These eggs hatch shortly after buds open in spring, and the young caterpillars move to the leaves. The caterpillars, which are pale green with a dark head, feed and develop over a period of a month or more, then crawl down the trunk or drop to the ground on silk strings. They pupate under bark flaps or other protected sites, forming a thin cocoon. The moths emerge in late spring and early summer, mate and lay eggs. There is one generation per year.

**Natural and Biological Controls:** Numerous parasites attack leafrollers, generally maintaining populations below levels that are damaging to the plants. Leafroller problems can become chronic in orchards when insecticide use destroys these natural controls. Leafrollers are susceptible to *Bacillus thuringiensis*.

**Chemical Controls:** Several insecticides control leafrollers. However, after leaves are tightly rolled, insecticides often cannot reach the caterpillars. Systemic insecticides, which can kill the caterpillars, cannot be applied to fruit trees or other edible plants. Dormant oil sprays can kill many of the eggs that overwinter on twigs.

**Miscellaneous Notes:** Leafrolling with silk is not done by many other garden inhabitants. Spiders will tie together leaves when they produce an egg sac. A few other caterpillars, such as webworms, may produce some silk on the plant leaves but do not tightly fold the leaves. Tent caterpillars and fall webworms produce large amounts of silk, feed in groups and do not roll leaves.

## Peach Twig Borer (*Anarsis lineatella*)

**Damage:** The peach twig borer primarily damages peach trees, causing most injury when larvae tunnel into the fruit. Apricots and plums are less common hosts of the insect. Less serious injury is produced earlier in the season when the caterpillars bore through the twigs of these trees, causing them to die back. The insect is most common in southern areas of the region, particularly west of the Rockies.

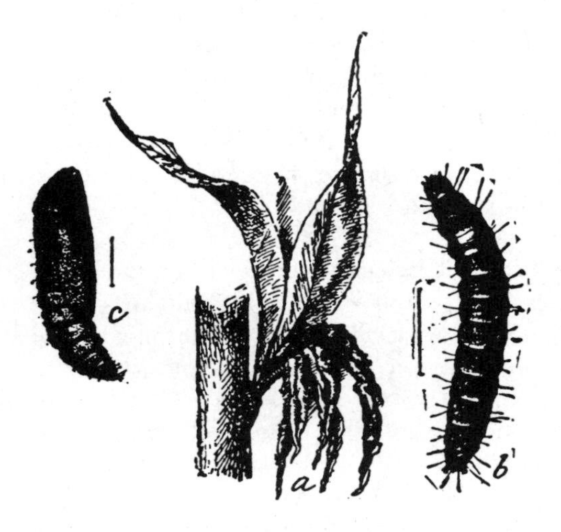


*Peach twig borer. A) Young shoot wilting from attack of borer; B) Adult larva enlarged; C) Pupa enlarged.*

**Life History and Habits:** The peach twig borer overwinters as a partially grown caterpillar, protected in a small, silk-covered cell on the bark of fruit trees. In early spring, the caterpillars become active, migrate to the twigs and tunnel the buds and emerging terminal growth. The damaged new growth typically wilts and dies (flagging). They then pupate in the tree and emerge as moths in May. The emerged moths lay eggs on twigs, small leaves and developing fruits. The caterpillars usually first feed on terminal twig growth, later moving to fruits shortly after the pits start to harden. After becoming full-grown, these caterpillars pupate on the trunk and larger branches. Moths that emerge from this second generation lay eggs on the fruit. Caterpillars restrict their feeding to the fruit at this time, causing most crop injury. (Apricots, which fruit earlier than peaches, are damaged more by the first generation.) After the caterpillars become full-grown, there is a third generation. This produces the overwintering larvae, which do not cause significant injury.

**Biological Controls:** Several different species of parasitic wasps commonly attack peach twig borers and may kill 50 percent or more of the caterpillars. In addition, caterpillars exposed on leaves or bark are susceptible to general predators, such as green lacewing larvae and damsel bugs.

**Chemical Controls:** Dormant oil or lime sulfur sprays used for control of other insects and diseases can kill many of the overwintering larvae on the tree. Sprays of several other insecticides, applied just prior to bloom, can kill the caterpillars as they begin to emerge from their winter shelters and move about on the tree. (Insecticides should never be applied during bloom to protect pollinating insects, such as honeybees.) The moths are small (about 1/4 inch) and ash-gray. They are rarely observed but can easily be captured in pheromone traps baited with the peach twig borer lure. This can help to better time sprays used for control of the insect later in the season. There is promising research indicating that the male confusion method, involving permeating the air with the female insect's sex pheromone to inhibit mating, might be useful in controlling peach twig borers in isolated orchards or plantings. These products are currently under development and have not yet been marketed.

## Diamondback Moth
## (*Plutella xylostella*)

**Damage:** Diamondback moth caterpillars feed on the leaves of cabbage and related plants. Although very common, the larvae are quite small and feed considerably less than other cabbageworms (imported cabbageworms, cabbage loopers, etc.), about 3 percent as much as cabbage loopers, for example. However, populations can be high and cause serious injury.

**Life History and Habits:** The diamondback moth overwinters in the adult stage. The small moth is one of the first insects to be active in the spring. Eggs are laid singly or in small groups on leaves of various cabbage family plants. The first, early-spring generation tends to occur on winter annual mustard plants. Larvae feed on the surface of the leaves and pupate in a loose cocoon attached to the plant. Generations are completed in approximately four to six weeks, with several occurring throughout the season.

**Natural and Biological Controls:** Parasitic wasps are common natural enemies of diamondback moth larvae. General insect predators (spiders, damsel bugs, etc.) are also important controls. It is susceptible to Bt.

**Chemical Controls:** Chemical controls are similar to those for other cabbageworms. The habit of diamondback moth caterpillars to remain exposed on the plant can simplify coverage. However, in several areas of North America this insect has become highly resistant to insecticides and is extremely difficult to control.

**Miscellaneous Notes:** The caterpillar stage of the diamondback moth is very active and wriggles vigorously when disturbed, an identifying characteristic different from that of the other cabbageworms. Caterpillars also may temporarily drop from the plant in response to disturbance. They later climb back up the silken thread spun while dropping.

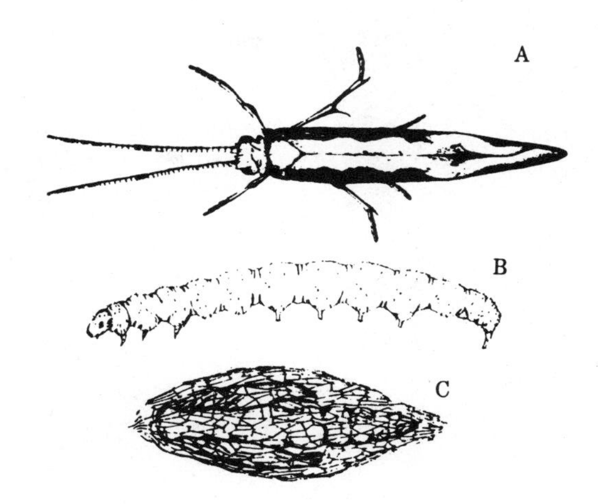


*Diamondback moth. A) Adult; B) Larva; C) Cocoon.*

## Peach Tree Borer
## (*Synanthedon exitosa*)

**Damage:** Larvae tunnel under the bark of peach, plum, cherry, apricot and other species of *Prunus*, usually near the base of the tree. Injuries weaken and may kill trees.

**Life History and Habits:** The adult moths emerge from mid-June through August. Females lay eggs on lower trunks or in nearby soil cracks, particularly favoring previously wounded trees. Eggs hatch in one to two weeks, and the small larvae tunnel into the outer layers of bark. They continue to feed through late spring, if temperatures allow, producing deep, gouging wounds that may destroy all the lower bark. Pupation occurs in a silken cocoon, mixed with wood fragments, usually on the surface of the lower trunk.

**Natural and Biological Controls:** Experimental work suggests that drenches of insect-parasitic nematodes around the base of trees during late summer or early spring can kill some larvae before they cause serious damage. However, it is not known if this can control larvae deep within masses of gum.

*Peach tree borer larva*

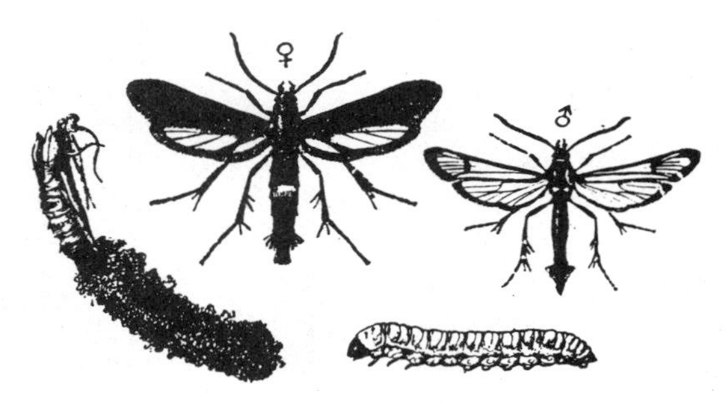


*Peach tree borer*

**Cultural Controls:** Egg-laying is concentrated around wounds. Practices that avoid wounding, and control of existing peach tree borer infestations, can reduce later infestations by the insect. White paint applied to the lower trunk can seal bark cracks used by female moths for laying eggs.

**Mechanical Controls:** Individual larvae can be dug out or killed by a sharp wire. However, this needs to be done with care to avoid excessive tree damage.

**Chemical Controls:** Larvae cannot be controlled by insecticide sprays after tunneling has progressed. Preventive treatments of insecticides should be applied to the trunk when moths are active and eggs are being laid, typically in early July and August. Borer crystals (paradichlorobenzene) can be used to fumigate larvae within the trunk. The crystals are applied around the base of the trunk and temporarily mounded with soil to retain the fumigant gas in the area of the borer infestation. The crystals should not directly touch the trunk. Application rates vary by trunk diameter and are indicated on label directions. Fumigation rescue treatments are best applied during warm periods in fall, after harvest. However, future registration of paradichlorobenzene fumigants for fruit trees may become more restricted, so always check labels to ensure that they are legal.

**Miscellaneous Notes:** Infestations of peach tree borers result in production of clear gum at wound areas, mixed with light brown wood fragments. Almost all injuries occur at, or slightly below, the soil line. This is different from other gumming on stone fruits caused by stress, mechanical injury or infection with Cytospora fungi (see page 186). These produce a clear, amber gum and can occur throughout the upper areas of the tree.

## Currant Borer (*Synanthedon tipuliformes*)

**Damage:** Larvae tunnel canes, particularly near the base of the plant. Leaves on infested canes are small and yellow, and canes often die back.

**Life History and Habits:** Currant borers spend the winter nearly full-grown in the base of canes of currant, gooseberry, sumac or black elder. They feed for a brief period in spring but cause little damage, then pupate and later emerge as adults in late May or early June. The bluish-black adult clearwing borer moths resemble wasps and can be seen resting or mating on leaves of the plants. Eggs are laid on the bark during June and early July, and the caterpillar stage larvae bore into the plant. They move downward, tunneling the pith and wood. These feeding injuries may girdle or weaken the plant, causing dieback of the canes in late summer.

**Natural and Biological Controls:** Insect-parasitic nematodes (species of *Steinernema*), applied as a drench to the crown area of the plants, can control currant borer larvae that are tunneling canes.

**Mechanical Controls:** Cut out and discard all canes that show evidence of wilting and dieback in spring, before adult moths emerge.

**Cultural Controls:** Increasing plant vigor through proper culture can reduce the severity of infestations.

**Chemical Controls:** Control of currant borers with insecticides has had only marginal success. If attempted, treatments should be applied to coincide with periods when the moths are flying and laying eggs, typically early June. However, very few insecticides are registered to permit use on currants and gooseberries.

## Raspberry Crown Borer (*Pennisetia marginata*)

**Damage:** Larvae tunnel into the base (crown) of raspberry plants, which causes wilting of the entire cane and can kill plants. Symptoms are most severe in midsummer.

**Life History and Habits:** The life cycle of the raspberry crown borer requires two years to complete. Adult moths emerge from the base of infested plants from mid-July through August. After mating, the females lay eggs around the canes. Eggs hatch in three to five weeks, and the young borer larvae tunnel into the lower canes. The first winter is spent in a small chamber cut into the side of the lower canes. The larvae feed throughout much of the following season, moving finally to the soft areas within the crown. They overwinter a second season within the plant and resume feeding in spring. Most damage is done by the larger, nearly full-grown larvae. During June and July, the larvae stop feeding, pupate and later emerge as adult moths.

**Natural and Biological Controls:** Drenches of insect-parasitic nematodes (species of *Steinernema*) around the base of plants can control larvae.

**Mechanical Controls:** Dig up and discard infested plants as soon as wilting occurs to kill larvae before they emerge.

**Chemical Controls:** Sprays of the insecticide diazinon, directed at the base of the plant during late summer, can kill the moths and newly hatched larvae. Drenches of insecticides applied after harvest in early fall can kill young larvae before serious injury occurs. However, registration of this treatment may be discontinued by the manufacturer. The raspberry crown borer has a life cycle that requires two years to complete, so controls need to be sustained over two to three years during outbreaks.

**Related Insects:** The bronze cane borer (page 100) and stemboring sawfly (page 127) are other regional insects that damage rose and raspberry.

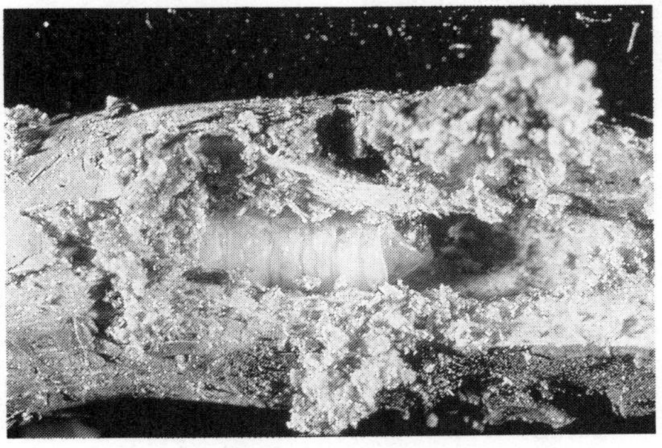

*Raspberry crown borer*

# *Hymenoptera* (*Bees, Ants, Sawflies, Etc.*)

## Small Carpenter Bees (species of *Ceratina* )

**Damage:** Several species of small carpenter bees (as well as several beneficial hunting wasps) commonly nest in the pith of rose, raspberry and other pithy plants. Most tunneling is confined to the dead areas of older canes. However, some tunneling may occur into live areas of the plant crown, causing dieback.

**Life History and Habits:** The adult bees are dark and metallic-colored. Like leafcutter bees, they are solitary bees that nest individually. After digging out a cavity in the plants for nesting, the mother sections it into small cells, packing each with pollen. The young bees feed on the pollen and become full grown in about 45 days.

**Mechanical Controls:** Cover exposed pith with wax, glue, shellac or a thumbtack to discourage the mother bees from nesting.

**Miscellaneous Notes:** Small carpenter bees show some unusually cooperative behaviors. After digging and provisioning the nest cells for her young, the mother bee usually waits in the tunnel until the new bees emerge. The developing bees, which are arranged in reverse order of distance from the nest opening, chew the opening above their head and then wait until the remaining bees have developed. The newly emerged adult bees then follow the mother out of the nest and often return to help clean the old nest so that it can be reused.

## Leafcutter Bees (species of *Megachile)*

**Damage:** Leafcutter bees cut characteristic semicircular notches out of leaves. Roses and ash trees are particularly favored by the bees. When very abundant, leafcutter bee damage can slightly reduce plant vigor and quality of rose blossoming. Gardens established in desert areas can serve as oases for leafcutter bees. In these instances, plantings can be very badly damaged.

*Leafcutter bees cut leaves to form nest cells.*

**Life History and Habits:** Adult leafcutter bees resemble small dark bumblebees. They are solitary bees; each female individually rears her young. When rearing young, the female bees cut leaves from rose, ash and other plants. They work quickly and leaf cutting is very rapid, occurring in as little as 10 seconds. The bees then take the leaf disks to nest sites excavated out of rotten wood, pithy plants or other hollows. The leaves are formed into thimble-shaped rearing cells, which are then packed with pollen on which the young develop.

**Biological Controls:** Leafcutter bees are attacked by many insects. Nests are frequently destroyed by blister beetles and velvet ants. Robber flies kill adult bees.

**Cultural and Mechanical Controls:** Usually, damage to a rose plant is little more than a curiosity and does not require control. Insect populations can be reduced by limiting breeding sites, such as the exposed pith of rose and caneberries, by sealing the opening with glue, shellac or a tack.

Where leafcutter bees are very numerous, protecting susceptible plants with netting is the only effective control.

**Chemical Controls:** Many insecticides can kill the adult bees but need to be reapplied frequently. Because of the beneficial habits of leafcutter bees as pollinators, insecticides should be used only when very serious infestations threaten. Insecticides are not effective where very high leafcutter bee populations occur.

**Miscellaneous Notes:** Most leafcutter bees (unlike honeybees) are native insects that are useful pollinators of several native plants and alfalfa.

## Gall Wasps (species of *Diplolepis*)

**Damage:** Several species of gall wasps are associated with *Rosa* species, producing bizarre growths on leaves and stems known as galls. These may take the form of balls, spikes or mossy growths. Species, rugosa and old garden roses are most commonly galled. Little plant injury occurs from these galls; they are primarily a curiosity.

**Life History and Habits:** Gall wasps overwinter in cells inside the galls, dropping to the ground to pupate in spring. Adults are small, inconspicuous, dark wasps that emerge and become active in spring. Eggs are laid on growing leaves and tissues, and subsequent feeding by the larval stage produces the gall. There is one generation per year.

*Galls on rose produced by gall wasps*

**Mechanical Controls:** Old galls can be handpicked and destroyed before adult insects emerge in spring.

## Stemboring Sawfly (*Hartigia trimaculata*)

**Damage:** A common cane borer infesting rose and raspberry plants is the stemboring sawfly. Damage is caused by the larvae, which tunnel into the stem, often girdling it. The top of the plant, beyond the injury, wilts and dies. Canes break easily at the injury point. Rose plantings near wild roses are more likely to be infested, since the wild plants may be a reservoir of this insect.

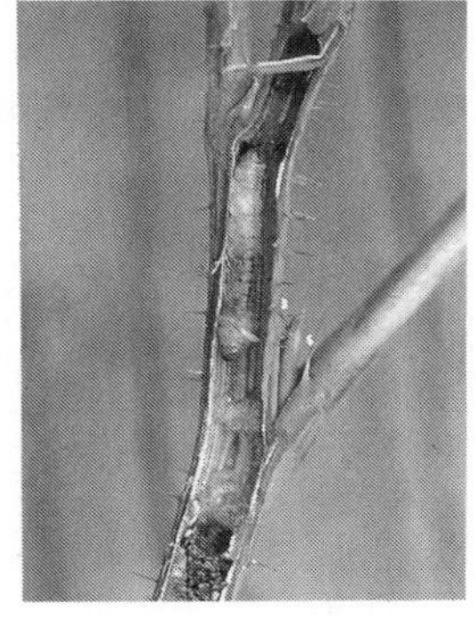

*Stemboring sawfly in raspberry*

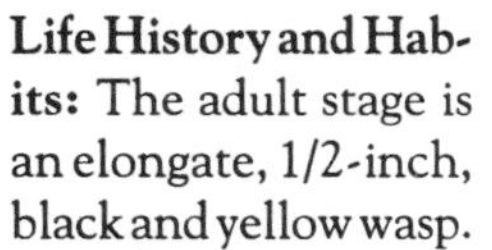

**Life History and Habits:** The adult stage is an elongate, 1/2-inch, black and yellow wasp. Adults emerge in late April and May and insert eggs under the bark at the tips of current season canes. Upon hatching, the larvae enter the stem to feed. They feed in the pith, eventually forming a small chamber in the upper stem during late June and July. There they pupate and a second generation of adult wasps gnaw their way out and emerge in late summer. These lay eggs, which hatch into larvae that eventually tunnel downward, spending the winter near the base of the plant.

**Natural and Biological Controls:** Parasitic wasps commonly attack and kill the full-grown sawfly larvae as they prepare to pupate.

**Cultural Controls:** Destroy the developing larvae by cutting and destroying canes when injury is first detected. Cut each affected cane below the wilted portion and examine the pith. Continue making short cuts until the pith is white.

**Related Insects:** Another stemboring sawfly, the raspberry horntail (*Hartigia cressoni*), attacks berry crops in the Pacific states.

## Imported Currantworm (*Nematus ribesii*)

**Damage:** The larvae chew the leaves of currants and gooseberries, often defoliating the plant extensively early in the season. Foliage in the interior of the bush is damaged first, but all leaves may be eaten. Yield and quality of fruit may be greatly affected by this injury.

**Life History and Habits:** The imported currantworm spends the winter in a cocoon in the soil around previously infested currants and gooseberries. The adult, a black and yellow wasp about 1/3 inch long, emerges early in spring. After mating, the female lays eggs in a row along the main veins of the leaf underside. The larvae hatch about 7 to 10 days after eggs are laid and first feed in groups, chewing small holes in the center of the leaves. Later they disperse throughout the plant and feed along the leaf margins, becoming full-grown in about three weeks. Young larvae are pale green, but develop distinctive dark spots as they grow and reach a size of about 3/4 inch. The full-grown larvae drop to the ground and form a cocoon. Some pupate and emerge in late June and July, producing a small second generation. The remainder become dormant and emerge the following year.

*Imported currantworm*

**Mechanical Controls:** In small plantings, the larvae can be controlled by handpicking or shaking. Careful examination of newly emerging leaves will reveal the eggs, which may be crushed. Most eggs and larvae are found in the interior of the shrub. At the end of the season, rake and remove all debris from the base of the plants. Most of the overwintering cocoons occur in this leaf litter and can be destroyed by this practice.

**Cultural Controls:** This insect is most damaging to shrubs in shaded sites. Planting in sun-exposed areas can reduce problems.

**Chemical Controls:** Chemical control options are limited. The larvae can be controlled with pyrethrins and malathion, some of the only insecticides permitted for use on currants and gooseberries. However, most garden formulations of malathion do not permit use on this crop. Sprays of irritants, such as soaps and wood ashes, have been reported to dislodge many of the larvae.

**Miscellaneous Notes:** The currant spanworm is another insect that often feeds on the leaves of currants and gooseberries. The larvae of this insect are also spotted, but are actually a type of inchworm with a distinctive looping walk. These are the immature stage of a small moth. The caterpillars can be controlled with *Bacillus thuringiensis*.

## Roseslug (*Endelomyia aethiops*)

**Damage:** The gray-green larvae of the roseslug feed on the upper leaf surface of roses and caneberries. Injury is characteristic since leaves are skeletonized, leaving the main veins and lower leaf surface intact. Old injuries often appear as small windowpanes through the leaf. Serious damage is extremely rare.

**Life History and Habits:** The roseslug overwinters in the pupal stage around the base of plants. The adult stage is a small, thick-bodied wasp. Female wasps insert eggs into the leaf. Emerging larvae feed on the upper leaf surface. After feeding is complete, they pupate in the ground. There is one generation per year.

**Mechanical Controls:** Hosing leaves with water can easily dislodge most feeding larvae.

**Chemical Controls:** Significant injury by roseslugs is extremely rare and controls are not often needed. The roseslug is also very susceptible to almost any garden insecticide.

**Miscellaneous Notes:** Occasionally, a second species, the bristly roseslug, may occur on rose or raspberry plants. This insect chews through the leaf but causes very little injury. The roseslug is also similar in habits to the pearslug, which feeds on fruit trees such as cherry, plum and pear, as well as some rose family shrubs, including cotoneaster.

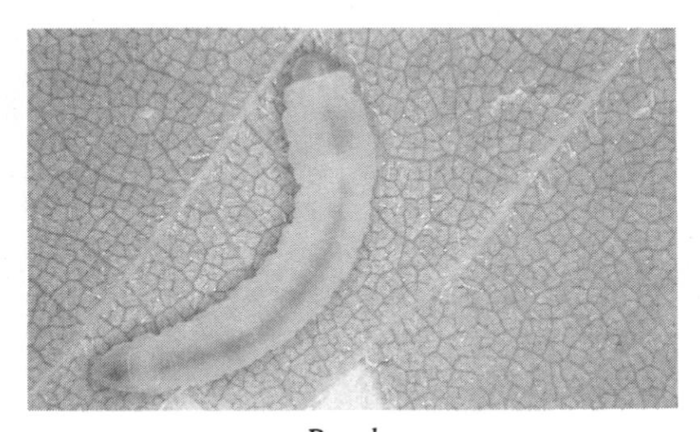

*Roseslug*

## Pearslug (*Caliroa cerasi*)

**Damage:** The sluglike larvae of the pearslug feed on the upper leaf surfaces of sweet cherry, plum, pear and several ornamental shrubs such as cotoneaster. The injury is characteristic, a skeletonizing defoliation where main veins and the lower leaf surface are not eaten. When abundant, pearslug injury can cause a reduction in fruit size and production the following season.

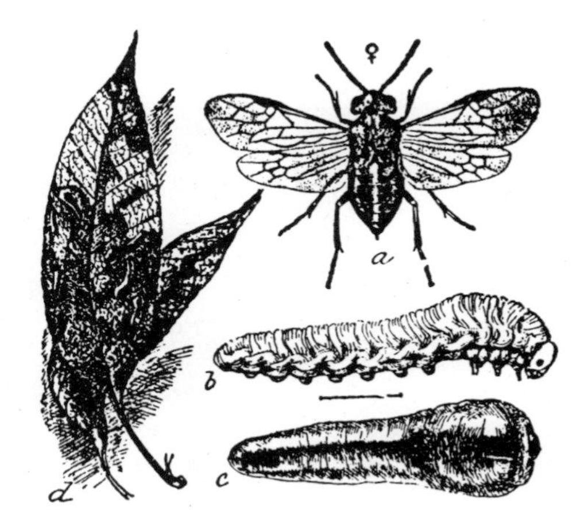


*Pearslug: A) Adult; B) Larva with slimy covering removed; C) Same as preceding in natural condition; D) Leaves showing slug injury.*

**Life History and Habits:** The pearslug overwinters as a pupa in a small cocoon at the base of plants attacked the previous season. The adult emerges in late spring, a small, black, nonstinging wasp that is rarely observed. After mating, the female wasps insert eggs into leaves, and the sluglike larvae hatch within one to two weeks. Young larvae are dark green, often turning more orange as they become nearly full-grown. They then drop from the plant and pupate in the soil. A second generation occurs in mid-August and September. The pearslug is also called the pear sawfly or cherry slug.

**Mechanical Controls:** Pearslug larvae can be washed off plants with a vigorous jet of water.

**Chemical Controls:** Dusting the plants with wood ashes can easily control pearslugs. The ashes stick readily to the body of the insect, and excess ashes can be removed immediately by shaking the plant. This treatment works most rapidly during hot, dry weather. The pearslug is also very easy to control with many garden insecticides. Insecticidal soaps are effective but may cause leaf spotting damage on some plums and cherries.

**Miscellaneous Notes:** Because of their fondness for cherries, pearslugs are often called cherry slugs. The closely related roseslug is a common pest of roses found in the region. Neither of these insects is related to true slug (see page 166), which are mollusks, a very different type of animal.

# *Diptera (Flies)*

## Seedcorn Maggot (*Delia platura*)

**Damage:** Germinating seeds of beans, corn, squash and most other garden vegetables can be destroyed or seriously scarred by the maggots. Damage is particularly severe where large amounts of decaying organic matter are present and when seedings are made in overly cool soils. Adult flies also can transmit the bacteria that cause soft rot of vegetables.

**Life History and Habits:** The seedcorn maggot spends the winter underground in the pupal stage. Adult flies emerge in early spring and lay eggs in soils abundant in organic matter. The newly hatched larvae feed on decaying plant matter, seeds or seedlings. Development is rapid and two to three generations may be completed when plantings are established in spring. The flies become dormant during periods of high temperatures and cause little injury after late spring.

**Natural and Biological Controls:** Dry soil conditions are very unfavorable to the survival of eggs and larvae. These soil stages are fed on by rove beetles and other predators. Adult flies commonly die from a fungus, which causes them to become stuck to the tops of plants and other high objects.

**Cultural Controls:** Plant in a manner that allows rapid germination to avoid the effects of maggot feeding. In particular, this involves use of adequately warmed and well-prepared seedbed, delaying planting of susceptible warm-season crops (beans, corn, squash) until soil temperatures are suitable.

Apply only decayed organic matter, such as compost, to reduce attraction to egg-laying flies. Till plant debris in fall and moisten it to allow some decay before spring.

**Mechanical Controls:** Yellow sticky traps can capture adult flies and help suppress populations. Fine-meshed cloth, such as cheesecloth or row cover, or screening can exclude flies from the planting.

**Chemical Controls:** Some insecticides can provide control when applied to the seed (seed treatment) or as a band over the plants during seeding.

**Related Species:** The onion maggot is a closely related fly that limits attacks to onions and related plants. The cabbage maggot, which rarely occurs in the region, attacks radishes, turnips and other cabbage family plants. Spinach is occasionally damaged by the sugarbeet root maggot (*Tetanops myopaeformis*), which tunnels the roots and causes wilting.

## Onion Maggot (*Delia antiqua*)

**Damage:** Onion maggot larvae tunnel into onion plants. Small plants may be killed and larger bulbs may become susceptible to bulb rots due to the wounding. Related alliums (chives, scallions) can also be attacked.

**Life History and Habits:** The onion maggot spends the winter underground in the pupal stage within the puparia, a stage that resembles a dark

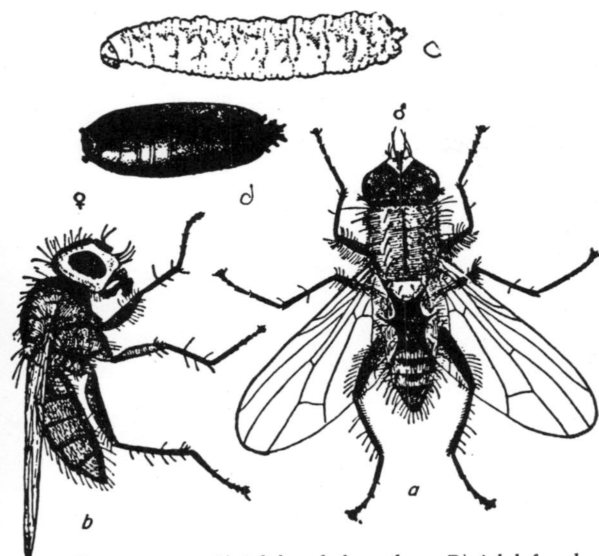

Onion maggot. A) Adult male from above; B) Adult female from the side; C) Larva; D) Puparium.

*Onion maggot*

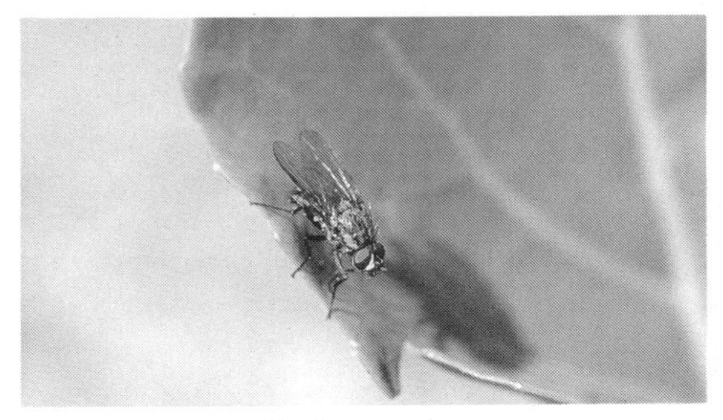
*Seed corn maggot*

piece of grain. Adult flies emerge over a period of several weeks in late spring. The flies lay eggs in soil cracks around the base of onion plants. The maggot-stage larvae hatch and tunnel into the plants, feeding particularly around the base. After finishing feeding, they move a short distance from the plant and pupate. Two or three generations may occur during the growing season.

**Cultural Controls:** Dispose of old, infested onions before maggot emergence in spring to reduce overwinter populations. Alternatively, you may wish to plant some rotting onions as a trap crop for the adult onion maggots, which prefer the larger onions early in the season. However, use of this technique requires that the trap crop onions be destroyed to kill developing maggots. High-moisture conditions favor survival of the eggs and maggots. Minimizing watering can assist in onion maggot control.

**Mechanical Controls:** In small plantings, floating row covers can be useful when combined with crop rotation. Adult flies can also be trapped on yellow sticky cards.

**Biological Controls:** Several predators of the onion maggot have been found in other areas of North America. In particular, larvae of rove beetles can be important natural controls. Attempts to control onion maggots with insect-parasitic nematodes (species of *Steinernema*) have been erratic, despite claims of some nematode producers.

**Chemical Controls:** Use of insecticides for control of the onion maggot in gardens has primarily involved treatment with diazinon at planting time. The future of this practice is questionable due to registration changes.

**Miscellaneous Notes:** The seedcorn maggot also commonly attacks onion bulbs that have been damaged by disease. Evidence shows that the seedcorn maggot does not directly damage healthy bulbs, as does the onion maggot.

**Related Species:** The seedcorn maggot (see page 130) is a closely related fly that damages germinating seeds and is found in decaying bulbs and tubers. The cabbage maggot, which rarely occurs in the region, attacks radishes, turnips and other cabbage family plants. Cabbage maggot controls are similar to those for onion maggots.

## *Leafminers*

Some of the most discriminating feeders among the insects are the leafminers. These insects tunnel between the upper and lower leaf surfaces, feeding on the soft inner tissue and avoiding the tough epidermis.

Many different types of insects share the leafmining habit, including the larvae of various flies, small moths, beetles and sawfly wasps. However, they are often classified as a group, based on the pattern of the mine they make. *Blotch* leafminers make irregular blotchy mines that are often black or brown. *Serpentine* leafminers make meandering mines that gradually become larger. Mines produced by *tentiform* leafminers pop up in the middle, like a pup tent.

Leafminer injury is often easily confused with leaf spotting caused by fungus infection. However, it can be diagnosed by pulling apart the leaf at the injury. Leaves damaged by leafminers have a distinctly separate top and bottom leaf surface and often will contain the insect or its droppings. Leaves damaged by fungi have collapsed and compressed cells, which makes them difficult to separate.

Adult stages of leafminers insert eggs into the leaves or lay them on the lower leaf surface. The young insects tunnel into the leaf, where they develop for about two to four weeks. Often the insects cut their way out of the mine when they have finished feeding, dropping to the soil to pupate. Most leafminers have about two generations in a season.

Because the mines are conspicuous, they often attract attention. However, leafminers rarely cause serious injury to plants, since plants easily outgrow the damage. Leafminers suffer from many natural enemies, so ignoring the problem will often allow it to go away. Spinach leafminers, common on beets, spinach and chard, can sometimes make these vegetables very unpalatable, however.

Leafminers are protected from most insecticides once they have entered the leaf. Only systemic insecticides are effective after the mining has begun. (This is *not* an option on food plants, such as spinach!) Nonsystemic insecticides are effective only if applied as eggs are laid, before tunneling occurs. Hand crushing eggs, picking infested leaves or use of row covers are other controls.

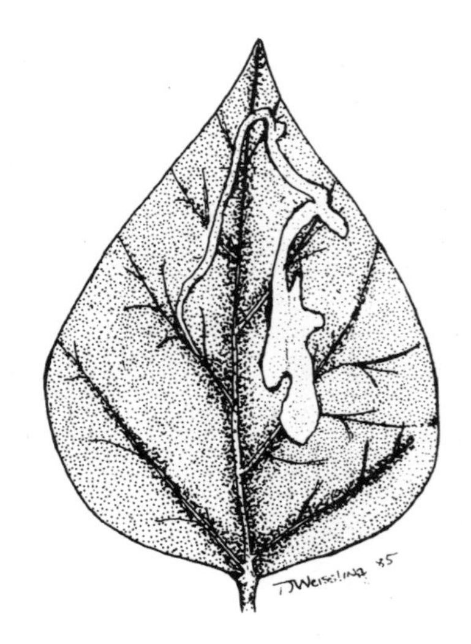

*Serpentine leafminer*

| COMMON REGIONAL LEAFMINERS AND THEIR HOST PLANTS | | |
|---|---|---|
| **Leafminer** | **Plants attacked** | **Description of mine** |
| Spinach leafminer | Spinach, beets, lambsquarters, chard | Dark blotchy mine |
| Cottonwood blackmine beetle | Cottonwood, poplars | Dark blotchy mine |
| Vegetable leafminer | Most vegetables (cabbage, melons, celery, etc.), some flowers, weeds | Small serpentine mine |
| Lilac leafminer | Lilac | Light blotchy mine, with some leaf folding |
| Tentiform leafminer | Apple | Light blotchy mine with spotting from sap sucking |
| Columbine leafminer | Columbine | Large blotchy mine |
| Grape leafminer | Grape, Virginia creeper | Very thin serpentine mine |
| Aspen/poplar leafminer | Aspen, cottonwood, poplar | Very thin serpentine mine |

## Spinach Leafminer (*Pegomyia hyoscyami*)

**Damage:** The immature maggot stage tunnels through older leaves of beet, Swiss chard, spinach and related weeds such as lambsquarters. The large, dark blotchy mines that are produced destroy the leaves for use as greens, although effects on plant growth appear to be minimal.

*Spinach leafminer larvae*

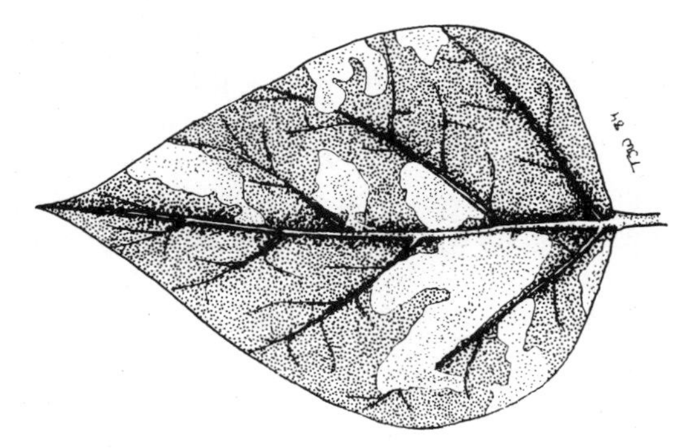

*Blotch leafminer*

**Life History and Habits:** The insect overwinters in the soil as a pupa, emerging in mid-spring. The adult flies lay small masses of white eggs on the underside of older leaves. Upon hatching, the

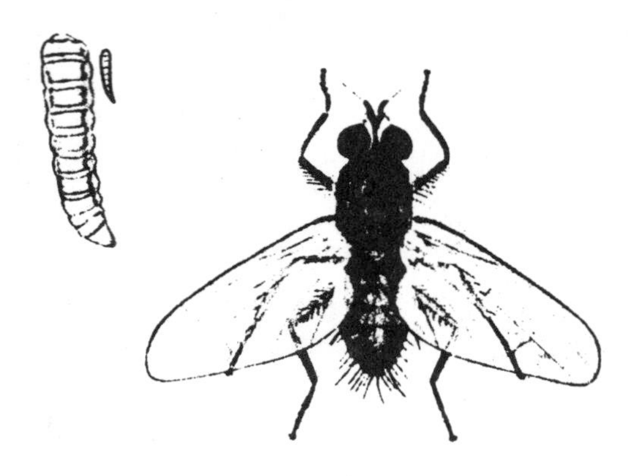

*Spinach leafminer larvae (left), adult fly (right)*

young maggots tunnel into the leaves, where they feed, typically for two to three weeks. Upon becoming full-grown, they cut through the leaf, drop to the ground and pupate in the soil. Several generations may be completed during the season, but almost all damage occurs during mid- to late spring.

**Mechanical and Physical Controls:** Floating row covers, cheesecloth or other screening can exclude the adult flies and prevent egg-laying. Regularly inspect leaves for newly laid egg masses on leaf undersides, which can then be crushed before tunneling begins. Infested leaves should be picked and destroyed to kill the larvae before they complete development. These infested leaves need to be removed from the planting, since the insects can continue to grow in the picked leaf.

**Cultural Controls:** Fall plantings and very early spring plantings can largely avoid infestations.

**Chemical Controls:** Weekly applications of insecticides to the foliage can kill adult flies and small larvae before they enter the leaves. No registered insecticides are available that will kill maggots already within leaves. Sprays used on spinach usually require that one to two weeks elapse before harvest.

## Western Cherry Fruit Fly (*Rhagoletis cingulata*)

**Damage:** The adult female flies "sting" the cherry fruit, producing small puncture wounds. Eggs are often laid in the punctures, and the immature maggots chew through the flesh of the fruit. Infested berries are misshapen, undersized and mature rapidly. In most areas, the western cherry fruit fly limits damage to cherries, including wild types. However, strains have developed in some regions (e.g., Utah) that attack and tunnel apples.

**Life History and Habits:** The western cherry fruit fly spends the winter in the pupal stage around the base of previously infested trees. In late spring, the flies emerge and feed on aphid honeydew and other fluids, including oozing sap from wounds made by fruit punctures. The females then insert eggs into the fruit. The immature maggots feed on the fruit, particularly the area around the pit. They become full-grown in two to three weeks and drop to the ground to pupate. In most areas there is one generation per year. However, a few flies may emerge and produce a second generation if susceptible fruit is available.

**Mechanical Controls:** Adult flies are easily trapped by yellow sticky cards or sticky red spheres. How-

*Apple maggot injury*

ever, it has not been shown that trapping alone can adequately control the western cherry fruit fly.

**Cultural Controls:** Prematurely ripening cherries, often infested with developing fruit flies, should be picked and destroyed before the insects drop to the soil. Control of aphids, which produce honeydew fed on by the flies, can reduce egg-laying. Where possible, tilling around the base of the trees in fall can kill many of the overwintering pupae.

**Chemical Controls:** The western cherry fruit fly can be controlled by several insecticides applied during periods when the adults are mating and beginning to lay eggs. Yellow cards or red spheres with a sticky coating can be used to determine when adult flies are present.

**Related Species:** The western cherry fruit fly is closely related to the apple maggot (*Rhagoletis pomonella*), a serious apple pest in the eastern United States. Although the apple maggot is sometimes found in the West, damage to apples is rare and it generally favors native plants, such as hawthorn. Controls are similar to those for the western cherry fruit fly. Another related species, the walnut husk fly (*Rhagoletis completa*), is widespread in the West, where it feeds on walnuts. It only rarely develops in peaches, although it never becomes abundant in these.

## Rose Midge (*Dasineura rhodophaga*)

**Damage:** The small maggot stage of the rose midge feeds by making small slashes in developing plant tissues to suck the sap. Developing flower buds are usually killed or distorted by this injury.

**Life History and Habits:** The rose midge overwinters in the pupal stage in the soil. The adult stage, an inconspicuous, small fly, emerges in late spring, sometimes after the first crop of blossoms. The adult stage lives only 1 or 2 days, but during this time the females lay numerous eggs under the sepals, in opening buds and in elongating shoots. Hatching larvae slash plant tissues and feed on sap. Eggs hatch in a few days, and the larvae feed for about a week before dropping to the soil to pupate. The complete life cycle can take about two weeks, with numerous generations occurring during a growing season.

**Mechanical Controls:** Infested plantings should be examined every few days and all damaged buds trimmed and removed.

**Chemical Controls:** The rose midge is difficult to control with insecticides. (Insecticides may be applied to the plants to kill maggot stages.) Treatments of soil at the base of the plant can further reduce populations by killing the insect as it attempts to pupate.

## Narcissus Bulb Fly (*Merodon equestris*)

**Damage:** The narcissus bulb fly tunnels into the base of narcissus, daffodil, hyacinth and many other bulbs during late spring and summer. Bulbs are killed completely or are greatly weakened and susceptible to rot.

**Life History and Habits:** The insect spends the winter as a large maggot inside the bulb. It pupates the following spring, and the adult flies emerge around mid- to late May. The narcissus bulb fly is yellow and black, resembling a bumblebee. Female flies lay eggs in soil cracks around the plant stems for one to two months. The newly hatched maggots crawl down along the plant stem and tunnel into the base of the bulb. They continue to feed throughout the summer, becoming full-grown in fall.

**Cultural Controls:** Carefully inspect all newly purchased bulbs and discard any that are soft and show signs of rot. Also inspect bulbs during fall digging and discard any that appear infested. Problems with this insect are reportedly less severe in windswept, versus more protected, sites.

**Chemical Controls:** Insecticides applied around the base of plants during late spring can kill the adult flies and newly hatched maggots. Egg-laying, however, occurs over an extended period and necessitates multiple treatments.

**Related Insects:** The lesser bulb fly (*Emerus tuberculatus*) is a small fly that has been found in narcissus, hyacinth, onion, shallot and iris. Its habits are little studied, but apparently there are two generations per season.

# *Thysanoptera (Thrips)*

## Onion Thrips (*Thrips tabaci*)

**Damage:** Onion thrips feed on the foliage of a wide variety of plants and can be damaging to several vegetables, including onions, beans and cabbage. Feeding injuries typically appear as scarring wounds in leaves resulting from destruction of the upper cell layers. Most damage occurs as reduced vegetable size related to the feeding wounds. Some varieties of cabbage react to injuries by producing warty growths, similar to edema, a physiological disorder related to moisture imbalance. (See section on edema on page 198.) Onion

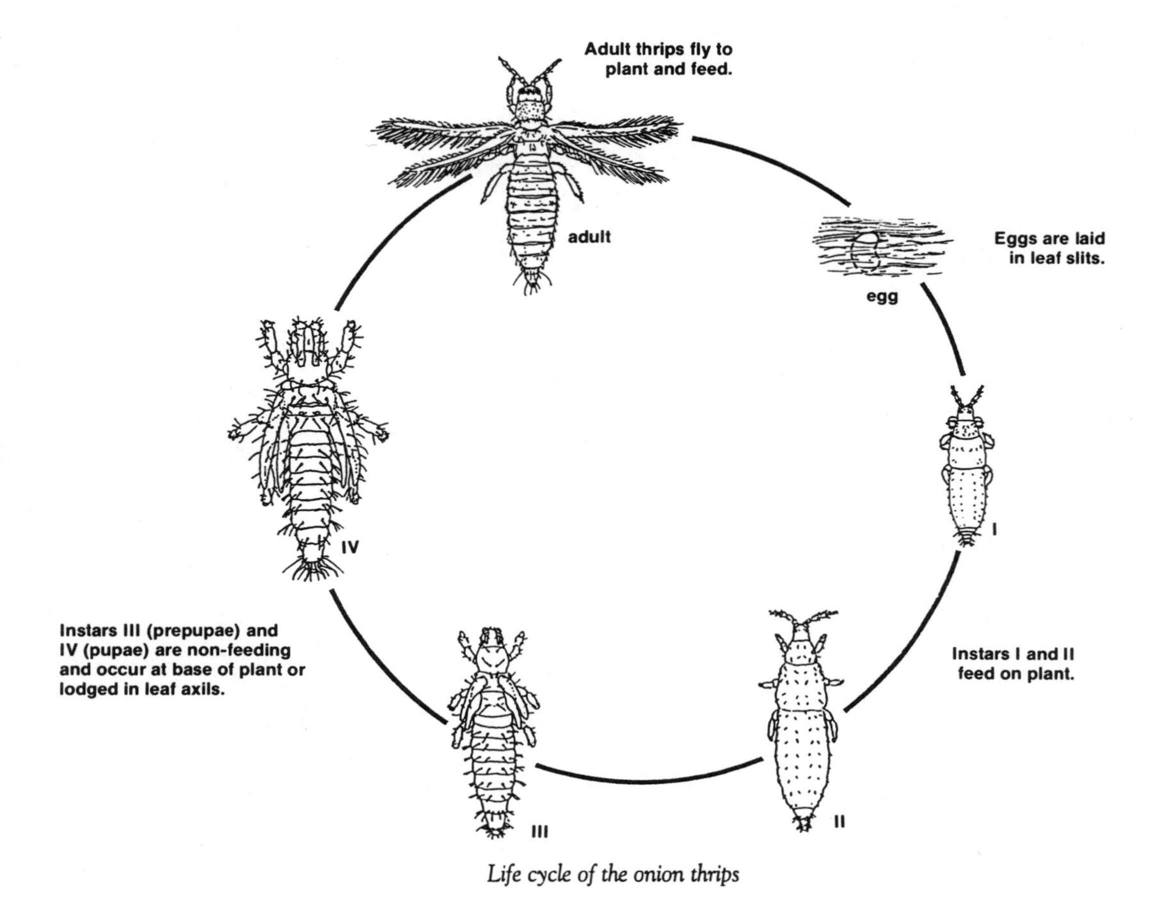


*Life cycle of the onion thrips*

thrips can transmit tomato spotted wilt virus, but they are much less important vectors of the disease than the western flower thrips.

**Life History and Habits:** Onion thrips overwinter in the adult stage throughout the region in protected sites and old plant materials. They may also be introduced into a field on infested transplants and are common on onion sets. Eggs are inserted into leaves and stems. They hatch in about one week, and two feeding wingless stages (nymphs) occur on the plant, lasting about two weeks. These are followed by two nonfeeding stages (pre-pupa and pupa), which occur in the soil or in crevices on the plant. The winged adult stage commonly disperses throughout an area and may fly long distances aided by winds. Several generations occur annually, and for most of the season all life stages may be present.

**Natural and Biological Controls:** Weather, notably rains that trap insects in the soil, can be very important in limiting thrips populations. Serious thrips infestations are typically related to extended dry weather. Small predators, such as minute pirate bugs, predatory thrips and predator mites, also commonly feed on thrips. However, thrips are often well protected in the plant and many predators may be unable to reach them.

**Mechanical Controls:** Overhead irrigation, or purposeful hosing of plants, is destructive to thrips and can help provide control.

**Cultural Controls:** Resistance of onions to onion thrips is associated with waxy leaves and a spreading habit of top growth. Cabbage varieties have a wide range of reactions to thrips feeding, with many varieties showing little effect from thrips.

**Chemical Controls:** Thrips exposed on leaves may easily be controlled with insecticidal soaps and many garden insecticides. However, thrips hidden within plant parts, such as inner leaves of cabbage or onions, can be extremely difficult to control.

**Miscellaneous Notes:** Onion thrips are extremely common on most produce, making them the most frequently eaten insect in North America.

## Western Flower Thrips (*Frankliniella occidentalis*)

**Damage:** Feeding injuries to vegetables and flowers by western flower thrips are minor, usually involving slight scarring. Halo spotting of fruits may also occur from egg-laying wounds. The importance of western flower thrips to vegetables has greatly increased in recent years because it is the most important vector of tomato spotted wilt virus. (See discussion of tomato spotted wilt on page 174.) Tomatoes and peppers, as well as many greenhouse crops, have sustained serious losses to this disease.

**Life History and Habits:** The western flower thrips is thought to overwinter in much of the region in the adult stage. Some may migrate in on winds from the south. Adults tend to seek out flowers for feeding and egg-laying, although feeding sometimes occurs on foliage. Eggs are laid in plant tissues, and the nymphs feed on the flowers and developing fruit. Nonfeeding pre-pupa and pupa stages occur in the flower, in crevices on the plant or in soil around the base of the plant. Generations are produced throughout the growing season at approximately three to four week intervals.

**Natural and Biological Controls:** Heavy rains and crusting soils are very important abiotic controls of flower thrips. Predatory thrips and minute pirate bugs are common natural enemies that reduce overall populations.

**Mechanical Controls:** Overhead irrigation can kill many thrips.

**Cultural Controls:** Darker flowers with blue coloration are more attractive to the adult thrips. The most important means of reducing tomato spotted wilt is to ensure that disease-free transplants are used. Although transmission of the disease between garden plants can occur, and is common in warmer areas of the country, most spread of tomato spotted wilt in the West is limited to greenhouses. Diseased plants should be removed and promptly destroyed (to kill the thrips), particularly early in the season.

**Chemical Controls:** Chemical control of flower thrips is very difficult and seldom exceeds 50 percent. Since the most important injury is transmission of tomato spotted wilt virus, it is far more important to concentrate on cultural controls that remove tomato spotted wilt–infected plants from the garden.

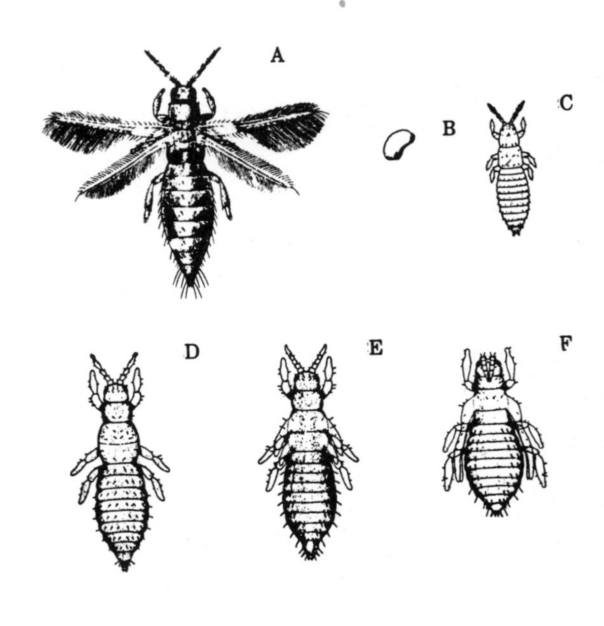


*Gladiolus thrips. A) Adult; B) Egg; C-D) Larvae; E) Prepupa; F) Pupa.*

## Gladiolus Thrips (*Thrips simplex*)

**Damage:** Gladiolus thrips are very small but can be extremely destructive to gladiolus. Feeding by gladiolus thrips results in scarring injuries caused when the insect punctures cells and removes plant sap. Infested leaves appear silvery, later browning and dying. Thrips damage to developing flowers causes flecking and distortion. Heavy infestations can prevent flower production. Gladiolus thrips may also damage the corms in storage, causing them to become sticky and dark from plant sap at wounds. Infested corms produce poorly when replanted. Gladiolus thrips may also feed on lily and iris, which rarely are seriously damaged.

**Life History and Habits:** In the region, gladiolus thrips cannot survive winters outdoors. Infestations are maintained by thrips that move to corms and hibernate on them in winter storage areas. When corms are replanted in spring and temperatures rise above 60°F, they begin to reproduce and infest the developing shoots. Several generations of gladiolus thrips occur during the growing season, each of which may be completed within a few weeks.

**Mechanical and Physical Controls:** Corms can be freed of thrips by dropping them in a hot water bath, between 112°F and 120°F, for 20 to 30 minutes. Spraying developing plants repeatedly with a vigorous jet of water can kill exposed thrips, although many remain protected in developing leaves.

**Chemical Controls:** Corms can be fumigated with naphthalene moth balls. The treatment requires exposure for a week and a half to three weeks at temperatures above 50°F. During the growing season gladiolus thrips can be controlled with insecticides, particularly those that are systemic. However, because gladiolus thrips tend to hide in tightly curled new growth, it is difficult to get effective coverage, and repeated applications are required.

# *Orthoptera (Grasshoppers, Crickets, Etc.)*

## Grasshoppers

Redlegged grasshopper
(*Melanoplus femur-rubrum*)
Differential grasshopper
(*Melanoplus differentialis*)
Twostriped grasshopper
(*Melanoplus bivittatus*)
Migratory grasshopper
(*Melanoplus sanguinipes*) and others

**Damage:** During outbreak periods, grasshoppers can be the most destructive insect pests of vegetable crops, flowers, rangeland and many field crops in the West. Pest species will chew on leaves and flowers, sometimes extensively defoliating and even killing plants. Although some garden plants seem to be preferred (e.g., lettuce, beans, corn, most flowers), essentially all plants are potential food. When very high populations are present, twigs and stems of woody plants may be chewed and girdled also.

**Life History and Habits:** Most of the pest species of grasshoppers in the region spend the winter as eggs. These are laid underground in late summer or early fall in the form of elongate egg pods containing 20 to 120 eggs each. Egg-laying is concentrated in dry, undisturbed areas, being very heavy along roadsides, in pasturelands and on native prairie. Relatively little egg-laying occurs in an irrigated yard or garden, and is largely limited to small dry areas in the yard, such as between sidewalk cracks. Eggs hatch in late May and June. The immature nymphs take two to three months to become fully developed. Adults are present during August and remain until they are killed by a heavy frost. Grasshoppers feed during the day, resting on shrubbery, tall plants or human structures during the late afternoon and night. Movement into yards accelerates as native vegetation becomes less suitable due to summer drying or defoliation. Light frosts that may kill many host plants in the fall further concentrate grasshoppers on remaining plants.

**Natural and Biological Controls:** Grasshopper outbreaks tend to occur in cycles of roughly eleven years, with serious populations often lasting for two to three years. However, localized occurrence can be much more erratic. A large number of natural controls can reduce outbreaks. Cool, wet weather in spring, particularly around the period of egg hatch, is one of the more common conditions associated with declining grasshopper populations. Grasshoppers can also suffer from several diseases. The fungal disease *Entomophthora grylli* can be very destructive to populations, although spread is dependent on adequate moisture. A very large nematode (*Mermis nigrescens*), several inches in length, also kills or sterilizes many grasshoppers. The microsporidian *Nosema locustae* can kill or weaken grasshoppers. It is also manufactured and sold as a type of microbial insecticide under trade names such as Semaspore, Grasshopper Spore and NoLo Bait. It is most effective against young grasshoppers and may kill many under optimum conditions. Older grasshoppers are much less sensitive and rarely die, although they may be less vigorous and lay fewer eggs the following season. Since *N. locustae* is slow acting, it is not effectively used within the garden when grasshoppers are moving into the yard. Instead, it should be applied to breeding areas outside the yard before migrations occur. Insects that feed on grasshoppers include the larvae of blister beetles (predators of the eggs), robber flies and parasitic flies. Grasshoppers are also common foods of many birds such as the horned lark, American kestrel and Swainson's hawk. Poultry, notably guinea hens and turkeys, feed readily on grasshoppers and can assist in controlling moderate grasshopper populations. Grasshoppers have even been used as a highly nutritious poultry feed.

**Cultural Controls:** Maintaining relatively lush areas of grass or other plants around the perimeter of a property can divert the feeding activities of many grasshoppers from the vegetable or flower garden. (This technique is further improved if this trap crop is treated with an insecticide.) Grasshoppers can also be largely excluded by floating row covers, although they can chew through lightweight fabrics and plastic screens. Tillage in the garden destroys most egg pods that are laid in these sites. During outbreaks, it becomes extremely difficult to grow many of the plants more favored by grasshoppers, so it is sometimes best to concentrate on growing early-maturing plants and plants that are less favored by grasshoppers, such as squash and tomatoes.

**Chemical Controls:** Since most breeding of grasshoppers occurs outside of the yard or garden, control of grasshoppers in these areas is most productive. Sprays of several insecticides can kill grasshoppers on rangeland or in pastures. In addition, various baits can be used (insecticide on a molasses-bran base) that allow the insecticide to be applied in a more selective manner, killing fewer beneficial insects. As grasshoppers pour over the fence, insecticide treatments in the yard and garden are of little effect. The relatively short duration of the insecticides (usually only a couple of days), plus the continuous immigration of new grasshoppers, means that sprays must be made very frequently to have any effect.

**Miscellaneous Notes:** Over sixty species of grasshoppers are native to the high plains and Rocky Mountain region. Most of these do not occur in large numbers and exist as destructive pests. Many also have selective feeding habits and feed only on grasses or native shrubs (forbs) that keep them out of yards and gardens. The life cycles of various grasshoppers can also vary, with many overwintering as nymphs or adults.

## *The Mystery of the Rocky Mountain Locust*

One of the gravest threats to the early European settlers of western North America was the Rocky Mountain locust. Periodically, huge swarms of these migratory grasshoppers would take flight from their breeding areas in Colorado, Wyoming and Montana and sweep south and east on prevailing winds. The intensity of the situation makes current grasshopper problems seem tame, as indicated in an account of the famous outbreak of 1874 by Nebraska settler Everett Dick, in his book *The Sodhouse Frontier*:

> They came like driving snow in the winter, filling the air, covering the earth, the buildings, the shocks of grain, and everything.... Their alighting on the roofs and sides of the houses sounded like a continuous hail storm. They alighted on trees in such great numbers that their weight broke off large limbs.... At times the insects were four to six inches deep on the ground and continued to alight for hours. Men were obliged to tie strings around their pant legs to keep the pests from crawling up their legs.

*Rocky Mountain locust, laying eggs in the ground. A) Females with their abdomens in the ground; B) An egg pod broken open; C) Scattered eggs; D) Egg packet in the ground.*

These flights of locusts had occurred for centuries, recorded in the famous Grasshopper Glacier near Granite Peak, Montana, where layers of grasshoppers had collected in bands. However, shortly after the new settlements began to spread and grow in earnest, the Rocky Mountain locust disappeared. No living specimen of this insect has been observed since 1902. Like the passenger pigeon, another animal that once was abundant beyond number, the Rocky Mountain locust appears to have become extinct!

What happened is one of the great mysteries of entomology. Some speculate that changes in the environment caused by settlement—farming, ranching and the slaughter of the bison—somehow destroyed the habitat required for the Rocky Mountain locust. Others feel that it is still roaming in the guise of the migratory grasshopper, *Melanoplus sanguinipes*, waiting for the proper conditions to allow its reemergence. Many grasshoppers, such as the Mormon cricket of the western United States and the desert locust of the Sahara, radically change their form and habits when overcrowded, periodically producing swarms similar to that of the Rocky Mountain locust. However, no one has ever been able to cause the migratory grasshopper, or any other grasshopper, to change into the Rocky Mountain locust form.

# *Dermaptera (Earwigs)*

## European Earwig (*Forficula auricularia*)

**Damage:** The European earwig is the most common earwig found within the high plains and Rocky Mountain region, having migrated into the area during the 1940s and 1950s. They have extremely wide food tastes and feed on several types of plant materials. Flower blossoms, corn silks (and some corn kernels) and tender vegetable seedlings are commonly fed on by earwigs. Their habit of crawling into tight, dark, moist places to spend the day makes them an unwanted presence in harvested fruits, vegetables and flowers. The European earwig also feeds on many insects and has even been used

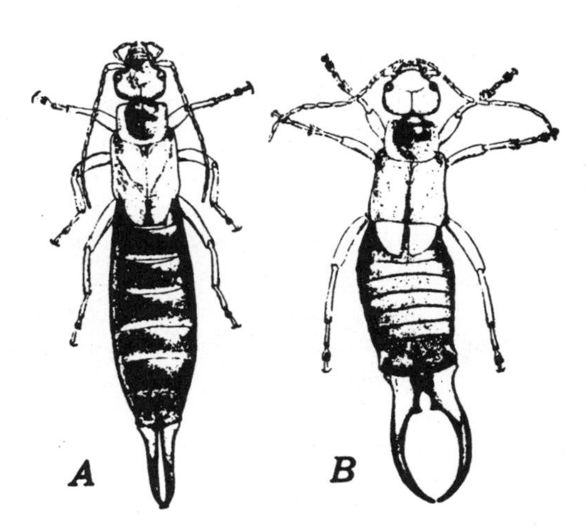


*European earwig. A) Female; B) Male.*

## *Earwigs*

Few garden insects are so universally disliked as the earwig. Its prominent pincers (more correctly known as forceps or cerci) have inspired numerous folktales. Earwigs have the habit of seeking out dark (preferably moist), tight spots to spend the day. This means they can get into almost everything and are often encountered where we least expect them.

The most common earwig found in the region is known as the European earwig (*Forficula auricularia*). An introduced species, the European earwig spread into the Rocky Mountain area shortly after World War II. Although they are capable of flying short distances, earwigs have largely "piggybacked" their way along with humans.

Earwigs can cause problems in vegetable and flower gardens. Tight flower petals and corn silks are favored hiding sites. Soft plant materials, such as flower petals and seedling plants, may be seriously damaged. However, earwigs are often blamed for injuries caused by other insects. For example, tunneling wounds produced by codling moths in fruit or corn borers in vegetables are often visited by earwigs as a daytime shelter. Although they did not cause the original injury, earwigs show up at the "scene of the crime."

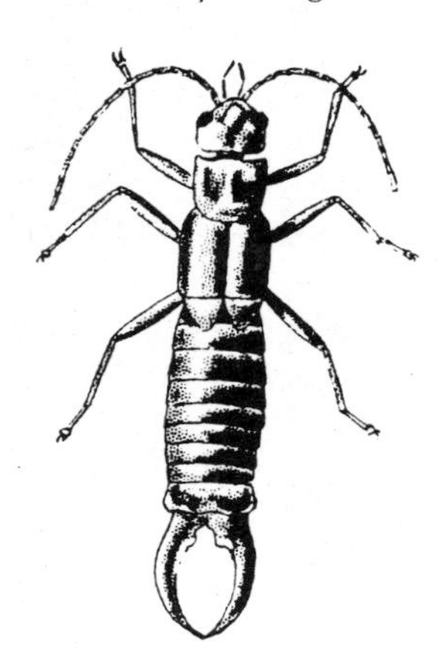

Earwigs also have several desirable habits that are often overlooked. Insects as well as plants are fed upon, and earwigs can be important predators of insects and insect eggs. Indeed, earwigs have even been successfully encouraged (by providing sheltering burlap bands) to control insect pests on fruit trees.

Earwigs provide an exemplary model of parental care, for an insect. Mother earwigs guard the eggs and tend the young earwigs for several weeks until they are able to forage on their own. From our perspective, this should compare very favorably to the "drop off the eggs and run" behavior of most insects.

to help control aphids in fruit trees. Soft-bodied insects, such as aphids and young insect larvae, as well as insect eggs are common foods of earwigs.

**Life History and Habits:** Earwigs overwinter in the adult stage. The female earwig produces an egg mass in late winter in a small cavity dug underneath a rock or in some other protected site. The mother earwig guards the eggs and tends to the young earwigs for several weeks until they have molted and are ready to leave the nest. The young then forage on their own, becoming fully grown in about one month. Often the mother lays a second smaller egg mass in May or June. There is only one generation per year, but developing earwigs are present throughout the growing season.

**Mechanical Controls:** The habit of earwigs to seek shelter during the day causes them to collect under boards, moistened newspapers or similar objects. They can easily be collected from these sites and destroyed.

**Chemical Controls:** Several types of baits containing an insecticide are available to control earwigs. These should be spread around the base of plants or under shelters where earwigs tend to concentrate their activity.

**Miscellaneous Notes:** Despite their appearance, earwigs are essentially harmless to humans. If handled or crushed, they can produce a moderately painful pinch from their jaws. The pincers, or cerci, which protrude off the hind end, are used during mating but have little force and do not cause a painful pinch. Males have more broadly bowed cerci; those of the females are slender and relatively straight.

# *Hemiptera (True Bugs)*

## Squash Bug (*Anasa tristus*)

**Damage:** Adults and nymphs feed upon most vine crops, particularly winter types of squash and pumpkins. During feeding they puncture cells and inject damaging saliva into the plant that cause areas of the stems and leaves to wilt and die off. Plants may be killed by midsummer. Late in the season squash bugs may also feed directly on fruits, producing wounded areas that allow rotting organisms to enter.

**Life History and Habits:** Squash bugs spend the winter in the adult stage in protected sites around previously infested plantings. They become active and first appear and feed in June, shortly after plant emergence. At this time they mate, and females lay masses of shiny brown eggs on leaf undersides. After hatching, the nymphs feed together in groups, usually on the shaded undersides of the plant. The nymphs of the first generation mature by early July, followed by a second generation that is often much more numerous and destructive than the first. Adults that develop late in the season do not lay eggs and leave the field for overwintering shelter. During warm seasons in southern areas, some of the second-generation adults may continue to develop and lay eggs, producing a third generation.

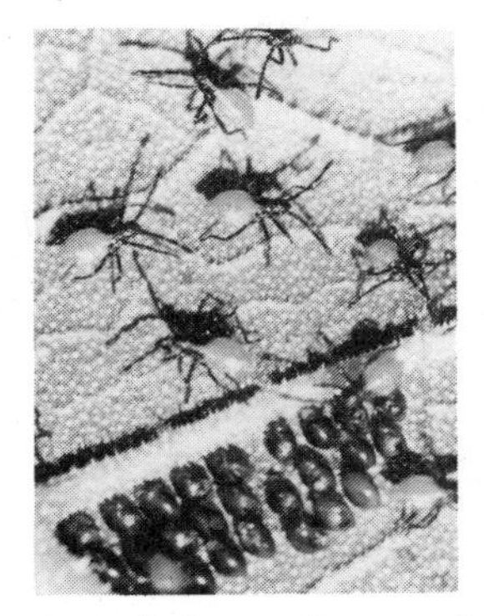

*Squash bugs. Eggs and hatched nymphs.*

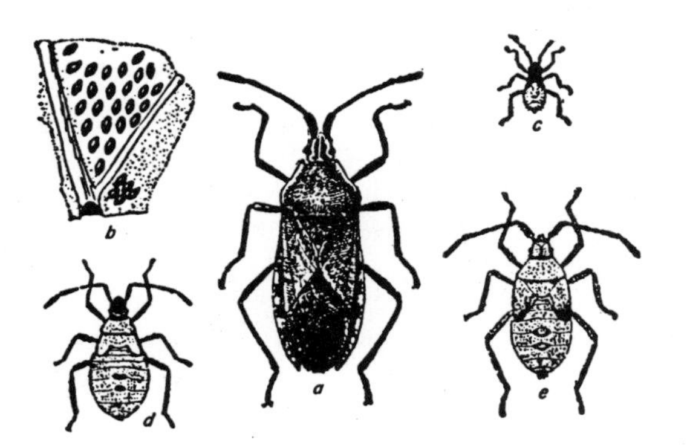


*Life stages of the squash bug. A) Adult; B) Eggs (about natural size); C-E) Various growth stages.*

**Biological Controls:** A wasp parasite of the eggs and a tachinid fly parasite of nymphs have been recorded. However, these do not usually provide adequate control in many areas. Northern areas of the region are rarely infested, perhaps due to erratic and killing spring frosts or other environmental conditions.

**Cultural Controls:** There are conflicting reports as to the feeding preference of squash bugs on different types of squash. Early studies indicated that winter squash (*Cucurbita maxima*) were more damaged than summer squash (*Cucurbita pepo*) and butternut squash (*Cucurbita moschata*). However, recent studies indicate that much egg-laying occurs on summer squash when the insect has a choice. Little egg-laying and development occur on cucumber and melons. Trap crop planting of the more preferred species can be used to divert feeding from less preferred cucurbits.

**Mechanical and Physical Controls:** Some protection early in the season is possible using floating row covers or other screening on small plantings. However, these covers must be removed prior to flowering to allow pollination.

Regularly inspect plants to locate egg masses on leaf undersides, which can then be crushed. Flat boards or shingles placed around plants attract many squash bugs. These can be collected and destroyed.

**Chemical Controls:** The squash bug has always been extremely difficult to control, since it is resistant to most insecticides and the leaf undersides on large plants are difficult to cover. The best control involves regularly checking plants to detect when egg masses are first laid. Application of insecticides at this time, followed by a second treatment approximately two weeks later, can provide fair season-long control. Undersides of leaves must be thoroughly covered during any applications.

## Harlequin Bug (*Margantia histrionica*)

**Damage:** Harlequin bugs feed on the sap of various cabbage family plants. Areas around the feeding site typically turn cloudy. When young tissues are fed on, growth develops in a distorted manner and may turn brown and die. Severe winters typically limit populations and subsequent damage caused by harlequin bugs in areas above about the fortieth parallel.

**Life History and Habits:** Harlequin bugs survive winters in northern regions only in the adult stage, which hides in protected sites, such as under older crop debris. The adults emerge from winter shelter in mid-spring and typically first feed on wild mustards and other weed hosts. By June they may be found in gardens feeding on cabbage, radish or other crucifers. Females then lay egg masses resembling rows of small, black and white banded barrels on the leaves of these plants. The immature nymphs usually develop in the same plants where eggs were laid and feed for about two months before becoming full-grown. If warm weather conditions occur, a second generation will be produced. Otherwise, adults will continue to feed without reproducing and move to winter cover at the end of the growing season.

*Harlequin bug*

**Mechanical Controls:** Individual bugs can be hand collected and egg masses crushed. Harlequin bugs tend to seek more shaded areas on the plant and are most easily found early in the morning.

**Cultural Controls:** Some plants resistant to harlquin bugs have been identified. These include cabbage varieties Copenhagen Market 86, Headstart, Savoy Perfection Drumhead and Early Jersey Wakefield; cauliflower varieties Early Snowball X and Snowball Y; and radish varieties White Icicle, Globemaster, Cherry Belle, Champion, Red Devil and Red Prince. Infested plants should be removed and destroyed immediately after harvest, such as by composting. This can kill stages of the insect that are developing and reduce local populations. In larger plantings, plowing or otherwise removing cover in fall can deprive overwintering bugs of shelter and reduce survival.

**Chemical Controls:** Harlequin bugs are quite difficult to kill with insecticides. If needed, sprays should be applied when adults are observed, before many eggs are laid or when nymphs are small.

**Related Insects:** Harlequin bugs are frequently confused with another brightly colored stink bug, the twospotted stink bug, *Perillus bioculatus* (page 19). However, the latter species is a predator of other insects, such as the Colorado potato beetle, and does not feed on plants.

## False Chinch Bug (*Nysius raphanus*)

**Damage:** False chinch bugs suck the sap from plants during feeding. The adults may appear in tremendous numbers, causing plants to wilt and die rapidly. Outbreaks are sporadic but can destroy

*False chinch bug*

plantings, particularly early in the year. Most garden plants can be damaged during outbreaks, but beets or cruciferous plants such as radish and cabbage are favored. Wild hosts include many weeds such as kochia, Russian thistle and sagebrush.

**Life History and Habits:** False chinch bugs spend the winter as eggs or small nymphs. They become active in early spring, feeding on various weeds. When adults develop, they often scatter over wide areas, sometimes concentrating in large masses on favored food plants. Migrations are most common when wild hosts are not available or become less suitable to the insects due to drought. Eggs are laid in loose soil, dirt clods or around flowers and seeds. Eggs hatch in about 4 days. The grayish nymphs feed for about three weeks, molting five to six times, before reaching the adult stage. The adults mate and females lay eggs over a period of two to three weeks before dying. There are several generations per year.

**Mechanical Controls:** The bugs can be brushed off plants and collected in water pans.

**Chemical Controls:** There is little information on effective controls for false chinch bugs. Insecticidal soaps and sabadilla dust are among the more effective treatments.

## Lygus Bugs (species of *Lygus*)

**Damage:** Lygus bugs feed on developing leaves, fruits and flowers, killing the areas around the feeding site. This can cause abortion of young flowers, seeds or buds. Older tissues may continue to grow but be deformed. Leaf curling and corky "catface" injuries to fruit are common distortions due to lygus bug feeding injury. Peach, apricot, strawberry and beans are among the garden plants most commonly damaged. Some flower abortion of most plants is normal, and lygus bug feeding injuries have little effect on plant yields unless the insects are abundant.

**Life History and Habits:** Lygus bugs spend the winter in the adult stage under the cover of piled

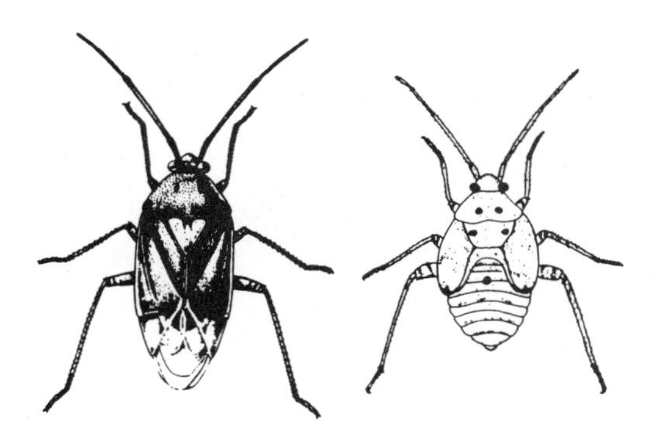

*Lygus bug. Adult (left), nymph (right).*

# *Plant Bugs*

This book focuses largely on insects and other "bugs" that damage plants. However, only one group of insects truly rates the distinctive moniker *plant bug*. Among entomologists only the insect family Miridae are considered true plant bugs. Lygus bugs, fourlined plant bugs and campylomma bugs are among the insects in this elite group that damage plants.

Plant bugs, or mirids, feed on plants with mouthparts designed to pierce plant tissues and suck fluids. However, plant bugs do not use these piercing-sucking mouthparts as do their relatives (aphids, psyllids and the like), which feed discreetly and cause little damage as they try to find the phloem. Instead, plant bugs thrust their stylet mouthparts into the plant, lace the area with copious amounts of saliva and then suck up the mixture. This lacerate-flush feeding causes extensive damage to the plant.

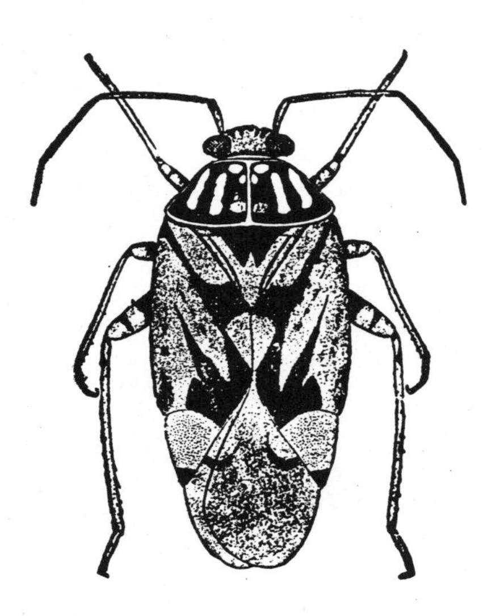

*Adult tarnished plant bug*

Obvious symptoms of plant bug injury usually develop several weeks after the insects have left. Areas of plants visited by feeding plant bugs often show dark spotting—cells killed by the feeding. Since plant bugs typically concentrate feeding in the new growth, this can result in deformities as the growing leaves and fruits overgrow the killed areas damaged by plant bugs. Curled leaves or puckered (catfaced) fruit develop following plant bug feeding.

leaves, bark cracks or other sheltered sites. They emerge in early spring and feed on emerging buds of trees and shrubs. Most then move to various weeds and other plants, and females insert eggs into the stems, leaves and buds of these plants. The young hatch, feed and develop on these plants, becoming full-grown in about a month. Several generations are produced during the year.

**Cultural Controls:** Legumes, particularly alfalfa, are important host plants for lygus bugs. If these plants are present around a garden, they should not be cut when fruits and vegetables are in susceptible stages, such as fruit set. Cutting can force migrations of lygus bugs. Clean up weeds and other debris in orchards to reduce overwintering shelter and survival of lygus bugs. However, most small yards have plenty of other sheltering sites.

**Mechanical Controls:** Lygus bugs can be trapped with sticky white cards. This is most useful for detecting when lygus bugs have moved into a planting, rather than for control.

**Chemical Controls:** Lygus bugs and most other true bugs are fairly difficult to control with insecticides. Since most injury occurs during early fruit development, insecticide sprays are best timed either immediately before flowering and/or immediately after petal fall. Insecticides should never

## *Leafhoppers*

One of the more common groups of insects found on plants are the leafhoppers (Cicadellidae family). These small (typically 1/8 inch), elongate insects vary in color from rainbow stripes to more pedestrian green or light brown. Adult leafhoppers can fly (and jump), sometimes flying hundreds of miles from southern areas to infest regional plantings. Immature leafhoppers (nymphs) are almost always found on the undersides of leaves, where the high humidity and shade is to their liking. Despite their name, they don't readily jump, but many can do a rapid "crab-crawl" when disturbed, moving rapidly either sideways (crablike) or forward and backward.

Leafhoppers have sucking mouthparts, but their preferred feeding spots vary. Many try to plug into the sugary sap of the plant phloem, excreting sticky honeydew as they feed and sometimes causing young leaves to curl. Other leafhoppers feed on the inner cells of the mesophyll, leaving white flecking wounds as evidence (on beets these wounds turn bright red). One group of leafhoppers, known as sharpshooters because of their pointed head, probe more deeply into the water-conducting vessels of the plants, or xylem.

Most leafhoppers cause little injury to fruit and vegetable crops. The most important species in the West are those that transmit diseases such as curly top virus (beet leafhopper), aster yellows mycoplasma (aster leafhopper) and X-disease of peaches. However, some leafhoppers (grape leafhopper, white apple leafhopper and rose leafhopper) become quite abundant and cause direct injury to plants by their feeding wounds. Occasionally, there are also problems with "hopperburn" injuries to plants such as pumpkins and winter squash in which the infested leaves progressively die back, appearing scorched.

be sprayed during flowering to avoid killing beneficial pollinating insects.

**Miscellaneous Notes:** Several species of lygus bugs occur in the region, including the tarnished plant bug (*Lygus lineolaris*) and the pale legume bug (*Lygus elisus*). In addition, other plant bugs (such as the campylomma bug), stink bugs and even box elder bugs occasionally damage fruit in a similar manner. Lygus bugs also occasionally feed on insects and can contribute to the biological control of aphids and other small, soft-bodied species.

# *Homoptera (Aphids, Leafhoppers, Whiteflies, Psyllids, Etc.)*

### White Apple Leafhopper (*Typhlocyba pomaria*)

**Damage:** The white apple leafhopper feeds on the leaves of apple trees, removing chlorophyll and sap. Feeding sites are white-flecked and heavily infested leaves look silvery. During early season outbreaks, the vigor and productivity of trees can be reduced. The white apple leafhopper excretes tiny tobacco juice–like droppings that can cover fruit, making it unattractive. It also occasionally damages leaves of roses, and peach, plum and cherry trees.

*Leaf flecking injuries produced by white apple leafhopper*

**Life History and Habits:** The white apple leafhopper spends the winter in the egg stage, inserted under the bark of twigs and branches. Eggs hatch in spring, generally about the time of bloom. The developing nymphs crawl to feed on the undersides of older leaves. They become full-grown by early June, producing winged adult stages. These first-generation adults mate, and female leafhoppers lay eggs that produce a second generation. Adults emerging from these produce the overwintering eggs. Peak feeding injury occurs in late July and August. There are two generations per year. The insect is easily observed on leaf undersides and will run away readily when disturbed. Even when it is no longer present on the leaf, the white flecking and old, shed skins can be used to diagnose white apple leafhopper injury.

**Natural and Biological Controls:** Several general predators, such as green lacewings and damsel bugs, commonly eat leafhoppers. Some hunting wasps specialize in gathering leafhopper prey to rear their young. Parasitic wasps and fungal diseases are other reported natural enemies of white apple leafhoppers. These biological controls usually keep leafhopper numbers low in unsprayed orchards.

**Mechanical and Physical Controls:** When possible, periodically hose off plants to dislodge many of the leafhoppers. Watering can also soften the droppings that cover fruit, making them easier to wash off. However, hosing treatments or overhead watering should not be done if there is danger of spreading fire blight.

**Cultural Controls:** Pruning removes overwintered eggs under the bark.

**Chemical Controls:** Dormant season applications of horticultural oils may provide some control of overwintering eggs, although most are too well protected by bark to be killed by these treatments. Newly hatched nymphs are susceptible to many insecticides, including insecticidal soaps and carbaryl (Sevin). However, insecticide-resistant strains are reported in the eastern United States. Sprays should be directed at leaf undersides where the nymphs feed.

**Related Species:** White flecking injuries on leaves of garden plants are characteristically produced by many different leafhopper species that feed on mesophyll cells of the leaf underside. One of the more common is the rose leafhopper, *Edwardsiana rosae*, which feeds on roses. The life history and control of this species are similar to those of the white apple leafhopper, which it resembles. Also common are various leafhoppers in the *Erythoneura* genus. These include the grape leafhopper (described on next page), a pest of grapes in warmer areas of the West, and the "zigzag" leafhopper, which commonly damages ivy and Virginia creeper. These spend the winter in the adult stage under loose bark and other protected sites. They are also attracted to lights and are an occasional nuisance.

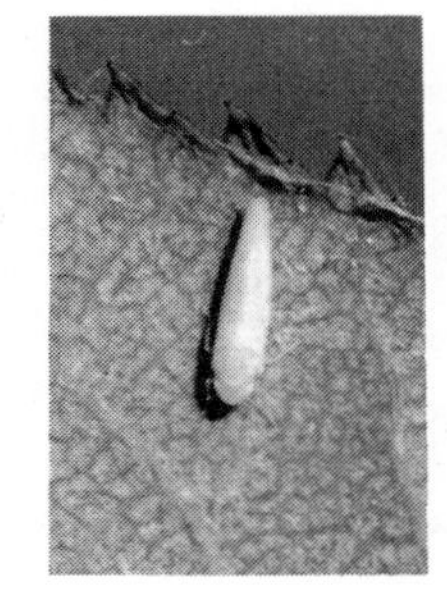

*Rose leafhopper*

## Grape Leafhopper (*Erythroneura comes*)

**Damage:** Grape leafhoppers feed by puncturing cells of the leaves and removing the sap. The damaged tissues around the feeding site turn white and later die. This loss of leaf area can reduce photosynthesis. Berries on heavily infested plants may have lowered sugar content and increased acids. Grape leafhoppers also excrete honeydew, which can stick to fruit, staining it and favoring the development of sooty mold fungi.

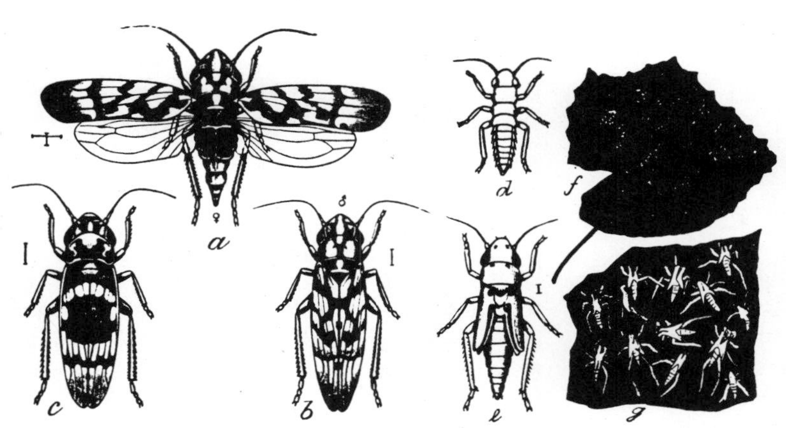

*Grape leafhopper. A) Adult female; B) Adult male; C) Another form of the species showing variations in markings; D) Newly-hatched nymph; E) Last-stage nymph; F) Appearance of injured leaf; G) Cast skins.*

**Life History and Habits:** The grape leafhopper feeds on a wide variety of plants, including Virginia creeper, maple, strawberry, burdock and various mint family plants. When temperatures reach the mid-60s, the overwinter adult leafhoppers move to these plants to feed and lay eggs. As new growth is produced on the grapes, leafhoppers begin to migrate into the vineyard. Adult females prefer to lay eggs in newly matured leaves. Eggs are inserted into the grape leaf. In about one to two weeks, the eggs hatch to produce the immature nymphs. Grape leafhopper nymphs are pale-colored and highly active. They feed on the underside of leaves, with populations particularly high on lower leaves. They develop over the course of a month, molting five times before reaching the adult stage. The cycle is repeated, with probably two to three generations occurring during a growing season. At the end of the season, leafhoppers move to weedy areas to spend the winter.

**Natural and Biological Controls:** Cold and wet weather conditions in spring and fall are reported to reduce populations of grape leafhoppers. Grape leafhoppers also have some natural enemies, including a small wasp (*Anagrus epos*) that parasitizes eggs. This wasp has been a very important natural control in California vineyards. Although it is also present in parts of the Rocky Mountain and intermontane region, it is unknown how effective it is for biological control.

**Cultural Controls:** Maintaining a weed-free area in and around the vineyard can reduce overwintering cover and grape leafhopper populations. Removal of lower sucker canes in late June can also eliminate many of the eggs.

**Mechanical Controls:** Adults can be trapped on sticky surfaces, such as a screen. The trap should be placed downwind of the vine and the vine then jarred. Escaping leafhoppers can be trapped on the screen in this manner.

**Chemical Controls:** The grape leafhopper is readily controlled by many insecticides, provided there is adequate coverage of the undersides of leaves. Sprays should be used only when sampling indicates a developing problem. Little berry spotting occurs unless leafhopper populations exceed an average of fifteen per leaf. However, grape leafhoppers are distributed unequally on the vine, with highest populations on more shaded parts of the plant. When sampling populations, it is suggested that mid-shoot leaves from the north or east sides of the plant be collected, averaging a count of at least fifteen leaves.

## Beet Leafhopper (*Eutettix tenellus*)

**Damage:** The beet leafhopper is the only insect capable of transmitting the virus that causes curly top disease. (See discussion of beet curly top disease on page 174.) Although feeding injury by the beet leafhopper is inconsequential, transmission of the virus can greatly stunt or kill tomatoes, peppers and beets. Curly top outbreaks tend to be very sporadic between seasons and are, in part, related to the severity of winter temperatures and the amount of rainfall. However, outbreaks can be a chronic problem in some regions.

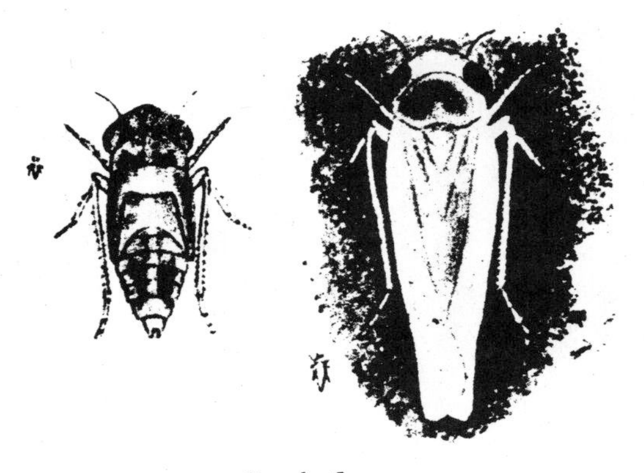

*Beet leafhopper*

**Life History and Habits:** The beet leafhopper overwinters in the region as an adult, feeding on wild host plants such as Russian thistle, wild mustards and salt bush. In early spring, eggs are laid and the first generation is produced. In late spring, large migrations are common, and the leafhoppers may fly for hundreds of miles. If migrating leafhoppers are allowed to feed on plants infected with the curly top virus, they may be able to transmit the disease for the remainder of their life, a period of several weeks.

**Cultural Controls:** The beet leafhopper thrives in open, arid areas and tends to avoid dense plantings. Higher plant populations and overhead irrigation will reduce colonization of the plantings. Damage may also be reduced by the use of earlier planting dates, since older plants are less sensitive to the effects of the disease.

**Chemical Controls:** Several insecticides can kill beet leafhoppers. However, insecticides are ineffective for controlling the curly top virus because it can be transmitted before the insect is killed, and leafhoppers continually migrate into plantings.

## Aster Leafhopper (*Macrosteles fascifrons*)

**Damage:** The aster leafhopper (also known as the sixspotted leafhopper) transmits the mycoplasma that causes aster yellows disease. (See discussion on aster yellows disease on page 196.) Aster yellows can affect over forty families of plants, including many cultivated vegetables and flowers. Infections of carrots are marked by hairy roots and bushy, off-colored top growth. Lettuce, particularly head lettuce, shows a distorted twisting of new growth. Several types of flowers (cosmos, marigold, petunia, zinnia, etc.) show distortion of flowering parts due to aster yellows. Feeding by leafhoppers that do not transmit the disease causes no serious damage. Infection of plants tends to occur in late July and early August.

**Life History and Habits:** The aster leafhopper cannot survive the harsh winter temperatures of the West. It survives the cold season in the southern United States along the Gulf of Mexico. Winged stages of the leafhopper annually fly northward, aided by storm patterns and winds. Eggs are laid in leaves and stems of plants. The immature

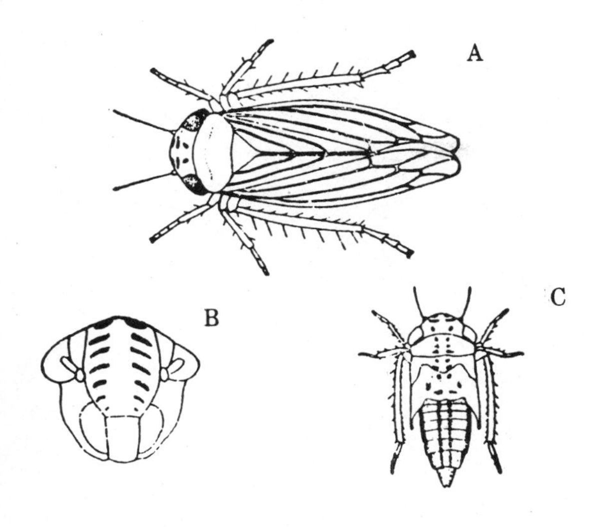


*Aster leafhopper. A) Adult; B) Front view of head; C) Nymph.*

nymphs take three to four weeks to complete development before becoming mature adults. Two or three generations typically are produced during the growing season. Leafhoppers that feed on plants infected with aster yellows acquire the mycoplasma that produces the disease. However, they cannot transmit the disease until the mycoplasma has circulated in the leafhopper and moved to its salivary glands, a process that typically takes 10 days to three weeks. Once this period has passed, they are capable of transmitting the disease organism for the rest of their lives. Typically, less than 3 percent of all leafhoppers carry the disease organism.

**Cultural Controls:** Carrot varieties have a range of susceptibility to aster yellows infection. Royal Chantenay, Scarlet Nantes, Toudo, Hi Color, Amtou, Charger, El Presidente, Six Pak and Gold King are among those that have the greatest tolerance. Head lettuce is much more susceptible to infection than are carrots. No highly resistant varieties exist. However, damage from aster yellows is much less severe on leaf-type lettuce than on head types. Aster yellows resistance has not been developed for susceptible flowers. Planting early avoids the midsummer period of peak infection.

**Mechanical Controls:** In small plantings, mulches of reflective aluminum or straw can repel leafhoppers.

**Chemical Controls:** When high numbers of infective leafhoppers are present, it is very difficult to control aster yellows infection. Under these conditions, leafhoppers continue to migrate into plantings, so repeated applications are needed, often at short intervals, to maintain control.

## Garden Leafhoppers (*Empoasca abrupta, Empoasca recurvata*)

**Damage:** Several related leafhoppers (species of *Empoasca*) may feed on the leaves of garden plants. Feeding injuries produced by the western potato leafhopper (*Empoasca abrupta*) on potatoes appear as white flecking. Another species (*Empoasca recurvata*) damages the leaves of pumpkin and squash, resulting in a "hopperburn" injury by which the leaves die back from the edge and appear scorched. Despite the appearance of these injuries, it is doubtful that they seriously affect yield.

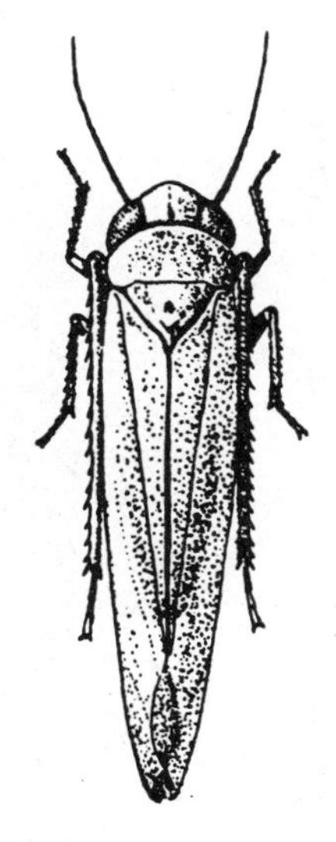

*Potato leafhopper*

**Life History and Habits:** *Empoasca* leafhoppers cannot survive in regions with extreme cold winter temperatures. Therefore, regional infestations occur from leafhoppers that fly northward, aided by winds, from more southern areas. Adult leafhoppers insert eggs into leaf mid-ribs and petioles.

The emerging nymphs feed on the plant, growing and molting several times before becoming full-grown. About two generations are produced annually in the West.

**Chemical Controls:** Controls are rarely needed, since damage tends to occur late in the season. Several insecticides (including insecticidal soaps) can kill leafhoppers if they become abundant.

**Related Species:** The potato leafhopper, *Empoasca fabae*, can cause great "hopperburn" injury to potatoes, beans, alfalfa and many other plants. However, the potato leafhopper is almost always limited to the central and eastern regions of the United States and Canada.

## Greenhouse Whitefly (*Trialeurodes vaporariorum*)

**Damage:** Nymphs and adults suck sap, weakening the plants. Honeydew is excreted during feeding, which contaminates fruit and vegetables.

**Life History and Habits:** The greenhouse whitefly originates from subtropical and tropical areas. In the northern United States and Canada, the greenhouse whitefly overwinters only indoors on houseplants and in greenhouses. The adult female whitefly lays eggs in a series of semicircles on the lower leaf surface. Eggs hatch in one to three weeks and immature nymphs emerge. The nymphs are pale yellow and difficult to detect. They rarely move, feeding for three to five weeks. After they have finished feeding, they change to a stage similar to the pupa stage, that is, a nonfeeding, immobile stage that lasts about one week. The adult whiteflies then emerge, and females may live for four to six weeks, laying up to 400 eggs. Generations continue throughout the season so long as temperatures allow.

**Natural and Biological Controls:** The greenhouse whitefly cannot survive freezing temperatures and dies out in unprotected locations. In high-temperature greenhouses (average above 72°F), the parasitic wasp *Encarsia formosa* can provide control. (See discussion of Biological Control Organisms in Chapter 2.)

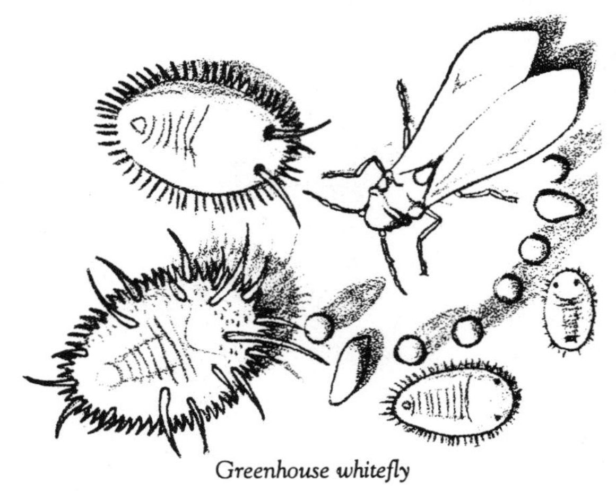

*Greenhouse whitefly*

**Cultural Controls:** Since the insect only survives winters indoors, carefully examine transplants and purchase only those that are free of the insect. ("Just say no!") When growing transplants it is important to eliminate all whitefly-infested plants several weeks before seedlings emerge.

**Mechanical and Physical Controls:** Yellow sticky traps can be used to capture adult whiteflies. Adult whiteflies are repelled by white or light-colored mulches.

**Chemical Controls:** Whiteflies are difficult to control, and use of most whitefly sprays (containing resmethrin, tetramethrin, etc.) is not permitted on edible vegetable crops. Insecticidal soaps have been among the more effective garden treatments, although they kill only immature stages. Repeated applications are necessary, since egg and pupal stages are often not controlled by any insecticides. Leaf undersides must receive coverage to kill nymphs. Oil sprays are among the more promising controls for nymphs, although such sprays are not widely registered for use on vegetables at present.

## *Aphids*

Aphids (Aphididae) are the most abundant family of insects found on plants in the West. Several hundred different species occur in the region, many as pests of vegetables, fruits, flowers and other plants.

Aphids reproduce during the growing season without males (asexually). Females are literally born pregnant, and the young may become full-grown about a week after birth. This extremely high rate of reproduction can allow them to become very abundant in a short time.

During the spring and summer, developing females may produce either a winged or wingless adult female stage. Winged aphids tend to be produced more often when the plants on which they are feeding become overcrowded and decline in food quality, or when days become shorter. When conditions are favorable, nonwinged females predominate. These females reproduce more than the winged stage, which tends to disperse.

However, the life cycles of aphids can be very complex and involve numerous different stages through the year. For example, many common aphids alternate between annual plants, such as vegetables, during the summer and a perennial tree or shrub host in the winter. In these types of aphids, males are produced at the end of the summer. The males—and special reproductive stages of the females—meet on plants on which the insects overwinter. After mating, specialized, egg-laying daughter aphids are produced, which lay overwintering eggs. Eggs hatch in spring. After a few generations, the aphids leave to search for summer plants.

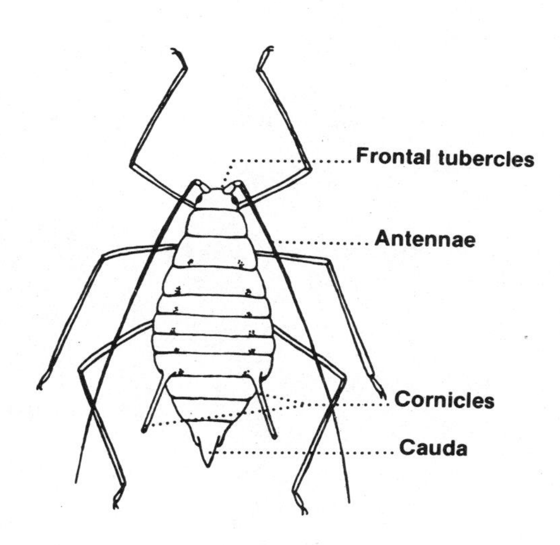


Aphids feed by sucking and removing sap from plants. When aphids are abundant, this can stress the plant, making it less vigorous. Some aphids also produce a saliva that is somewhat toxic to the plants, causing discoloration or curling of developing leaves during infestations.

Perhaps the greatest injury aphids cause is by transmitting viral diseases. Dozens of viral diseases of vegetables, flowers and berry crops are spread by aphids. Although the aphids themselves may cause little damage, the effects of viral infections they spread can be very destructive.

COMMON REGIONAL APHIDS AND THEIR HOST PLANTS

| Aphid | Plants attacked |
|---|---|
| Asparagus aphid (*Brachycorynella asparagi*) | Asparagus* |
| Cabbage aphid (*Brevicoryne brassicae*) | Various wild and domestic cabbage family plants (Cruciferae)* |
| Turnip aphid (*Hyadaphis erysimi*) | Various wild and domestic cabbage family plants (Cruciferae)* |
| Green peach aphid (*Myzus persicae*) | Peppers, cabbage, many garden plants, peach,* plum,* apricot* |
| Currant aphid (*Cryptomyzus ribis*) | Currant,* motherwort, marsh betony |
| Black cherry aphid (*Myzus cerasi*) | Sweet cherry,* sour cherry, plum, wild mustards |
| Crescent-marked lily aphid (*Neomyzus circumflexus*) | Columbine, aster, lily, dock, violet |
| Rose aphid (*Macrosiphum rosae*) | Rose* |
| Potato aphid (*Macrosiphum euphorbiae*) | Rose,* potatoes, tomatoes and many other garden plants |
| Rose grass aphid (*Metalophium dirhodum*) | Rose,* grasses, corn |
| English grain aphid (*Sitobium avenae*) | Apple,* grains |
| Corn leaf aphid (*Rhopalosiphum maidis*) | Corn, other grains |
| Pea aphid (*Acyrthosiphum pisum*) | Peas, beans, clover, alfalfa* |
| Lettuce aphid (*Acyrthosiphum lactucae*) | Lettuce |
| Rosy apple aphid (*Dysaphis plantaginea*) | Apple,* pear, plantain, mountain ash |
| Melon aphid (*Aphis gossypii*) | Rose, melons, squash, many other garden plants |
| Bean aphid (*Aphis fabae*) | Euonymus,* viburnum,* beans, beets, cucumber, carrot, lettuce, etc. |
| Apple aphid (*Aphis pomi*) | Apple,* pears, mountain ash, hawthorn |
| Western aster root aphid (*Aphis armoraciae*) | Aster, dandelion, milkweed, sagebrush, corn, dock, many garden flowers |

| Aphid | Plants attacked |
|---|---|
| Leafcurl plum aphid (*Brachydaudus helichrysi*) | Plum,* various aster family plants, clover, vinca, thistle |
| Sunflower aphid (*Aphis helianthi*) | Dogwood, sunflower, pigweed, four o'clock, ragweed |
| Chrysanthemum aphid (*Macrosiphoniella sanborni*) | Chrysanthemum |
| Carrot aphid (*Caveriella aegopodii*) | Carrot, parsley, dill, willow* |
| Wild parsnip aphid (*Aphis heraclella*) | Carrot, celery, parsnip, cow parsnip |
| Sugarbeet root aphid (*Pemphigus populivenae*) | Beets, many garden plants, narrow-leaf cottonwood* |
| Woolly apple aphid (*Eriosoma lanigerum*) | Apple, crabapple, mountain ash, elm* |

* Primary overwintering host plant in the West. Where no plants are indicated, the overwintering host plant in the region is unknown, or the aphid is thought to rarely survive winters in the region and occurs primarily as an annual migrant.

## Asparagus Aphid (*Brachycorynella asparagi*)

**Damage:** The asparagus aphid sucks the sap from asparagus plants. This can distinctly deform fern growth, producing stunting and bushiness. Continued feeding causes the premature release of dormant buds that are formed for the subsequent crop. Yields can be severely reduced and young plants killed by these injuries.

**Life History and Habits:** Asparagus aphids overwinter as eggs on and around the debris of asparagus plants. The eggs hatch in spring and the aphids move to the tips and bud scales of developing spears. There they reproduce, having several generations during the growing season. By late spring, some winged forms are produced that infest new plantings. Mating occurs in late summer, and overwintering eggs are laid on lower areas of ferns,

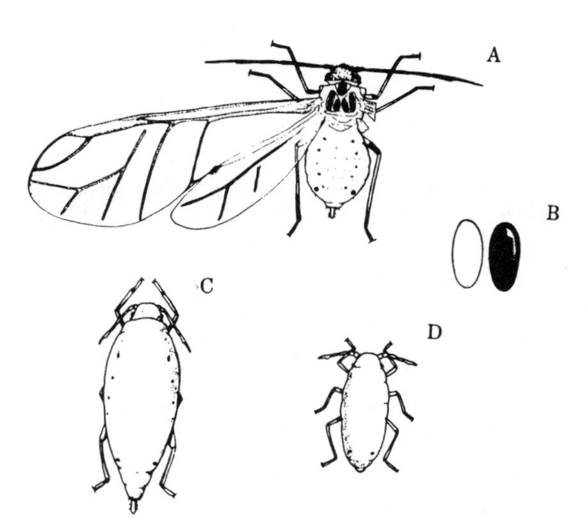


*Asparagus aphid. A) Winged adult; B) Eggs; C) Wingless adult; D) Nymph.*

particularly on new growth produced from released dormant buds.

**Natural and Biological Controls:** Parasitic wasps and general aphid predators (lacewings, syrphids, lady beetles) provide some control. However, the asparagus aphid is an introduced insect that has only recently moved into the region, and effective biological controls are often lacking.

**Cultural Controls:** Remove old ferns to eliminate overwintering eggs. Control is further improved by carefully tilling the soil over the dormant crowns in spring. Regular picking of asparagus shoots in spring can remove many of the early-hatching aphids, although plants continue to be infested after harvest. Some varieties of asparagus show better tolerance to injury, although no varieties are highly resistant.

**Chemical Controls:** Chemical control options for aphids are very limited on asparagus. Repeated applications of insecticidal soaps, malathion or nicotine sulfate may provide some control during early stages of infestation. However, when dense fern growth has been produced, the aphids are largely protected from insecticides.

## Green Peach Aphid (*Myzus persicae*)

**Damage:** The green peach aphid feeds on over 200 species of plants and is the most important aphid to transmit viral diseases to vegetables. Diseases spread by the aphid include potato virus Y, potato leafroll, watermelon mosaic and bean mosaic. Green peach aphids also feed on many vegetables but rarely are abundant enough in gardens to cause serious injury by feeding injuries alone. On overwintering plants (peach, apricot, certain plums), green peach aphids can produce leaf-curling injuries during early spring. Green peach aphids are also the most common aphid found in greenhouses or on houseplants. They are highly resistant to most insecticides and to insecticidal soaps, so problems are increasing.

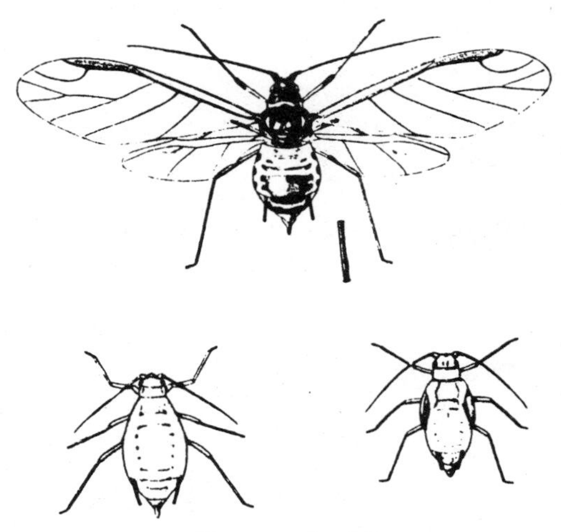

*Green peach aphid*

**Life History and Habits:** Green peach aphids spend the winter in the egg stage around the buds of peach, apricot and certain types of plum and cherry trees. Eggs hatch in April and May, and the aphids feed on developing leaves, often producing leaf-curling injuries. After one to two months, winged forms begin to be produced that fly to summer host plants, including many vegetables and flowers. The developing nymphs can become full-grown in as little as a week and live for a month or more. Numerous generations occur on these summer hosts, which produce both wingless and migratory winged forms. In fall, they fly back to the overwintering plants, mate and a special female form lays eggs. Indoors, the green peach aphid reproduces continuously and does not produce special overwintering forms and eggs.

**Natural and Biological Controls:** Green peach aphids are preyed upon by many general aphid predators, including lady beetles, lacewings, syrphid flies, predator midges, minute pirate bugs and damsel bugs. Various parasitic wasps also kill aphids. High populations of green peach aphids rarely persist on vegetables unless these biological controls are disturbed by pesticides or interference with ants. (See discussion on ants and aphids on page 25.)

**Cultural Controls:** Most of the viral diseases spread by green peach aphids are carried for short distances. Therefore, remove all "off"-looking plants that may be infected with the virus to help prevent further spread to healthy plants. Purchase virus-free seed and transplants to reduce later problems with virus spread by aphids. Flights of green peach aphids increase steadily through the summer. Early planting can avoid much of the infection of vegetables susceptible to viruses spread by aphids.

**Mechanical and Physical Controls:** Mulches made of reflective materials, such as aluminum foil or straw, can repel winged aphids from plants.

**Chemical Controls:** On the winter host plants (peach, plum, apricot, cherry), green peach aphids are easily controlled with applications of dormant oil. Insecticides are ineffective in preventing the spread of aphid-transmitted viral diseases. Spread of most diseases can occur within a few seconds after landing on the plant, far more rapidly than insecticides work. Some studies have even shown increases in virus spread by aphids following use of insecticides, since they can increase aphid activity. Sprays of highly refined stylet oils and oil-containing materials such as milk are capable of preventing transmission of many aphid-transmitted viruses to plants.

## Cabbage Aphid (*Brevicoryne brassicae*)

## Turnip Aphid (*Hyadaphis erysimi*)

**Damage:** Cabbage and turnip aphids feed on cabbage/mustard family (Cruciferae) plants, including cabbage, Brussels sprouts and broccoli. Most serious damage occurs when high populations of aphids feed and damage developing cabbage heads, causing permanent distortion. Cabbage aphids also can become very abundant during fall on Brussels sprouts and contaminate the sprouts.

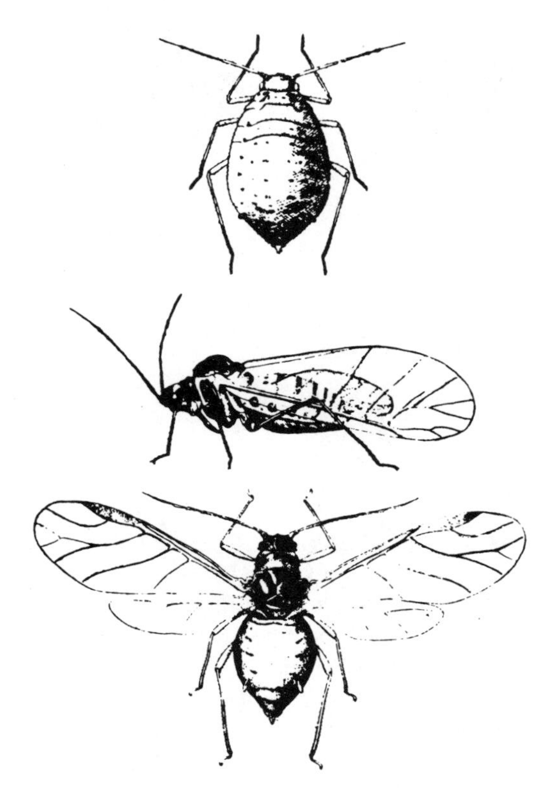

*Cabbage aphids*

**Life History and Habits:** Cabbage and turnip aphids spend the winter as eggs on wild or cultivated cabbage/mustard family plants. Eggs hatch in early spring and development can require two to six weeks, depending on temperature. Numerous generations are produced, and the aphids are active throughout the year. As with other common aphids, during most of the season, pure female populations are produced, with one sexual generation (males and females) developing at the end of the season. Winged individuals are occasionally produced, particularly following overcrowding, and these fly and colonize new plants. Cabbage aphids tend to make tight colonies on the new growth of the plant. Turnip aphids are more evenly distributed throughout the plant.

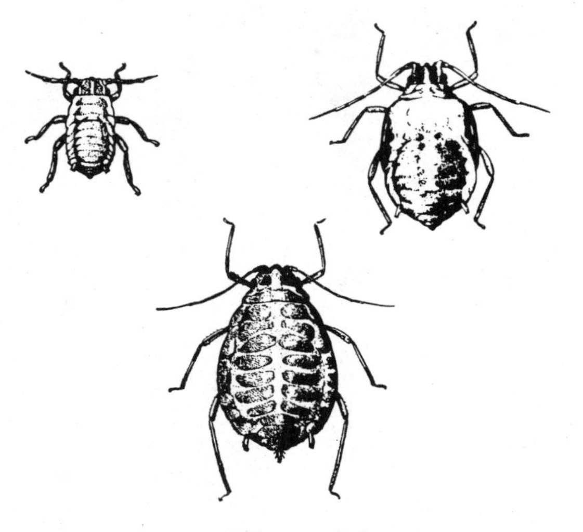

*Turnip aphids*

**Natural and Biological Controls:** Cabbage and turnip aphids are fed upon by most general aphid predators, such as syrphid flies and lady beetles. However, their waxy coating makes them less preferred than other aphids by lady beetles. Parasitic wasps are also common biological controls. Unfortunately, parasitized aphids often stick tightly to the foliage and can increase contamination problems.

**Cultural Controls:** Before spring, destroy or remove cabbage family plants in and around the garden to kill overwintering stages.

**Chemical Controls:** Several insecticides can control cabbage aphids. However, because the plants and the aphids have a waxy covering, soap, detergent or some other wetting agent should be used to provide better coverage. Control can be very difficult when infestations occur in protected areas within leaf curls, particularly with Brussels sprouts.

## Pea Aphid (*Acyrthosiphum pisum*)

**Damage:** Pea aphids suck sap from various legumes such as peas and beans, particularly damaging the new growth. During heavy infestations plants may wilt and new growth may die back. Plants can often tolerate large pea aphid populations, but plants that are stressed by drought are much more sensitive to feeding injury.

**Life History and Habits:** The pea aphid overwinters as an egg on perennial legumes, such as clover and alfalfa. After hatching and reproducing on the winter host plant, migratory winged stages begin to be produced in mid-spring. These colonize new plantings, including gardens. Reproduction is continuous throughout the growing season, with generations being completed in about two weeks. As with most aphids, the young are born

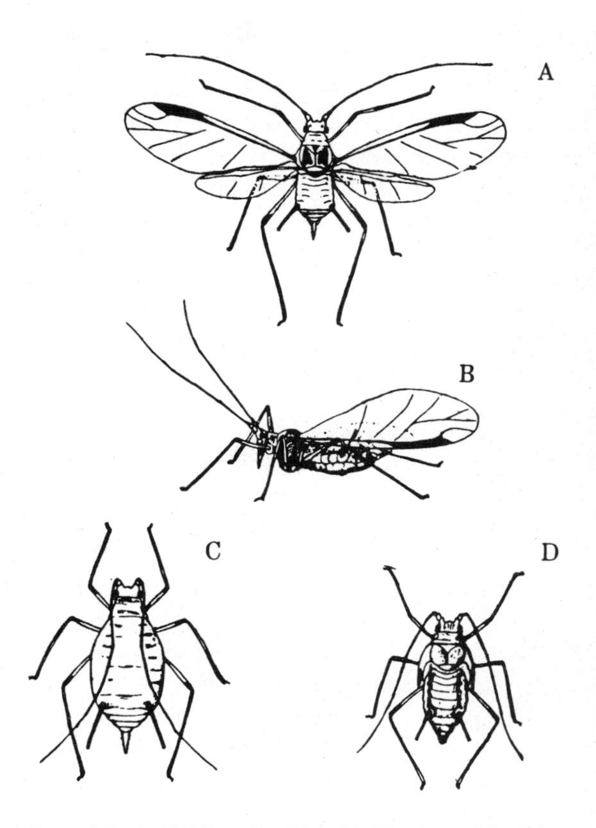


*Pea aphid. A-B) Winged adults; C) Wingless adult; D) Nymph.*

live for most of the year, with eggs being produced only at the end of the season.

**Natural and Biological Controls:** General aphid predators, such as lady beetles and syrphid flies, as well as parasitic wasps commonly feed on pea aphids, usually keeping populations under control. Most garden problems occur when ants interfere with biological controls or when ineffective insecticides are used that destroy the natural enemies of pea aphids. (See discussion on ants and aphids on page 25.)

**Mechanical Controls:** Pea aphids are quite delicate, so hosing plants with a strong jet of water (syringing) can kill and dislodge many aphids.

**Chemical Controls:** Most general-use insecticides are usually effective against pea aphids, although insecticide-resistant strains have developed in some areas where the insect is routinely sprayed in alfalfa. Soap or detergent sprays can also help to control aphids.

**Miscellaneous Notes:** The bean aphid, *Aphis fabae*, is the other common aphid that attacks legumes in the West. It is easily distinguished from the pea aphid by its dark color and smaller size.

## Woolly Apple Aphid (*Eriosoma lanigerum*)

**Damage:** The woolly apple aphid feeds on the bark of apple, crabapple, pear, mountain ash and related plants. Large colonies of the wax-covered aphids may develop around tree wounds above and below ground. The soft callous tissues of healing wounds are favored by the aphid, preventing proper healing and causing knotlike growths to form at sites repeatedly attacked. These cankers girdle, weaken and ultimately can kill heavily infested areas of the tree. Woolly apple aphids produce waxy threads that cover their body, and colonies have a conspicuous cottony appearance.

*Woolly apple aphid*

On elms, their alternate host, the aphid causes a leaf-curling injury in late spring. These "elm" aphids are covered with powdery wax, although they do not produce the conspicuous threads.

**Life History and Habits:** The woolly apple aphid has an unusual life cycle, alternating between rose family plants (apple, pear, etc.) and elms. Most survive the winter as eggs on elm trees. Aphids emerging from these eggs feed on the developing leaves, causing a thickened folding of the leaves. The aphids reproduce and often fill the leaf before they begin to produce winged forms that leave the tree. The winged aphids fly to and colonize the summer hosts, feeding in bark cracks and around old wounds. In early fall, many of them again produce winged stages, which move back to elms and later produce overwintering eggs. Some remain on the summer host and may survive the winter in protected areas, including the roots.

**Natural and Biological Controls:** During cold winters, the aphids are unable to survive over winter on apple trees. In these areas, infestations originate annually only from aphids that survived as eggs on elm trees and later fly to apple or other fruit trees. Parasitic wasps are the most effective biological controls of the woolly apple aphid. Mummified aphids killed by these wasps are dark-colored with exit holes cut out of their backs.

**Mechanical Controls:** Woolly apple aphids concentrate around recent wounds and can be easily crushed in small plantings.

**Chemical Controls:** In areas where the aphids overwinter on apple and other fruit trees, dormant applications of oil or lime sulfur before buds open can provide control of colonies exposed on branches. Several insecticides applied directly to the bodies of aphids can provide control, although the addition of soap to help "wet" the insects is useful. Insecticidal soaps can also control exposed aphids.

## San Jose Scale (*Quadraspidiotus perniciosus*)

**Damage:** San Jose scale feeds on the sap of apple, pear, currant and many other fruit and shade trees. Heavy infestations weaken the branches and cause them to die back or produce poor fruit crops. During the growing season, some scales may also infest the fruit, particularly around the stem and blossom ends. On fruit and twigs the scales cause a small reddish spotting that gives it a mottled look. Infestations of San Jose scale can be diagnosed by the small, circular wax coverings it produces. These are generally gray, with a lighter-colored central nipple. Old scale coverings remain on bark for several years before weathering away.

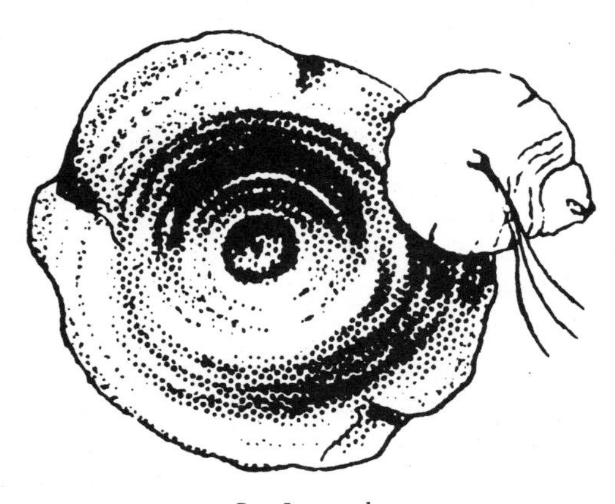

*San Jose scale*

**Life History and Habits:** San Jose scale spends the winter attached to bark as a dormant, partially grown nymph. As sap flow begins in spring, it resumes feeding and becomes full-grown around the time that apple trees bloom. The small male scales fly and mate with the stationary females, which shortly thereafter begin to produce eggs.

Eggs hatch over a period of weeks, after petal fall, producing tiny yellow *crawlers* (newly hatched scales). The crawlers move over the bark until they find a suitable site to feed, then insert their mouthparts. Within a few days after settling, they molt their skin and start to secrete a waxy covering. San Jose scale repeats the life cycle throughout the growing season. Typically three to four generations are produced per growing season, the most of any scale insect found within the region.

**Natural and Biological Controls:** San Jose scale is attacked by parasitic wasps. Some birds, such as nuthatches and brown creepers, will also feed on scales.

**Mechanical Controls:** Infested twigs and branches should be pruned and removed.

**Chemical Controls:** Sprays of oils or lime sulfur before spring bud break are the most effective means of controlling San Jose scale. During the growing season, settled scales are not easily controlled by insecticides, since they are protected by their waxy covering. However, crawler stages are susceptible to most orchard sprays. Crawler treatments are most effectively applied during the two to four weeks the first generation is present. This tends to occur about one month after apples first bloom. Timing of crawler emergence can be best determined by regularly shaking infested branches over a sheet of dark paper to detect the newly hatched crawlers.

## Pear Psylla (*Psylla pyricola*)

**Damage:** Pear psylla feed on the leaves of pear trees, excreting conspicuous droplets of honeydew. These droplets soil the leaves and fruit of the plant. Honeydew also allows sooty mold fungi to grow, which discolors fruit. High numbers of psyllids on trees can reduce plant vigor and cause it to yield poorly. Trees suffering from psylla shock may take several years to recover.

Pear psylla transmit the mycoplasma that produces the disease of pear decline. Fortunately this disease is uncommon in the region. Pear psylla occur throughout the region but are most common west of the Rockies.

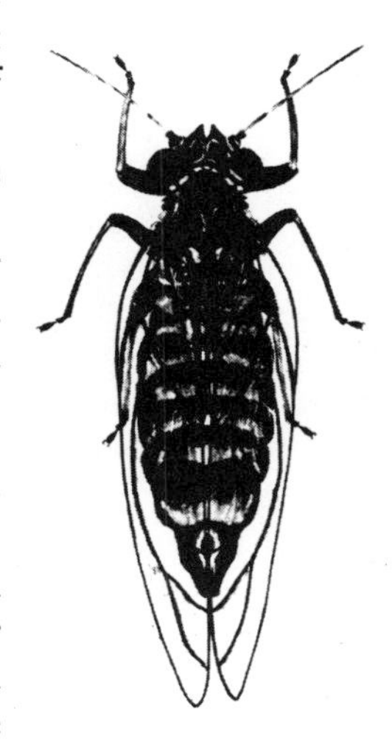

*Pear psylla*

**Life History and Habits:** Pear psylla overwinter in the adult stage in protected areas (under bark, plant debris on soil or other cover) in the vicinity of previously infested trees. They become active in late winter or early spring and move to pear trees, laying yellow-orange eggs as pear buds begin to swell. The emerging nymphs then move to feed on the tender new growth. As they feed, they are first covered with the honeydew droplets they produce. The final stage of the developing insect has conspicuous pads where wings are developing and does not live in the honeydew. They then molt to the adult stage. Later generations lay eggs on the new leaves, often concentrating on sucker sprouts late in the season. Two to three generations are normally produced during a season. At the end of the year, dark-colored winter adult forms move to shelter.

**Natural and Biological Controls:** During very dry weather, honeydew droplets covering the nymphs may dry and crust to kill the insects. Minute pirate bugs, lady beetles and other general predators also feed on pear psylla. A chalcid wasp is an important parasite of the pear psylla in many areas.

**Cultural Controls:** Pear psylla thrive on tender new growth. Remove sucker shoots during midsummer to deprive the insects of favored food sources after the main leaves have hardened.

**Chemical Controls:** Pear psylla are very resistant to most insecticides and difficult to control. The best control is to treat for the insect two to three times before bloom, killing the overwintered adult stages before they have laid eggs. Oils can be used for application. Such treatment is most effective if other pears in the area are also treated, limiting reinfestations.

## Potato (Tomato) Psyllid (*Paratrioza cockerelli*)

**Damage:** Potato (tomato) psyllids inject saliva into plants during feeding that can be damaging to tomatoes and potatoes. Damaged plants often stop growing and typically show leaf curling and color changes. The disease caused by potato psyllid injury is sometimes described as psyllid yellows. Effects on potato tubers include reduced tuber size, premature sprouting and rough skin. Tomatoes damaged by potato (tomato) psyllids produce small fruits that are soft and of poor quality. Yields of potatoes and tomatoes can be greatly reduced by potato (tomato) psyllid injuries. Infestations during early stages of plant growth cause much greater injury than later infestations. Potato (tomato) psyllids infest a wide variety of other plants, including pepper, nightshade, morning glory, bindweed and even beans. However, damage only appears to occur to tomatoes and potatoes. Potato (tomato) psyllid damage can occur throughout the western United States and parts of northern Mexico. Colorado, Wyoming and Utah are the areas most commonly infested by psyllids. However, outbreaks have been recorded throughout all areas of the region, except the Pacific Northwest.

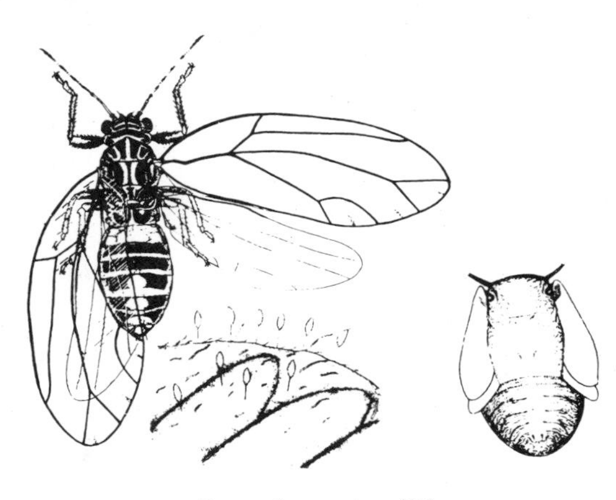

*Potato (tomato) psyllid*

**Life History and Habits:** The potato (tomato) psyllid does not overwinter within the region, dying out after hard freezes. Annual infestations originate from populations that survive on native plants in the southern United States and northern Mexico. As temperatures in these areas warm in spring, the adult psyllids migrate north on winds. If food plants are available and temperatures are not excessively high, the psyllids may settle and lay eggs. Otherwise they may continue the northward migration. Females tend to lay eggs in groups, so the insects tend to be unevenly distributed on the plant. Eggs are small and yellow-orange. These hatch in about 5 days, producing a scalelike nymph stage. Nymphs are yellow when young and often turn light green as they become older. They can move slowly and tend to concentrate on undersides of leaves in more shaded areas of the plant. After three to four weeks, the adults emerge, repeating the cycle. During a season, three to four generations may be completed.

**Natural and Biological Controls:** High temperatures (at least 90°) greatly reduce egg-laying and survival and may temporarily sterilize females. Predatory bugs, such as big-eyed bugs, minute pirate bugs and damsel bugs, appear to be among the more common predators. Occasionally, parasitic wasps (species of *Tetrastichus*) attack and kill nymphs of the potato (tomato) psyllid. However, none of these controls appears to provide reliable control during outbreaks.

**Cultural Controls:** Early plantings avoid much injury. There is a range in susceptibility to the effects of psyllid injury among both potatoes and tomatoes, but this is still being researched. Peppers, although commonly infested, do not appear to be very susceptible to psyllid injury.

**Chemical Controls:** Infestations are irregular in occurrence and scouting can help to identify periods when plants might benefit from treatment. Check leaf undersides for nymphs or their characteristic droppings, which resemble granulated sugar or salt. Pepper, although not a plant that is damaged by the insect, is a good indicator plant of local psyllid populations early in the season. Sulfur and lime sulfur are highly effective for psyllid control. Some common garden insecticides, such as diazinon and malathion, are also effective, but carbaryl is not. Insecticidal soaps can give some control when applied directly to psyllids, although coverage is often difficult.

**Miscellaneous Notes:** Potato (tomato) psyllids also infest a wide variety of other plants, including peppers and eggplant. However, they do not appear to be damaging to these. Psyllid injury on the leaves and growth of potatoes and tomatoes can resemble that caused by infection with various viral diseases. However, the symptoms of reduced tomato fruit size, sprouting and rough potato (tomato) skin are unique to psyllid yellows. Unlike many related insects that feed on plant sap and produce sticky honeydew, potato (tomato) psyllids excrete a granular material known as psyllid sugar. This pelleted excrement resembles salt or sugar and fluoresces under ultraviolet light. The presence of psyllid sugar is a very good indicator of infestation by potato (tomato) psyllids.

# *Arachnids (Spider Mites and Relatives)*

## Twospotted Spider Mite (*Tetranychus urticae*)

**Damage:** The twospotted spider mite pierces plant cells and feeds on the sap. Small white flecking injuries are typical around feeding sites. More generalized discoloration, typically bronzing, occurs as infestations progress. Vigor of plants may be seriously reduced. Premature leaf drop occurs on heavily infested plants. Most garden plants are fed upon to some extent by twospotted spider mites. Bean, cucumber, raspberry, rose and marigold are among the plants most commonly damaged seriously. The twospotted spider mite is also the most common species that damages greenhouse crops and interior plants.

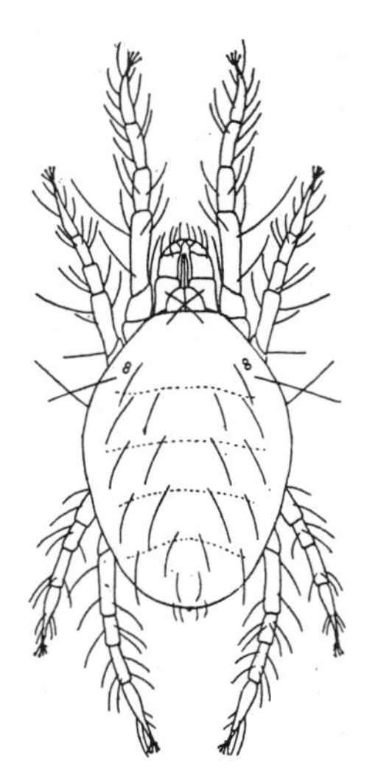

*Twospotted spider mite*

**Life History and Habits:** The twospotted spider mite is not an insect and is more closely related to other garden habitants, such as spiders and ticks—the arachnids. Life stages include the egg, an early, six-legged immature stage, an eight-legged immature stage and the adult stage. Under conditions of warm temperature and low humidity, the generations are completed in as little as 10 days. Adult females typically lay up to five eggs per day over the course of two to three weeks. During periods when temperatures are unfavorable or food is absent, twospotted spider mites temporarily change to a less active, nonfeeding stage, turning red or reddish-brown.

**Natural and Biological Controls:** Several organisms prey on spider mites in field settings. Minute pirate bugs and predatory species of mites are among the most important. The banded-winged thrips are also sometimes important. When adequate humidity is available, spider mites are sometimes killed by fungal disease.

**Cultural Controls:** Watering and water management are very important in controlling twospotted spider mites. Overhead watering—and purposeful hosing of plants with water in a garden setting—can dislodge and kill many spider mites. Providing adequate water for plant growth needs is also important in managing spider mites. Drought and fluctuating wet/dry soil conditions can stress plants, causing spider mite populations to increase. Water is also important in limiting the rate at which aphids feed, since they have a limited ability to evaporate excess water. Higher humidity rates can decrease feeding by twospotted spider mites.

**Chemical Controls:** Spider mites are very difficult to control with pesticides, and many commonly used insecticides (e.g., Sevin) can aggravate problems by destroying their natural enemies. In addition, some miticides, such as malathion and Orthene, are often ineffective because spider mites have developed resistance to them. Insecticidal soaps, sulfur dusts and nicotine sulfate are registered on most vegetable crops for spider mite control. Dicofol (Kelthane) has been a standard miticide for crops such as cucumbers, squash and beans, although availability has been limited in recent years.

# *Eriophyid Mites*

One unusual and little studied group of "bugs" are the eriophyid mites (Eriophyidae family). They differ greatly from their more familiar cousins, the spider mites (Tetranychidae family), in many ways. For one, these are extremely small mites, usually not visible to the human eye, that make the common spider mite a garden giant by comparison! Eriophyid mites also have only two pairs of legs (instead of four) and the general body shape of a little walking carrot.

Injuries by eriophyid mites also differ. Several species live on the leaf surface (called leaf vagrants), usually causing little observable injury. However, some (the rust mites) may cause leaves or fruit to turn a rusty brown color.

Most interesting are the gall-making mites, which poke and probe the plants during feeding to produce strange growths. This may result in simple blisters on the leaves (blister mites) or bizarre swellings of leaves (finger galls), flowers (flower galls) and buds (bud galls). Eriophyid mites are most common on trees and shrubs in the yard and garden.

**COMMON REGIONAL ERIOPHYID MITES AND RELATED PLANT INJURIES**

| Eriophyid mite | Plant injury |
|---|---|
| Pear leaf blister mite | Rusty blister on leaves of pear |
| Apple leaf blister mite | Rusty blister on leaves of apple, crabapple |
| Pear russet mite | Russeting discoloration of pear fruit |
| Chokecherry finger gall mite | Green/red finger galls on the upper leaf surface of chokecherry |
| Pear finger gall mite | Green finger galls on the lower surface of wild pear |
| Dryberry mite | White, dry sections on raspberry and other caneberry fruit |

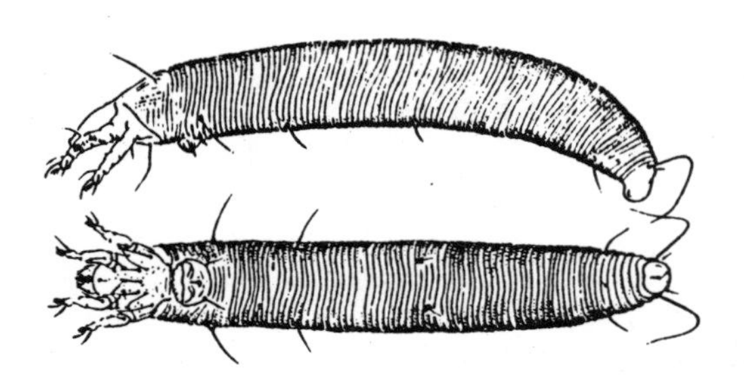

*Pearleaf blister mite, an eriophyid mite*

Eriophyid mites develop high populations and cause injury only to developing leaves and fruits. Plant parts that have already formed are not susceptible to injury.

They are controlled by many natural enemies, including predator mites, predator thrips and minute pirate bugs. Eriophyid mites are often an important alternative food for these predators, which feed on them when other foods are scarce, allowing the predators to survive and develop so they can later feed on more serious pests.

On fruit trees, eriophyid mites can be controlled with dormant oils, since they spend the winter around buds of the trees. They are also very susceptible to carbaryl (Sevin) and moderately susceptible to insecticidal soaps.

# *Isopods (Sowbugs and Pillbugs)*

## Sowbugs and Pillbugs ("Roly-polys")

**Damage:** Sowbugs and pillbugs have small mouth-parts and feed on tender plant materials. Usually they only feed on decaying plant matter and fungi. However, tender seedlings of garden plants and plant roots may also be chewed. When abundant, they can be a problem in newly seeded gardens.

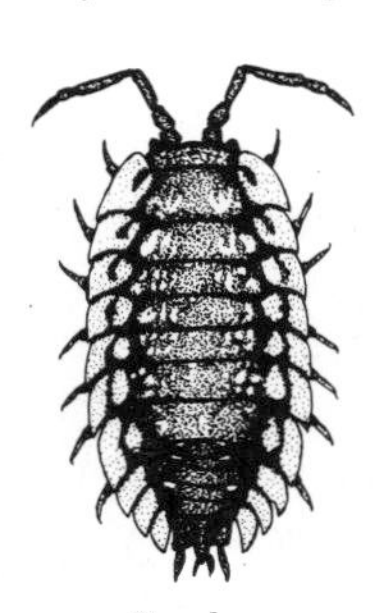

*Sowbug*

Sowbugs and pillbugs are classified as crustaceans, more closely related to lobsters, shrimp and crayfish than to insects. Both are covered with a hard shell that allows them to live on land. However, they can be separated from each other by their ability to curl protectively when disturbed. The dark gray, heavily armored pillbug (*Armadillidium vulgare*), also known as the roly-poly, can roll into a hard, protective "pill" form; the sowbugs (species of *Porcellio* ) can not.

**Life History and Habits:** Sowbugs and pillbugs are rather long-lived, but slow-growing. Eggs and the newly hatched young remain inside the mother for several months, protected by the pouchlike marsupium of the mother. They then leave and undergo a series of molts as they develop. Young stages resemble the adults, except for size. They become full-grown after about one year. Most of the young fail to survive after leaving the mother, usually due to drying. However, adults can live for two or more years. Pillbugs and sowbugs usually feed at night, spending the day under cover. However, they can be seen during the day, particularly after rains or during overcast conditions.

**Natural and Biological Controls:** Most sowbugs and pillbugs die from problems related to water—either too little or, less commonly, too much. They are also food for many animals, such as birds, spiders, ground beetles and centipedes.

**Miscellaneous Notes:** Sowbugs and pillbugs are common nuisance pests in homes in the West. Following periods of extended wet weather, they are often seen climbing walls and may accidentally enter homes. Within a home they rarely survive more than one to two days due to low humidity.

# *Diplopods*

## Millipedes

**Damage:** Ripening strawberries on the ground may be tunneled and destroyed by millipedes. Occasionally, millipedes may damage tender vegetable seedlings and root crops such as carrots.

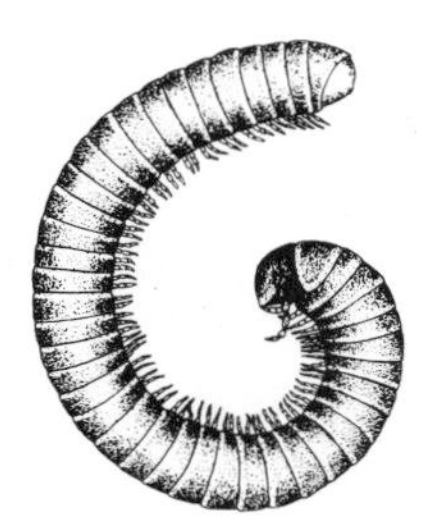

*Millipede*

**Life History and Habits:** Millipedes reproduce by laying clusters of sticky eggs in the soil or moist areas of the soil surface. Eggs hatch in a few weeks, and the pale-col-

*Millipede*

ored, young millipedes move about and feed. They grow slowly, increasing in size and number of body segments with each molt. As they get older, they increasingly resemble the adult form, becoming darker and thicker.

**Cultural Controls:** Millipedes are intolerant of dry conditions, so reducing moisture in the planting can reduce the numbers of millipedes. Problems are most severe in ripe or overripe fruit, so regular picking can also limit infestations. Surface-dry mulches that keep the berries off the ground can reduce injury. Decaying vegetable and animal manures are particularly favorable to millipedes and should be reduced when problems with millipedes are severe.

**Mechanical and Physical Controls:** Millipedes, like slugs, can concentrate on overripe fruits, under fruit rinds or in moistened newspapers. Daily removal of millipedes from these sites will reduce populations.

**Chemical Controls:** Baits containing the insecticide carbaryl (Sevin) can control millipedes, sowbugs and pillbugs.

**Miscellaneous Notes:** Millipedes sometimes make mass migrations, usually following wet weather. They are a common nuisance invader of homes in spring and fall.

*Slug eggs*

# *Gastropods (Slugs and Snails)*

## Gray Garden Slug (*Deroceras reticulatum*)

**Damage:** Slugs chew on the foliage of many plants, preferring soft plant tissues such as strawberries, seedling lettuce or beans and many other garden plants. Young plants may be killed or plantings retarded by damage. Fruits and harvestable parts of leaf and tuber crops may be destroyed by slug infestation.

**Life History and Habits:** The gray garden slug reproduces by laying masses of round, clear eggs in soil cracks and around the base of plants or other organic debris. The young slugs feed on organic matter, including soft plant parts, becoming full-grown in three to six months. Egg-laying may occur during both spring and fall, with development suspended during much of the hot season. Slugs feed at night, strongly avoiding sunny, drying conditions. During the day they move to sheltered areas under debris and in soil cracks. Occasionally, slugs are also active during rainy or overcast periods of the day.

**Natural and Biological Controls:** Several organisms feed on slugs, including toads and snakes. Larvae of certain species of fireflies and parasitic flies attack and parasitize immature stages of slugs.

**Cultural Controls:** Slugs are highly sensitive to dry conditions, having a high water content and producing large amounts of protective mucous. Consequently, practices that reduce humidity within a garden planting are fundamental to limiting populations of slugs. To eliminate potential shelter for slugs, remove surface debris in and around the garden and avoid organic mulches such as straw and grass clippings. Also, increase air movement around plants and reduce high-moisture conditions with trellises and wider row spacing. Drip irrigation, soaker lines or other irrigation techniques that use limited amounts of water can control slugs by decreasing humidity. If overhead irrigation is used, it should be applied early in the day to allow more time for leaves and soil surfaces to dry before the nightly activity of slugs.

**Mechanical Controls:** Slugs are attracted to chemicals produced by many fermenting materials. These materials can be used to make traps baited with the attractant materials. For example, pans of beer or sugar water and yeast can effectively attract, trap and drown slugs. A single baiting with these materials can remain effective for several days, as long as sufficient liquid remains. However, the attractive range of these traps is often only a few feet, so they must be placed throughout a planting. Effectiveness of the traps is also reduced if highly desirable plants are nearby. (*Note*: Alcohol is not an attractant to slugs or useful in capturing slugs.) Selective use of trap boards, moistened newspaper or inverted fruit rinds placed on the soil surface can also be used to concentrate slugs that seek shelter. These traps should be checked early every morning and any slugs collected should be destroyed. If traps cannot be regularly inspected, they should not be used, since they provide favorable shelter for the slugs to spend the day. Some plants are particularly attractive to slugs, such as lettuce, beans, hosta, calendula and okra. Purposeful planting of these trap crops can divert some feeding by slugs on other garden plants. Slugs often avoid traveling over materials that are highly acidic, alkaline or abrasive. Diatomaceous earth, wood ashes and similar materials placed around plants provide some protection. However, the effect of these treatments is often reduced following rains or other moisture that cakes the materials. Salt is also toxic to slugs and direct application of table salt is very lethal. However, this technique has limited effectiveness, since slugs are often not visible on plants, and excessive applications of salt can aggravate salinity problems in area soils. Certain metal ions are also highly repellent to slugs. Barriers of copper foil exclude slugs from greenhouse benches and raised bed plantings. Other copper-based materials, such as copper sulfate, are also useful as slug repellents.

**Chemical Controls:** Pesticides effective against slugs and snails are known as molluscicides. These are typically very different types of chemicals than those used to control insects and other garden pests. Slugs are not susceptible to poisoning by most insecticides. Metaldehyde has been the most commonly used molluscicide and is sold in a variety of bait or liquid formulations. Current labeling allows its use around fruit and vegetable crops. Metaldehyde kills slugs by dehydration through secretion of large amounts of mucous. Successful use of metaldehyde for slug control requires care in application and favorable weather conditions. Slugs often recover from metaldehyde poisoning if high-moisture conditions occur. Also, control is reduced during very cool or very warm periods, since slugs are relatively inactive during these times. Metaldehyde is most effective if applied during evenings when warm, moist conditions that favor slug activity are then followed by drying weather. Metaldehyde breaks down on exposure to sunlight, so baits should be placed under leaves or in harborage areas used by the slugs.

CHAPTER 7

# Management of Common Plant Diseases

Conditions that disturb the normal healthy growth of plants produce plant disease. Sometimes these are caused by microorganisms (plant pathogens), such as fungi, bacteria and viruses that attack the plant. Others are caused by adverse environmental or cultural conditions.

Because of the harsh environments of the West, *abiotic* problems predominate. Dry air and soil conditions often cause perennial plants, such as raspberries and evergreens, to die from winter drying. Plants that survive and tolerate the cold get teased out by warm winter weather and suffer when spring frosts return. High temperatures and drying weather stress many plants in summer.

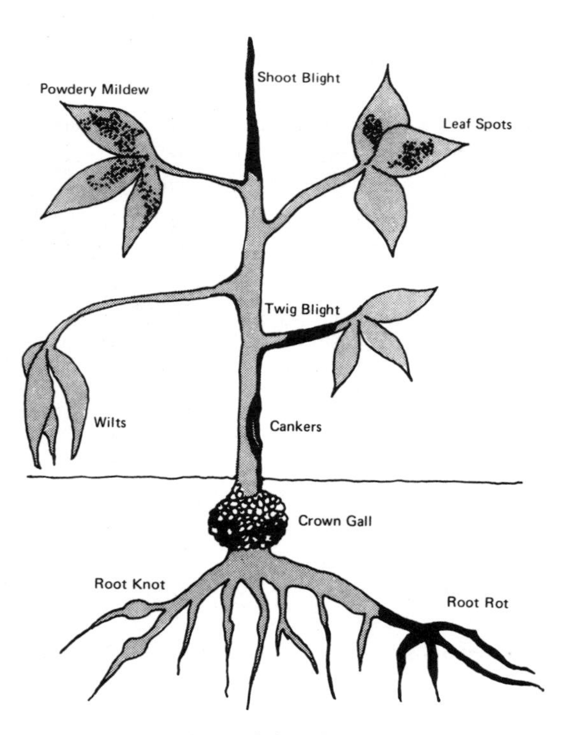


*Signs of plant disease*

Many garden problems are soil-related. "Tight" heavy clay soils restrict root growth and drainage, suffocating roots. Nutrient imbalances, such as iron chlorosis, are a common result of excess pH. High levels of salts sometimes damage garden plants.

However, the most important plant diseases often result from poor gardening practices. Plant varieties that do well in other areas but are poorly adapted to the region fail to thrive or die. Planting at the wrong time may also doom a crop to failure. Overzealous gardening habits, such as inadvertently cutting roots while weeding (cultivator blight) and overwatering, are other common problems in the garden.

Some problems are due to *parasitic* diseases caused by organisms such as fungi, bacteria, viruses and nematodes. Examples of these *biotic* diseases include powdery mildew, a common fungal disease of roses and melons, and fire blight, a devastating bacterial disease of apples and pears.

## Fungal Diseases

Fungi are the most important plant pathogens in the garden. Evidence of fungal disease includes rotting/decay of vegetables, powdery mildew on leaves, wilting and leaf spotting.

Fungi reproduce primarily by *spores*. Tremendous numbers of spores may be released by a fungus

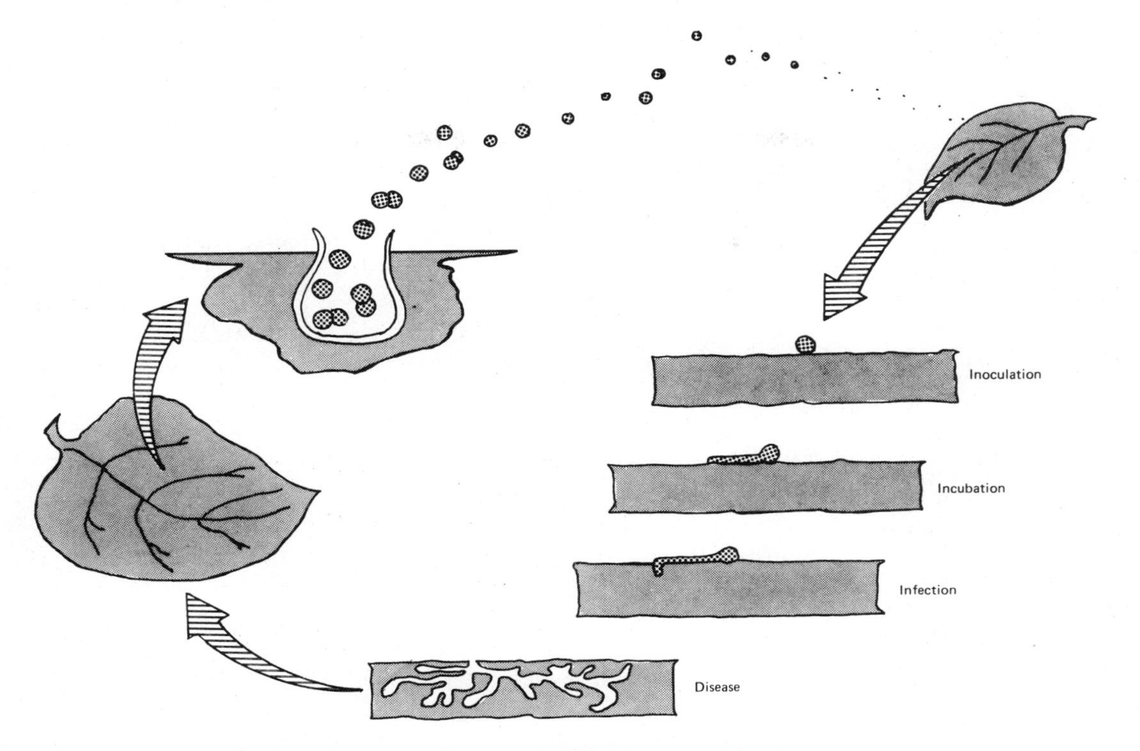


*Fungal diseases usually start with spores. After the spores land on the plant, they may grow into and infect the plant after the incubation period. The fungus then grows inside the plant, causing disease. Later new spores are produced that may cause new infections.*

that is actively growing and fruiting. For example, the fine dust released by a mushroom consists of millions of spores, which are spread over large areas by wind and splashing. When spores of fungi land on a susceptible plant, they germinate if surface conditions are proper and free moisture is available for extended periods. After germination, the growing fungi penetrate and grow in the plant, reducing its ability to grow normally.

Some fungi that cause plant disease can survive only on living plant tissues. Other fungi live primarily on dead plants and may only rarely invade living tissues. These latter fungi are weak pathogens since they may only injure plants that are in a weakened condition.

As they develop, fungi form reproductive parts to produce the spores. The familiar mushroom is the spore-producing structure of some fungi, such as fairy ring in lawns. However, most fungi that attack garden plants, such as powdery mildew, are less conspicuous, and spores may be produced on what appears as a fuzzy area on the plant.

Many of the important fungi that attack the roots of garden plants can also produce resistant stages that allow the fungus to remain alive for several years in the soil.

## Bacterial Diseases

Bacteria are very small, one-celled organisms. They reproduce by dividing (binary fission)

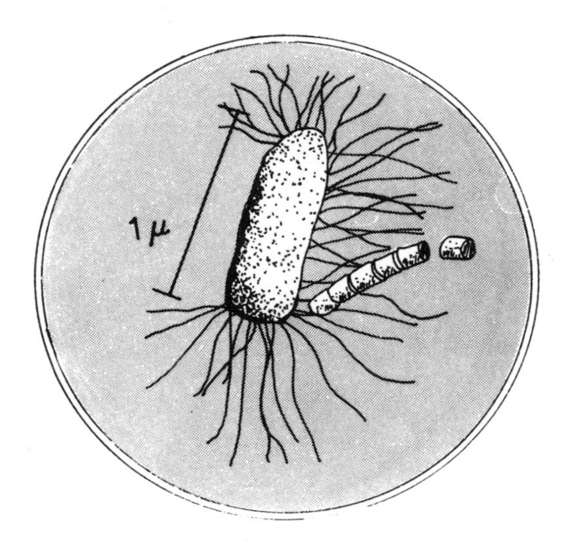


*Typical bacteria*

and are capable of rapidly increasing in number under favorable conditions. Bacteria that cause plant disease dissolve plant tissues with enzymes, or plug the sap stream of the plant, causing it to wilt. Fire blight of apples and pears, soft rots, halo blight of beans and crown gall are common bacterial diseases.

Bacteria have a limited ability to move by themselves. Most spread of bacteria occurs through splashing water or infected seed. Insects transmit some bacterial diseases. High-moisture conditions are an important requirement for development of all bacterial diseases.

One special type of bacteria is known as the *mycoplasmas* (more properly, the mycoplasma-like organisms, or *MLOs*). The mycoplasmas infect cells of the sap stream (phloem) of plants and produce symptoms such as bushy or greatly distorted new growth. Aster yellows is a very important mycoplasma disease in the West, affecting many vegetable and flower crops. Pear decline and X-disease of peach trees are serious mycoplasma diseases of orchard crops. All mycoplasma diseases in the region are transmitted by leafhoppers or psyllids that feed on the plant.

## Viral Diseases

The presence of viral infection is usually recognized by symptoms such as leaf streaking, mosaic-patterned discoloration of leaves, yellowing of leaves, leaf curling or crinkling and ring spots on leaves or blossoms. Viruses do not reproduce like other plant disease agents, but instead multiply by causing the infected cells of their host plant to form virus particles. By changing the normal function of the plant cells, the plant is injured. Of all the plant disease–producing organisms, viruses are the most dependent on spread by other organisms. Viruses, which are simply constructed, minute particles of protein-covered nucleic acid, have no means of movement on their own. They must be physically moved by other organisms into plant cells to cause an infection. Many viruses are spread mechanically as plants rub together, or by hands and tools as gardeners work between diseased and healthy plants. Other viruses are spread only by certain insects, usually aphids or leafhoppers. A few viruses are also spread by mites, nematodes and even fungi. Potato leafroll, beet curly top, tomato spotted wilt and mosaic diseases of melons and cucumbers are among the more important viral diseases spread by insects.

*Virus-infected (left) and healthy (right) peppers*

| VIRAL DISEASES AND AFFECTED CROPS | | |
|---|---|---|
| **Virus** | **Crop affected** | **Means of spread** |
| Tobacco mosaic | Wide host range including peppers and tomatoes | Mechanical (hand, rubbing of leaves, cultivation), maintained on crop debris |
| Potato virus X | Primarily potatoes | Mechanical (hand, rubbing of leaves, cultivation) |
| Cucumber mosaic | Wide host range (34 plant families) including cucurbits, beans and tomatoes | Mechanically transmitted (hand, cultivation, machinery, etc.), aphid transmitted (green peach aphid, cotton aphid), maintained on seed of some plants |
| Common bean mosaic | Beans | Aphid transmitted, mechanically transmitted |
| Watermelon mosaic | Cucurbits | Aphid transmitted only |
| Potato virus Y | Peppers, potatoes, tomatoes and related plants | Aphid transmitted only |
| Potato leafroll | Potatoes, tomatoes | Aphid transmitted only |
| Corky ringspot (tobacco rattle) | Potatoes, many other plants | Nematode transmitted |
| Tomato spotted wilt | Wide host range including tomatoes, peppers, lettuce, chrysanthemums, field bindweed, aster | Thrips transmitted (onion thrips, flower thrips), maintained on infected perennial plants and overwintered thrips, can be mechanically transmitted |
| Beet curly top | Tomatoes, beets, peppers | Beet leafhopper |
| Tobacco ringspot | Potatoes, peppers | Stubby root nematodes, through seed of some weeds (purslane, hairy night shade, common cocklebur), seed potatoes |

# Nematodes

Nematodes are a type of animal sometimes known as a roundworm or eelworm. Most are microscopic in size and wormlike in general form. Thousands of different types of nematodes occur in healthy soils and water, and some nematodes are important for controlling pest insects. However, some species (e.g., root knot nematodes, sting nematodes) are known to attack plants. Because of their small size, the plant problems they cause are generally considered plant diseases.

On their own, nematodes can move only short distances in the soil. However, nematode problems may originate and become established in an area by human movement of infested soil or plant materials. This is particularly common with nematodes that produce eggs inside a protective cyst (cyst nematodes) or a gelatinous sack (root knot nematodes), which allow them to survive for several years.

Fortunately, nematode problems are uncommon in the West. Soil drying, clay soils and severe winter temperatures are all unfavorable to most nematodes. Although a few species affect regional agriculture, such as the sugarbeet cyst nematode and alfalfa stem nematode, they rarely damage garden plants.

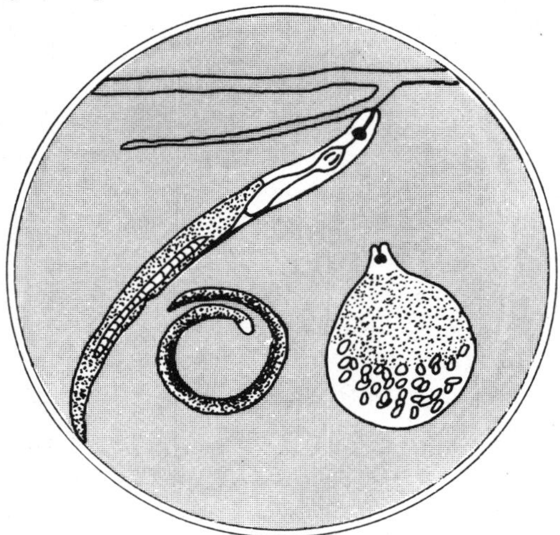

*Plant parasitic nematodes*

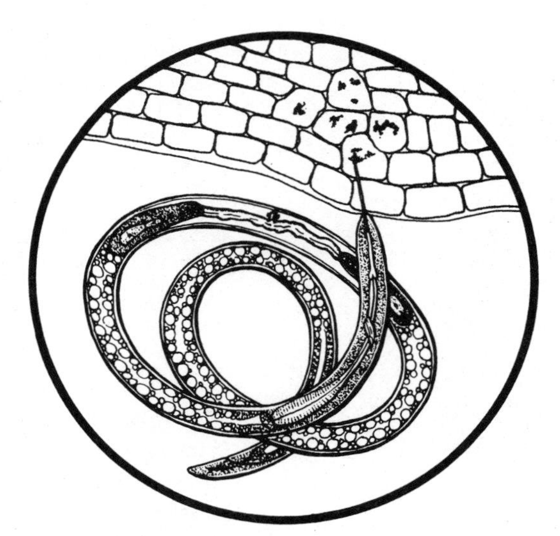

*Nematode feeding on plant*

# Common Viral Diseases

## Cucumber Mosaic

**Damage and Symptoms:** Cucumber mosaic can affect several garden vegetables but is particularly damaging to cucurbits, such as cucumber, squash and melons. Symptoms include a general stunting of plant growth, yellowing or mottling (mosaic) patterns and wrinkling of leaves. Yellow mottling and reduced size are common effects of the disease in fruit of cucumber and summer squash. Symptoms are most pronounced when plants are infected during early growth. Effects of the disease can also be increased by moisture stress.

**Disease Cycle and Development:** Cucumber mosaic is caused by infection of plants with the cucumber mosaic virus (CMV). Because it has an extremely wide range of hosts, overwintering sources often include weeds that are infected in the vicinity of the planting. A few plants are reported to allow infection of CMV between generations through infected seeds. However, spread of the disease through seed occurs rarely. Cucumber mosaic virus is spread between plants either by feeding of aphids and by chewing insects such as the striped cucumber beetle, or mechanically by practices such as handling of infected plants and cultivation. Mechanical transmission is much less likely than with common mosaic, and aphid transmission is generally considered to be most important in disease spread. Both of these methods of spread involve rapid infection (within minutes) after plant contact. The virus does not persist for long outside the plant host. Once inside the plant cell, the virus is reproduced at the expense of other plant growth processes.

**Cultural Controls:** Since infections can occur during greenhouse propagation, closely inspect and discard "off"-looking plants during purchase. During the growing season, also remove and destroy plants that are suspected of showing symptoms of viral infection to reduce its spread. If potentially diseased plants are retained in the garden, they should be handled last, after working around other susceptible plants. Higher plant populations can be useful for reducing losses to cucumber mosaic. Larger numbers of plants often result in a lower percentage being infected. Infected plants should be removed as soon as possible. Light-colored mulches, particularly those that reflect light such as aluminum foil, are very effective at repelling winged aphids. This can delay infection of plants until after they have overgrown the mulch. By this time, plants are much less sensitive to effects of the virus. Resistance to cucumber mosaic virus has been bred into some cucumber lines, but resistant varieties are not widely available to gardeners. Highly silver-leaved varieties also show potential resistance because of reduced landing by the aphids that spread the disease.

**Chemical Controls:** Cucumber mosaic virus is transmitted extremely rapidly (in seconds or minutes), usually by winged migrant aphids that rarely remain on the plant for long. Insecticides applied against the insect vector are ineffective for preventing spread of this virus. Oil- or fat-containing sprays, including milk solutions, can reduce transmission of the virus to plants by visiting aphids. However, these treatments have short-term effects and need to be reapplied at very short intervals (1 to 3 days), particularly when plants are growing vigorously.

## Watermelon Mosaic

**Damage and Symptoms:** Watermelon mosaic infects all the cucurbit (squash family) plants. It has been reported most damaging to watermelon and winter and summer squash. The symptom of the disease on leaves is a variable green mottling. Blistered or puckered malformations also may be present. Fruits from infected plants may be abnormally shaped and colored, often with knobs and other outgrowths. Watermelon mosaic causes obvious stunting of watermelons and muskmelons. Symptoms are most severe when plants are infected early. Watermelon mosaic may also occur with cucumber mosaic, causing a general loss of plant productivity.

**Disease Cycle and Development:** Watermelon mosaic is caused by infection with the watermelon mosaic virus (WMV). It is spread to new plants by winged aphids, which can acquire the virus from diseased plants and transmit it to healthy plants within a few seconds. The aphids that spread the disease are migrants that often do not remain on the plants for long or leave their young.

The virus is not transmitted to progeny via infected seed, nor is it mechanically transmitted by handling and wounding during the growing season. Overwintering sources of the virus in the region are not known, but in more southern regions, wild species of perennial cucurbits are hosts. New infestations may spread annually from aphids migrating from these southern areas. Once the disease has become established in the field, aphids can continue to spread it to new plants.

**Cultural Controls:** Light-colored mulches, particularly aluminum foil, can prevent aphid landing and retard watermelon mosaic infection. Intercropping of susceptible plants with nonhost plants (plants not related to squash and melons) can also reduce disease incidence. Since the disease is most serious to young plants, cultural practices that hasten plant development can avoid loss. Obviously diseased plants should be removed and destroyed. Breeding to develop plant varieties that resist effects of watermelon mosaic virus is being actively pursued, but no resistant varieties are yet available to gardeners.

**Chemical Controls:** Insecticides are ineffective for control of watermelon mosaic virus, since the disease is spread too rapidly. However, special lightweight oil sprays that can prevent aphids from transmitting the virus during feeding have been developed. These specialty products are not generally available to gardeners, but other oily materials (milk, horticultural oils, etc.) may be effective.

## Beet Curly Top

**Damage and Symptoms:** The most typical symptom of beet curly top infection is dwarfing. On beans, leaves turn yellow, curl downward and often become very brittle. Plants infected when young may be killed, while plants later infected may show few symptoms. Beet curly top can infect dozens of plants, including tomato, pepper, beet and bean. In addition, numerous weeds (Russian thistle, peppergrass, etc.) are hosts of the beet curly top virus. The beet leafhopper, which transmits beet curly top (see page 150), is a desert species, native to the West. Both the leafhopper and the beet curly top virus it transmits are most common in dry regions during periods of high temperatures.

**Disease Cycle and Development:** Beet curly top is caused by infection with the beet curly top virus. The virus is transmitted only by the beet leafhopper. Leafhoppers that have acquired the virus from infected plants can transmit it for the rest of their lives. Transmission to healthy plants can occur about 15 to 30 minutes after feeding.

**Controls:** See beet leafhopper (page 150).

## Tomato Spotted Wilt

**Damage and Symptoms:** Tomato spotted wilt can infect a very wide range of garden vegetables and flowering plants, including peppers, tomatoes and lettuce. However, over 200 species of plants in more than thirty families are potential hosts of tomato spotted wilt, including many common weeds, such as field bindweed, nightshade and pigweed. Characteristic symptoms of tomato spotted wilt are concentric ring spots on the leaves and fruit, although these are not always evident. Infected plants may be stunted and yellow, similar to symptoms produced by other viral diseases. Often the disease produces brown streaks or spots on stems and leaf petioles that cause parts of the plant to wilt and die back. In recent years, tomato spotted wilt has spread rapidly throughout North America and has become an extremely serious problem of greenhouse production. Spread of tomato spotted wilt outdoors has been infrequent in the West. Most infections in garden plantings are related to purchase of diseased transplants.

**Disease Cycle and Development:** Tomato spot-

*Tomato spotted wilt virus infection; symptoms on tomato fruit.*

ted wilt is caused by infection with the tomato spotted wilt virus. It is spread by thrips, primarily the flower thrips (species of *Frankliniella*) and onion thrips (*Thrips tabaci*). Young thrips developing on diseased plants acquire the virus during feeding and later, during the adult stage, can transmit it to healthy plants. The virus typically requires two to three weeks of reproduction within the plant before it can be acquired by thrips and spread to new plants. No other insects are capable of transmitting the virus. However, tomato spotted wilt virus can be mechanically transmitted by wounding plants during gardening operations.

**Cultural Controls:** Carefully inspect all transplants for evidence of tomato spotted wilt symptoms. Discard and destroy all plants that are suspected of harboring the disease. If tomato spotted wilt becomes established in a greenhouse, a vigorous, sustained effort must be made to eradicate all plants with the disease. These sanitation efforts can be complemented with, but not replaced by, chemical control of the thrips.

**Mechanical Controls:** Very fine mesh screens can exclude thrips from new plantings.

**Chemical Controls:** Control of the thrips vector can sometimes reduce spread of the disease, when combined with sanitation efforts to eliminate disease sources. However, both flower thrips and onion thrips are resistant to most insecticides.

## Common Mosaic (Tobacco Mosaic Virus)

**Damage and Symptoms:** Common mosaic produces a wide variety of symptoms in susceptible garden plants. Leaves typically show some yellow and green mottling (mosaic), and new leaves may show puckering and distortions. Stunting and poor growth often result from common mosaic infection. The host range of common mosaic is extremely large, affecting many vegetables (e.g., peppers, tomatoes, spinach, beets) and flowers (e.g., phlox, snapdragon, zinnia), as well as several weeds.

**Disease Cycle and Development:** Common mosaic results from infection of plants with the tobacco mosaic virus (TMV). It is an extremely persistent virus that can remain on seeds, crop debris from infected plants and clothing for several months or even years. The virus is spread mechanically, entering plants through wounds. This may occur during transplanting, pruning or other handling of plants. Roots growing through crop debris infected with the virus may also be infected. Transmission by chewing insects, such as grasshoppers, may occur but is considered of little importance. TMV is not spread by aphids, leadfhoppers or other insects that transmit many other plant viruses, but it can be transmitted to healthy plants through tobacco products, particularly cigar wrapper leaves.

**Cultural Controls:** Carefully inspect all garden transplants before purchase and avoid planting those that appear diseased. Diseased plants observed in the garden should be discarded. If plants suspected of being infected are retained, the plants should be handled last to prevent subsequent spread. Most seeds are free of tobacco mosaic virus, although the virus can survive on plant tissues that remain on the seed. Therefore, avoid collecting seed from "off"-looking plants that may be TMV-infected. Breeding work has identified several sources of resistance to tobacco mosaic virus. TMV-resistant plants are most commonly available in some tomatoes and peppers (e.g., VFTN).

**Chemical Controls:** Tobacco mosaic is not spread by insects, so insecticides are ineffective for control. Spread of the virus can be reduced by spraying plants with skim or reconstituted milk. This treatment can prevent new infections but does not affect the disease in plants that are already infected. Dipping hands in milk while handling plants can also prevent accidental transmission.

# Common Fungal Diseases

## Phytophthora Root Rot (species of *Phytophthora*)

**Damage and Symptoms:** Phytophthora root rot can affect most crucifers (cabbage, cauliflower, Brussels sprouts, etc.) as well as potatoes, spinach, beets and carrots. Original symptoms are a purplish discoloration along the edges of leaves, followed by dying of the leaf. Wilting and death of all leaves usually occurs on young plants, while older plants may only die back along one side. A watery root decay is present, and infected plants can be easily pulled from the ground.

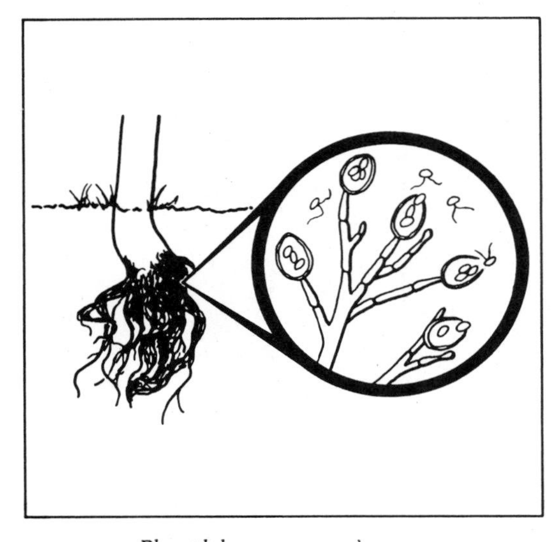

*Phytophthora crown and root rot*

**Disease Cycle and Development:** Phytophthora root rot is caused by the fungi in the genus *Phytophthora*, most often *Phytophthora megasperma*. The spores of the fungus can swim short distances if wet conditions are present and enter susceptible root or stem tissues. The fungus grows within the plant and later produces stages that can survive for over a year in the soil.

**Cultural Controls:** Poorly drained soils that allow standing water greatly favor spread of the disease organism. Improving drainage through techniques such as soil amendment with organic matter and raised beds can control the disease. Crop rotation involving non-susceptible plants for more than two years can also provide control.

## Phytophthora Blight (species of *Phytophthora*)

**Damage and Symptoms:** Phytophthora blight is most serious on peppers, but can affect related plants such as eggplant and tomatoes as well as all the commonly grown cucurbits (squash, cucumbers, melons). Marigolds are also susceptible. Symptoms can extend to all parts of the plant. Infection of stems can result in wilting and death of the plant due to girdling. When infections occur on leaves, there is an irregularly shaped spotting with a scalded appearance. Fruits are infected from the stem end and rot.

**Disease Cycle and Development:** Phytophthora blight is caused by a fungi in the genus *Phytophthora* (usually *Phytophthora capsaci*) that survives on seed and in soil. Conditions of high soil moisture allow the swimming spores to infect new plants. After plants become infected, other types of spores are produced that can be wind-blown for long distances. Warm, wet conditions are particularly suitable to infection of plants.

**Cultural Controls:** Crop rotations with nonsusceptible plants are very important to reduce the survival of the organism in the soil. Also, improving drainage around plants can prevent some of the original infections that are dependent on standing water. Later infections of leaves and fruit can be reduced by increasing air circulation around plants and watering in a manner that prevents the plant from remaining wet for long periods.

**Chemical Controls:** Repeated applications of copper-based fungicides or Bordeaux mixture can help control infections of leaves and fruit. However, these fungicides may cause damage to plants, particularly cucurbits, during high temperatures.

**Related Species:** *Phytophthora infestans*, or late blight, is a highly destructive disease of potatoes and related plants that caused the spectacular Irish potato famines in the mid-1800s. However, late blight requires extended periods of rain or fog and is rarely found in the West.

*Verticillium infection in eggplant. Note the dark streaking.*

## Verticillium Wilt (*Verticillium albo-atrum, V. dahliae*)

**Damage and Symptoms:** Plants infected with Verticillium wilt first show a yellowing of leaves, with leaves dying progressively from the margins. Stunting also may occur. On established plants, the disease progresses upward, first affecting older leaves. Infections on new canes of raspberry, rose or other perennial plants may show the reverse, with dieback occurring first at the tip. Symptoms on all plants can occur rapidly following periods when plants are under stress from drought or heat. Under favorable growing conditions, little yield reduction may be observed. Dark streaking of the vascular system (xylem) is often a symptom of infection with Verticillium wilt. These symptoms are similar to those of Fusarium wilt, although streaking by the latter often occurs higher up on the plant.

*Verticillium wilt*

**Disease Cycle and Development:** Verticillium wilt results from infection of plants by fungi, either *Verticillium albo-atrum* or *V. dahliae*. The fungus survives in the soil and infects plant roots that contact various stages of the fungus, generally entering through wounds. The fungus then grows inside the root, destroying the vascular system. As the infected roots die, the fungus reproduces by forming spores. Verticillium wilt can infect a very wide range of plants, including various flowers, trees, fruit crops and most vegetables. Grasses and sweet corn are not hosts of the disease-producing organism.

**Cultural Controls:** Verticillium wilt is introduced into gardens on infected plants or soil. These materials should not be brought into a garden unless the source is free of the disease. Many plant varieties are resistant to Verticillium wilt. In particular, resistant tomato varieties are widely available, as indicated by the "V" in VFN (or VFTN, etc.) designation on seed packs or labels. Crop rotations involving a cycle of three to four years or more can be very useful for limiting the disease. In the vegetable garden, these rotations should include nonsusceptible plants, such as sweet corn, as well as plants that are rarely infected such as onions, spinach, most cabbage family plants, beans and peas. In areas with high summer temperatures and sunlight, soil solarization can kill many of the surviving stages of the Verticillium wilt fungus in soil. This method should be used in coordination with other cultural controls, such as crop rotation, for best effect. Infected plants should be removed and destroyed by disposal. Composting or tillage often will not kill the more resistant stages of the fungus. Providing favorable growing conditions of adequate soil moisture, moderate temperatures and nutrients can reduce effects of infection. It is particularly important to avoid overwatering early in the season.

## Fusarium Wilt of Tomato (*Fusarium oxysporum, lycopersici* strain)

**Damage and Symptoms:** Fusarium wilt attacks the root system of plants, causing wilting of the top growth. First symptoms often include yellowing of lower leaves followed by general wilting. Sometimes leaves will die without wilting symptoms. Effects of the disease may be confined to only one branch of the plant. Brown streaking of the interior of stems and leaf petioles is characteristic. Fusarium wilt of tomato infects tomatoes, eggplant and several weeds such as pigweed.

**Disease Cycle and Development:** Fusarium wilt is caused by infection with a fungus, *Fusarium oxysporum, lycopersici* strain. Spores of the fungus are introduced into a garden on infested seeds, soil or decaying plant parts. The spores germinate when susceptible plants grow roots adjacent to them. The germinated fungus invades the plant root system, often through wounds. Once inside the roots, the fungus grows through the xylem of the plant. On susceptible plants, it may grow up the stem and even into the petioles of the leaves. In some resistant plants, infections are restricted to the roots. Growth of the fungus is favored by temperatures of about 80°F and slows greatly in very cool or very warm soils. The disease can progress rapidly in actively growing, succulent plants.

**Cultural Controls:** Use of resistant varieties is fundamental to managing problems with Fusarium wilt of tomato. Resistant varieties are indicated in descriptions of plant characteristics and can include resistance to both Race 1 (VFN) and Race 2 (VFTN) of the fungus. Several soil and growing conditions can retard development of the disease. Near-neutral soil pH (6.0–7.2) reduces disease incidence. Fertilization should also be moderate, since high nitrogen (particularly ammonium nitrogen) favors the disease. Cultivation around plants should be minimized to prevent root wounding. Crop rotations are of minimal value, since the spores are extremely persistent. Extra effort should therefore be made to prevent introducing the disease on infested soil and plants. In areas with high summer temperatures and sunlight, soil solarization can kill many surviving stages of the Fusarium wilt fungus in soil. This method should be used in coordination with other cultural controls, such as crop rotation and resistant varieties, for best effect.

**Miscellaneous Notes:** Several different species and strains of the Fusarium wilt fungus occur in garden soils. Most are weak pathogens that damage primarily young plants grown under stressful conditions. Species of *Fusarium* also cause problems with seed piece decay of potatoes and storage rot of onion.

## Anthracnose (*Colletotrichum orbiculare*)

**Damage and Symptoms:** Anthracnose is a common disease of most cucurbits, particularly cucumbers, muskmelons and watermelons. First symptoms of the disease are small yellow leaf spots that later turn brown-black. These irregular spots of dead tissue often drop out, leaving holes. The spotting can also spread, killing leaves, stems and developing fruit. Anthracnose symptoms also occur on fruit. Dark, sunken spots pit the skin of infected fruit. Although anthracnose does not penetrate into the fruit flesh, other rotting organisms can enter the wound and further destroy the fruit.

**Disease Cycle and Development:** Anthracnose results from infection with the fungus *Colletotrichum orbiculare*. The fungus can survive on non-decomposed plant debris and on seed from infected plants. Spores are splashed or mechanically moved about by insects or during handling. Once on the plant, spores may germinate following a favorable period of high humidity and enter the plant cells. External symptoms may be visible in about a week. The fungus continues to grow within the plant, periodically producing new spores that spread the infection. Plants may be infected at any stage of growth, including after harvest.

**Cultural Controls:** Because the fungus survives on old debris of infected plants, plants and old fruit should be destroyed by tilling or composting. Crop rotations out of cucurbits of three or more years are also of benefit, although this option is less useful in small garden plantings than in commercial fields. Since high humidity and water splashing are important for disease spread, trickle or furrow irrigation can reduce disease incidence. If gardens are watered with sprinklers, watering should be done early in the day to allow leaves to dry rapidly. Several cucumber varieties are resistant to strains of anthracnose including Dasher II, Slicemaster and Pixie. Resistant varieties of watermelon include Allsweet, Crimson Sweet, Jubilee, Sugarlee and Triple Sweet.

**Chemical Controls:** Fungicides used preventively can control infections. Treatments should be made during periods when leaf wetting favors the disease. (*Note*: Sulfur fungicides should not be used since they can burn leaves of sulfur-shy plants such as cucurbits.)

## Early Blight (*Alternaria solani*)

**Damage and Symptoms:** Early blight primarily causes a leaf spotting characterized by a series of dark, concentric rings often surrounded by a slight yellow area. As the disease spreads, the entire leaf may be killed and drop from the plant. Older leaves are most susceptible and the disease typically progresses upward in the plant. Infections may also occur around the calyx end of fruit. These appear as dark spots that radiate from the stem. Often the fruit spots may also have a distinctive concentric ring pattern. Infections also occur on potato tubers, appearing as dark spotting on the surface. Early blight infects tomatoes, potatoes, eggplant and several weeds in the nightshade family.

**Disease Cycle and Development:** Early blight is caused by infection with the fungus *Alternaria solani*, which survives on old plant debris in the soil or on the soil surface. When favorable conditions occur, the fungus continues to grow on these plant parts, producing spores that are spread by wind, rain and insects. The fungus spores germinate after a few hours at very high humidity. They then enter the plant through natural openings or may penetrate directly through the plant cuticle if the surface remains wet for several hours. Symptoms of spotting may become visible 2 to 3 days after infection and grow rapidly for about a week. As the fungus continues to grow, it periodically produces spores that can cause new infections.

**Cultural Controls:** Vigorously growing plants can resist much infection by early blight. Older plants, or plants carrying heavy fruit loads, are more susceptible to the disease. Cultural conditions that provide optimal fertilization, watering and temperatures can greatly retard development of the disease. Infected plant remains should be

Alternaria *infection: "target" spots, lesions coalesce.*

handled in a way that hastens decomposition, since the fungus requires plant parts to survive. Composting or rototilling into the soil can reduce the amount of fungus that will remain the following season. Crop rotation is also useful, particularly in larger gardens. Watering should be done in a manner that minimizes time when leaves are wet. Trickle types of irrigation can greatly reduce humidity around the plant, which favors infection. When overhead irrigation is used, it should be applied during periods when leaves will dry quickly.

**Chemical Controls:** Protectant fungicides such as sulfur can prevent spores from infecting plants.

**Related Diseases:** Several species of *Alternaria* fungi occur on vegetable crops, including Alternaria leaf spot/blight of cucumbers and other squash family plants (*A. cucumerina*), purple blotch of onion (*A. porri*) and Alternaria leaf spot/black spot of cabbage (*A. brassicae*). Cultural controls for these diseases are similar to those for early blight.

## Botrytis Blight/Gray Mold (*Botrytis cinerea*)

**Damage and Symptoms:** Botrytis is a common disease of vegetables, grapes and many flowers and attacks various plant parts. On roses and peonies, Botrytis produces a fuzzy, gray mold that kills flower blossoms. Damage to grapes occurs as a bunch rot of the developing fruit. Leaf blight symptoms occur from Botrytis infection of lettuce, tomatoes and onions. Botrytis also causes a common gray mold on the fruit of tomatoes, peppers, carrots and other vegetables kept in storage.

**Disease Cycle and Development:** Botrytis blight, also known as gray mold, is caused by infection with the fungus *Botrytis cinerea.* Spores survive the winter on previously infected plant parts around the garden. In addition, Botrytis is an extremely common organism and is annually transported long distances as wind-blown spores. Infections often originate on dead or damaged tissues before invading healthy parts of the plant. Favorable conditions for germination of the spores are extended periods of very high humidity (90–100 percent) and temperatures of 68°F–76°F. However, the disease can infect stored vegetables at quite low temperatures.

**Cultural Controls:** High humidity is very important for infection with Botrytis. Susceptible plants should be provided with good air circulation to

*Botrytis fruit rot*

encourage leaf drying. If possible, watering should be done in a manner that does not allow leaf wetting or prolonged high humidity around the plants. Where Botrytis occurs as a leaf disease, cultural conditions that favor vigorous plant growth reduce susceptibility. Also, since Botrytis typically enters wounds, use care when working around plants when infection is likely. Because the disease organism is so common and spreads easily, crop rotations are not very useful for control.

**Chemical Controls:** Several fungicides are available that can help control Botrytis on many garden plants.

**Related Species:** *Botrytis squamosa* produces leaf blighting and dieback of onion leaves, a much more severe injury to this plant than is produced by *B. cinerea. Botrytis fragariae* can be a common problem on strawberries, giving them a moldy taste.

## *Powdery Mildew*

Powdery mildew is one of the more common types of fungal disease in the yard and garden. Infected plants usually have a white powdery covering on leaves, tender stem tissues and developing flower buds. This covering is made of spores that the fungus produces after growing within the plant. Infected plants show reduced growth, and parts of the plant may die. A great many plants in the yard and garden are susceptible to various types of powdery mildew. Among the most commonly damaged plants are rose, cucurbits (squash, cucumbers, melons), apple, grape and flowers such as zinnia. Grass is also susceptible to a type of powdery mildew. However, each of these diseases is caused by a different species of fungus. Although they may look the same side by side in the garden, the fungus that produces powdery mildew infection of roses is different from the kind that infects cucumbers, and those on bluegrass differ from powdery mildew of peas.

Unlike most other fungi that attack garden plants, powdery mildew can thrive during the heat of the summer. Germination of spores does not require that leaves be wet, only that humidity be high (greater than 95 percent relative humidity) for several hours. This situation can often occur on nights when temperatures drop, so increasing air circulation around plants to lower humidity is an important powdery mildew control. Powdery mildew–resistant varieties are available for some types of plants. Fungicides can also be effective controls for most powdery mildew.

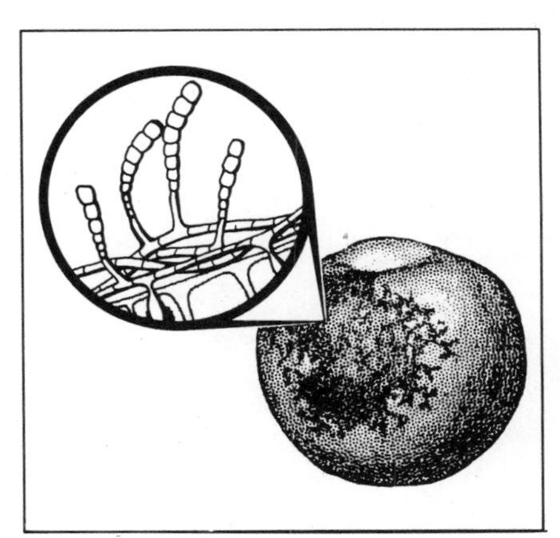

*Powdery mildew*

Powdery mildews should not be confused with downy mildews (e.g., species of *Peronospora* ). The latter are uncommon in the region because of their need for sustained damp conditions. Outbreaks of downy mildew (page 188) can occur on plants such as onion, spinach and cabbage family plants if unusually wet, cool periods of weather provide favorable conditions for this fungus. The symptom of downy mildew infection is typically a dark leaf spotting, and the furry mildew growth is only visible when plants are damp.

## Powdery Mildew of Cucurbits (*Erysiphe cichoracearum*, *Sphaerotheca fulginea*)

**Damage and Symptoms:** A fine, powdery growth forms on the leaves, often starting on the more shaded areas of the plant. The growth may spread and cover much of the leaf surface of the plant. Affected leaves often shrivel, dry and die, exposing the fruit to sunburn. Although fruits are rarely infected directly, damage to the leaves can cause small fruit size, reduce sugars that develop fruit flavor and produce fruit of poor texture. Problems are most common with squash and pumpkins, since many varieties of resistant cucumbers and muskmelons are currently available.

**Disease Cycle and Development:** Powdery mildew of curcurbits is caused by infection with one of two species of fungi, *Erysiphe cichoracearum* or *Sphaerotheca fulginea*. The fungus requires a living plant to continue to develop and dies out annually after frost kills the plants. However, the spores are produced in enormous numbers and, surviving in more southern regions, they are easily spread northward in winds. When conditions are favorable, the spores can germinate and infect leaves. Unlike with other fungi, very high humidity is not needed to infect plants—infection has occurred at relative humidity as low as 46 percent. Optimal temperatures for infection are around 80°F. Leaves are most susceptible two to three weeks after unfolding.

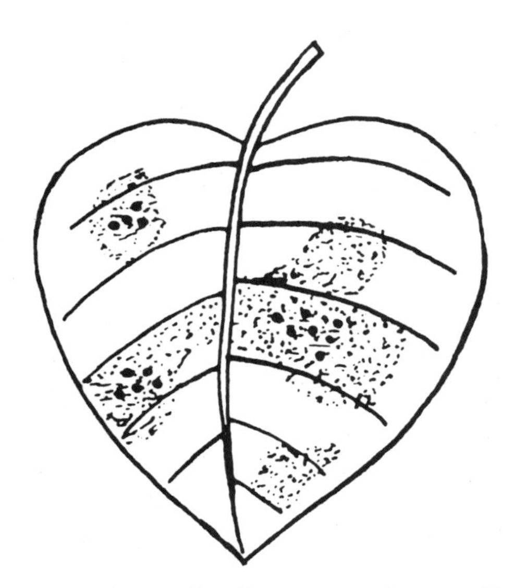

*Powdery mildew. Affected leaves, stems, flowers and fruits have surface coating of gray-white mycelium that can be rubbed off. Black specks may later develop in gray-white powder. Tissue under mycelium may remain green while surrounding leaf tissue is yellow (green-island effect).*

**Cultural Controls:** Development of plants resistant to powdery mildew has been the most effective control. Breeding has been most successful with cucumbers and muskmelons, but has lagged with pumpkins and squash.

**Chemical Controls:** Some fungicides are useful for control of powdery mildew. Copper-based fungicides are probably most widely available for garden use at present, although benomyl is also effective. Sulfur, and sulfur combinations like lime sulfur, should not be used, since cucurbits are notoriously sulfur-shy and easily burned by these treatments.

**Related Diseases:** *Erysiphe cichoracearum* also infects many garden flowers, such as zinnia, dahlia, chrysanthemum, phlox and sunflower. A related species, *Erysiphe pisi*, is common on garden peas.

*Powdery mildew on cucumbers*

| CUCURBITS RESISTANT TO POWDERY MILDEW | | |
|---|---|---|
| **Cucumbers** | | |
| Ashley | Atlantic | Belle Aire |
| Bravo | Burpless F1 Hybrid | Burpless Tasty Green |
| Burpless Green King | Cherokee | Commanche |
| Dasher | Early Set | Early Triumph |
| Gemini 7 | Marketmore 76 | Pickmore |
| Pixie | Poinsett | Polaris |
| Quick Set | Shamrock | Slicemaster |
| Sprint 440 II | Supersett | Sweet Slice |
| Tablegreen 65 | Tex Long | Victory |
| **Muskmelons (Cantaloupe)** | | |
| Ambrosia | Classic | Dixie Jumbo |
| Gold Star | Gulfstream | Imperial 4-50 |
| Luscious | Magnum .45 | Mainstream |
| Planters Jumbo | PMR 45 | Saticoy |
| TAM Uvalde | | |
| **Winter Squash** | | |
| Table King | | |

## Powdery Mildew of Rose (*Sphaerotheca pannosa*)

**Damage and Symptoms:** Powdery mildew of rose often first develops on young stem tissues, later spreading to leaves and flower buds. Infections reduce plant photosynthesis, resulting in reduced growth. The overall appearance of plants and flower quality are damaged by infection. Apricots, peaches and plums (all members of the rose family) are also attacked by this fungus.

**Disease Cycle and Development:** Powdery mildew of rose overwinters behind bud scales and other protected sites on the canes. In spring, the fungus resumes growth and produces spores when warm temperatures return. These spores are wind-blown or splashed onto leaves or other parts of the plant and may germinate if conditions are favorable. Very high humidity (but not leaf wetting) and temperatures above 65°F are required for germination of spores. The spores then produce a penetration tube that forces the fungus into the upper cell (epidermis) of the plant, causing infection. After growing within the plant cells for a few days, the fungus begins to grow along the leaf surface and invade new cells. Periodically, the mass of powdery mildew on the leaf surface produces and releases new spores (conidia) which further spread the infection. Infections are most common on newer growth, since powdery mildew cannot infect old growth. During periods when new growth increases, such as after flowering, infections increase. Daily temperatures ranging between 65°F and 80°F, with high humidity at night, favor outbreaks.

**Cultural Controls:** Since germination of powdery mildew spores requires high humidity, planting in locations with good air circulation can reduce humidity and infections. Air circulation can also be improved by use of raised beds and reducing the density of plantings. Interestingly, powdery mildew of rose does not need wet leaves to allow spores to germinate and infect the plant. This was demonstrated during the late 1930s and early 1940s when frequent water sprays were used as the primary spider mite control. Under these conditions, problems with powdery mildew decline, since the fungus is delicate and easily damaged by these water sprays (although other fungal diseases can increase with watering). Because the fungus overwinters on shoots, pruning and removal of infected shoots can help provide control. It may also be useful to remove fallen leaves, which under some conditions allows the fungus to survive. Resistance to powdery mildew varies among the several types of roses. Hybrid teas and climbers are generally highly susceptible, whereas wichurianans and all shrub roses are more resistant. Although many new varieties are resistant to powdery mildew, this resistance often breaks down as new races of powdery mildew develop.

**Chemical Controls:** Several fungicides can help control powdery mildew. Systemic materials, such as benomyl, are useful for control on flowers, roses and other nonfood plants. Protectant fungicides, such as sulfur and triforine, can also be used to prevent new powdery mildew infections. (*Note*: Sulfur can injure plants when temperatures are above 90°F.) It is most important to control the disease during periods when the disease spreads most rapidly. Sprays should be applied to protect the newly produced tissues, which are susceptible to infection. Antitranspirants, which are sometimes used to retard water loss during transplanting or over winter, may also reduce spread of powdery mildew. This action results from physically thickening the leaf surface, inhibiting infection.

*Powdery mildew on chokecherry*

## Powdery Mildew of Grape (*Uncinula necator*)

**Damage and Symptoms:** Powdery mildew of grape infects new plant tissues and produces a powdery covering. Leaves infected may be distorted and stunted. Infections of fruit clusters may cause poor fruit set. Infected berries are discolored and often split. Wines produced by the infected berries have off-flavors. The disease is much more serious on the European wine grapes, *Vinis vinifera*, than on the species native to North America. Ivy and Virginia creeper are additional hosts.

**Disease Cycle and Development:** Powdery mildew of grape is caused by the fungus *Uncinula necator*. In areas of cold winter temperatures, the fungus survives between seasons in a protected stage (cleistothecia) in bark cracks. In spring, spores are produced, which may blow or splash to developing grapes.

When environmental conditions are favorable, the spores germinate and may infect the plant. The fungus grows into the epidermis of the plant and spreads along the surface. Spores may be produced a week after infection, allowing further spread of the disease. Temperatures from 68°F to 77°F are optimal, although the fungus can develop in a wide range of conditions. Relative humidity from 40 to 100 percent is needed for spore germination, but free water, such as rainfall,

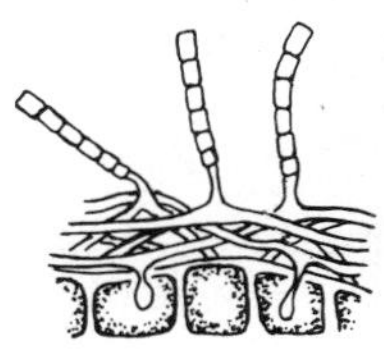


*Close-up of the visible stages of powdery mildew infection*

is damaging to the spores. The fungus is also favored by shading from direct sunlight. Infections occur only on new tissues. Grape berries are only susceptible until they reach a sugar level of about 8 percent.

**Cultural Controls:** Most grape varieties containing parentage of the native grapes of North America, *Vinis labrusca,* are resistant to powdery mildew. This includes Concord, Freedonia, Valiant and many of the more commonly grown grapes in the West.

**Chemical Controls:** Several fungicides can control powdery mildew of grape. These are best applied early in the season to protect the plant when susceptible new growth is being abundantly produced. Early treatment also prevents the production of new spores, which can cause damaging outbreaks. Sulfur is commonly used for powdery mildew control. However, some grape varieties, such as Concord, are damaged by sulfur. On wine grapes, sulfur should not be used late in the season to avoid production of hydrogen sulfide during fermentation.

## Common Smut (*Ustilago maydis*)

**Damage and Symptoms:** Common smut can infect any of the aboveground tissues of corn and is particularly noticeable when ears are involved. Attacked cells become greatly enlarged and fill with the fungus. This produces very conspicuous spongy swellings with a thin gray cover. As the fungus matures, it becomes black with formation of huge numbers of spores.

**Disease Cycle and Development:** Common smut is produced by the fungus *Ustilago maydis*. Spores of the fungus overwinter in infected parts of the plant that have not decomposed. Spores are produced in spring and are splashed or blown to new plants. Corn smut usually invades the plant through wounds, although it can also directly penetrate the epidermis. Corn is particularly susceptible to corn smut infection when it is only a couple of feet

*Common corn smut*

*Corn smut*

tall and before tassels form (whorl stage). High-moisture conditions are also necessary for the spores to germinate and infect the plant.

**Cultural Controls:** Crop rotation and tilling decrease the amount of surviving corn smut in the area of the planting and can reduce infections. Corn plants are also less likely to be infected if conditions allow steady, normal growth.

**Miscellaneous Notes:** The young stages of the fungus, before dark spores have been produced, are edible and considered by many to be a delicacy. They are widely used in cooking in Mexico and Latin America. They may be cultured by purposely wounding the young plants and exposing the wound to the spores.

## Coryneum Blight (Shothole Fungus) (*Stigmina carpophilum*)

**Damage and Symptoms:** Coryneum blight primarily infects peaches, although apricots and nectarines can also be attacked. On leaves, small red-purple spots develop, which later drop out of the leaf, producing a characteristic shothole injury, similar to flea beetle damage to vegetables and flowers. More serious damage results from infection of fruit. Early infections cause rough, scab-like spots on the fruit. Late-season infection causes slightly sunken, grayish areas on the fruit.

**Disease Cycle and Development:** Coryneum blight is caused by infection with the fungus *Stigmina carpophilum* (=*Coryneum beijerinckii*). The fungus overwinters on buds of dormant leaves and flower buds or in infections on twigs. Spores are produced during wet periods and are splashed by rain or blown by wind. Sustained wet conditions are required for the spores to successfully germinate and infect new plant parts. Once infections occur, the fungus reproduces and new infective spores are released. Infection is most rapid under optimal temperatures of 70°F to 80°F; however, germination can occur at very low temperatures. Once established in an orchard, Coryneum blight can be difficult to eliminate.

**Cultural Controls:** Early identification and control of infestation can limit the need for control practices. Pruning and removing infected twigs and raking infected leaves can help limit identified infestations. Most damage occurs on lower leaves, which remain wet longer. Improved aeration can reduce infections.

**Chemical Controls:** Several fungicides can help control Coryneum blight. Treatments are most important when early-season wet weather favors outbreaks of the disease.

**Related Species:** Shothole infections also occur on cherries. These are caused by another fungus (species of *Coccomyces*).

## Cytospora Canker (Gummosis) (*Cytospora leucostoma*)

**Damage and Symptoms:** Cytospora canker kills the wood underneath the bark of peach, cherry, plum and apricot trees. This injury weakens the trees and often kills branches. Newly transplanted trees are at particular risk of infection if replanted in areas where the disease has previously occurred. Large amounts of amber gum ooze through the bark around infected areas (gummosis).

*Cytospora canker*

**Disease Cycle and Development:** Cytospora canker is produced by infection with the fungus *Cytospora leucostoma*. Spores of the fungus are widely distributed by wind and air and can be found on the bark of trees year-round. The spores germinate and enter the tree through wounds caused by pruning, borers, cold temperature injury or other conditions. The fungus then grows under the bark, killing the wood and producing the canker symptoms. Growth of the fungus is greatest in spring as temperatures first warm. During periods of active tree growth, fungus growth is temporarily inhibited by natural tree defenses.

**Cultural Controls:** Because Cytospora is a wound parasite, cultural practices that avoid stress and injury are very important in managing the disease. This includes proper fertilization and watering so that trees grow vigorously in late spring and early summer but become suitably hardened off to survive fall freezes. Pruning wounds and borer injuries should be minimized to avoid infection at these sites.

**Mechanical Controls:** Infected branches can be pruned. This, and all other pruning, is best done in late winter. In areas of intense sun, use of white reflective paint on trunks can reduce damage by sunscald injuries, thus avoiding later problems with Cytospora canker.

**Chemical Controls:** Protectant fungicides can reduce infection if applied immediately after wounding by pruning or hail injury.

**Miscellaneous Notes:** Cytospora canker and other stresses commonly cause stone fruit trees to ooze clear, amber gum. Damage by an insect, the peach tree borer, also causes gumming, but the gum is characteristically mixed with bark fragments chewed from the trunk.

## Damping-off (species of *Pythium*, other fungi)

**Damage and Symptoms:** Seeds may decay before emergence and be killed or weakened. Tender stems of seedlings that emerge from the soil may be infected, causing plants to fall over and die. Surviving plants may show darkened depressions on the stems and grow poorly.

**Disease Cycle and Development:** Problems with seed decay and damping-off are caused by infection with various fungi, primarily species of *Pythium*. These are common in garden soils, generally surviving in decaying plant debris. These fungi can also survive on seeds, although most commercial seed is treated to be disease-free. However, when conditions of moisture and temperature are suitable, fungi can successfully invade and reproduce inside seeds and tender seedling tissues.

**Cultural Controls:** Seeds should be planted under conditions that favor rapid seed germination and seedling development, since older plants are more resistant to infection. However, excessive moisture greatly favors the disease. Watering should be done carefully to avoid prolonged soil wetting, such as in early morning, to allow drying by sun. Planting in a way to allow good air circulation can also reduce humidity near the soil surface. Excessively heavy seeding rates can increase spread of the disease, although some increase in seeding can help compensate for lost plants. This recommendation is typically included in seed package instructions that suggest seeding at a high rate and later thinning. When growing seedlings in a cold frame or greenhouse, many problems with damping-off can be avoided by use of disease-free soil. Some, but not all, artificial potting mixtures do not contain damping-off fungi. When garden soil is used to grow seedlings, it should be treated by high temperatures to kill the fungus spores.

**Chemical Controls:** Several fungicide seed treatments (e.g., thiram, captan) are available and widely used to control damping-off. Many seed suppliers routinely provide seed that is already treated with these fungicides, or they may be purchased separately if desired.

## Downy Mildew/Blue Mold (*Peronospora effusa*)

**Damage and Symptoms:** Blue mold can be the most important disease of spinach in the region. Infections appear as yellowish patches on the leaves, which later turn black and die. Late in the season, infected spinach may survive but leaves become yellow and plants are stunted. During periods of wet weather, which favors the fungus that produces the disease, leaves rot rapidly and plants may be killed or severely weakened.

**Disease Cycle and Development:** Downy mildew of spinach, also known as blue mold, is caused by infection with the fungus *Peronospora effusa*. It survives winters on living spinach plants, in the soil or on seeds. New spores that can infect plants are first produced in spring. Germination of spores requires free water on the leaves for several hours, provided by sources such as dew, rain or overhead irrigation. This is a cool-weather fungus (optimum around 46°F).

**Cultural Controls:** Since leaf wetting is required for infection with the blue mold fungus, watering should be done in a manner that does not allow leaf wetting or prolonged high humidity around the plants. If possible, plantings should be made in a way to increase air circulation, such as by use of raised beds. The life cycle of the fungus can be broken by crop rotations. In a garden, eliminating spinach for a year usually should be sufficient to cause the disease to die out. Growing winter crops of spinach, which allow the fungus to survive the following season, will increase problems with blue mold. After harvest, spinach plants should be removed or spaded into the soil as soon as possible. Downy mildew reproduces well on dying spinach leaves and survives well on seeds. Blue mold has several different races, and resistant plants have been developed for some of the most common ones. Among the spinach varieties resistant to the common races of blue mold are Melody, Norgreen, Marathon, Bonus, Duet, Mount St. Helens, Highpak and Badger Savoy. However, new races of the fungus that are able to damage these varieties are present in the region.

**Related Species:** Although the generally arid climate of the region does not favor infection with any downy mildew fungus, some plant damage can occur. Most important is *Peronospora destructor*, which sometimes damages the leaves of onions and related crops during seasons with cool, wet weather. Downy mildew can also occur on lettuce (caused by the fungus *Bremia lactucae*) and some cabbage family plants (caused by the fungus *Peronospora parasitica*).

## Rhizoctonia (*Rhizoctonia solani*)

**Damage and Symptoms:** Rhizoctonia causes a wide variety of disease symptoms in various vegetable crops. It is a member of the damping-off gang that causes seedling plants to collapse and die. Surviving plants affected by the disease may be weakened and grow poorly because of wirestem, caused by destruction of much of the plant stem. On lettuce the disease produces bottom rot. Leaf mid-ribs touching the soil are usually first affected. Later the rot grows from leaf to leaf through the lower head. Infected tissues are often invaded by secondary organisms, which produce a slimy rot. On potatoes, the most common symptom of the disease is black scurf, sometimes described as "the dirt that won't wash off." Rhizoctonia may also cause small darkened areas on the stem (cankers) but rarely has much effect on plant growth. If much of the stem area is damaged, aboveground (aerial) tubers may form.

**Disease Cycle and Development:** The fungus *Rhizoctonia solani* produces the various Rhizoctonia diseases. Several strains of the fungus are known, each attacking separate types of plants. Rhizoctonia is easily transported on infected soils and plants and is commonly present in garden soils. It can persist for very long periods in its resistant sclerotia form (e.g., the black scurf on potatoes). When susceptible plants are grown, the fungus is stimulated to grow and infect crops. Although wet soil is needed for infections to occur, favorable temperatures vary. For example, the strain of Rhizoctonia that causes black scurf on potatoes needs an optimum temperature of about 54°F; the optimum for bottom rot on lettuce is about 78°F degrees.

**Cultural Controls:** Disease is favored by high humidity. Increasing air circulation around the plants to avoid prolonged wetting of the base of the plant can avoid some infection of lettuce. Growing on mulches that prevent soil/leaf contact can also avoid this problem. Rhizoctonia prefers dead plant materials and, when soils are rich in organic matter, may gradually show reduced ability to infect healthy plants. Increasing soil humus can help reduce problems with disease caused by the fungus. Obviously diseased plants should be removed and discarded to prevent further development of the fungus. Crop rotations can reduce incidence of some strains, such as the one that infects potatoes, although most forms of Rhizoctonia are too persistent to be affected by garden rotations. Crop rotations involving nonsusceptible plants, such as sweet corn and beans, must be of several years' duration to reduce the disease. The black scurf condition of potatoes develops rapidly in cool, wet soils. Prompt harvesting can reduce incidence of this problem.

## Hollyhock Rust (*Puccinia malvacearum*)

**Damage and Symptoms:** Hollyhock rust damages the leaves, stems and vigor of hollyhock. Common mallow (cheeseweed) and okra are also susceptible. First symptoms are yellow-orange spots on the upper leaf surface. These expand and ultimately form rusty-brown swellings (pustules) on the underside of the leaves.

**Disease Cycle and Development:** Hollyhock rust is caused by infection by the fungus, *Puccinia malvacearum*, which survives on the old plant debris infected the previous year. It begins to regrow in spring, producing spores that can infect the plant. The spores germinate and grow following exposure to leaf wetting for several hours. The fungus grows into the plant, ultimately producing

*Hollyhock rust on mallow*

## *Rust Fungi*

The rust fungi are some of the most notorious plant diseases, periodically devastating wheat, coffee and other crops worldwide. They usually attack leaves and stems, and the germinating spores directly penetrate the leaf surface or move through natural plant openings such as stomates. Areas that are also infected die or function poorly, and plants may yield poorly. Fruits and berries of some crops are also infected. In the course of infection, most rusts produce yellow, white or, more commonly, rusty red or orange spots, each packed with spores (urediospores).

One unusual aspect of the rust fungi is that many require two different types of host plants to complete their development. For example, juniper-hawthorn rust alternates between juniper and hawthorn. It spends the winter on juniper, producing a jellylike, horned growth that appears following wet weather in spring. Infections caused by spores from this growth occur during the summer on hawthorn, where it produces rust-colored growths on the leaf underside. Cedar-apple rust (cedar/apple), wheat stem rust (wheat/barberry), white pine blister rust (white pine/currants or gooseberries) and white rust of spinach (spinach/saltgrass) are other examples of these two-host rusts. In the absence of both hosts, the disease is generally less important, being dependent on spread of the spores for long distances by winds. (White pine blister rust is such an important disease of white pine in the Northeast that culture of its alternate host—currants and gooseberries—is restricted in some areas.)

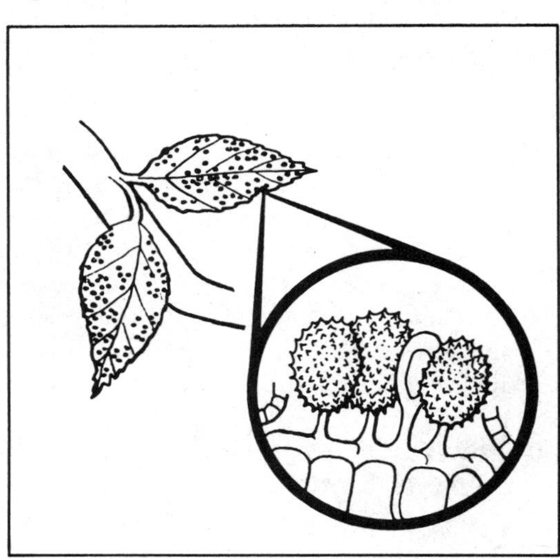

*Rust*

Much work in the control of rust fungi has involved breeding resistant plants. However, the rusts have shown tremendous ability to overcome these defenses as new races of the fungi overcome plant defenses. For very important rust diseases, such as wheat stem rust, there are ongoing sophisticated programs to detect new races early and breed resistant varieties.

Rust fungi can also be controlled by many fungicides. Rusts are susceptible to triforine (Funginex), copper fungicides and maneb fungicides.

the rusty-brown pustules on leaves, which are packed with summer-stage spores (urediospores). These can produce later infections during the growing season. More resistant overwintering stages of the fungus are formed in late summer.

**Cultural Controls:** Old plant parts should be removed and destroyed to kill overwintering stages of the fungus. During the growing season leaves showing symptoms of infection should also be promptly picked and removed to prevent spread.

A new cover of mulch placed around the base of the plant can further prevent spread of overwintered stages of the fungus. Planting in sites that allow good air circulation will reduce problems with infection by hollyhock rust.

**Chemical Controls:** Application of sulfur or other preventive fungicides can prevent spores from infecting plants. These are most useful during periods of high humidity.

### Rose Rust (species of *Phragmidium*)

**Damage and Symptoms:** Rose rust infects the leaves and occasionally the young stems and sepals of roses. The disease is most noticeable when orange or red-orange spots (pustules of aeciospores) are formed on the leaf underside. As the spots enlarge, brown spots may be observed on the upper leaf surface. Infected leaves are less vigorous and sometimes will drop following infection.

**Disease Cycle and Development:** Rose rust is caused by infection with *Phragmidium* fungi. Nine different species of this fungus have been found on roses. They differ from other rust diseases that infect turfgrass, hollyhock, corn, beans and other garden plants. Spores of rose rust blow in the air and infect leaves through natural openings, such as stomates. Continuous moisture for 2 to 4 hours and favorable temperatures (65°F to 72°F is optimal) are needed for successful infection. Hot, dry conditions limit the disease. The fungus grows through the leaf tissues. They later produce clumps of spores in characteristic pustules on the leaf underside. The orange or orange-red urediospores that are produced cause new infections throughout the growing season. At the end of the year, the fungus forms black, resistant spores (teliospores) that survive the winter on fallen leaves and stems. The following spring, these stages of the fungus renew growth and produce spores that can repeat the cycle.

**Cultural Controls:** Reducing periods of leaf wetting, particularly when temperatures are favorable for infection, is fundamental to managing rust on roses. Water early in the day so leaves dry quickly. Plantings pruned to allow increased air circulation also allow leaves to dry rapidly and reduce infection. Rose varieties vary greatly in their susceptibility to rose rust. Many are quite resistant; others are highly susceptible.

**Chemical Controls:** Several fungicides are effective for control of rose rust. These are most effective when applied during periods when environmental conditions favor infections (e.g., moderate temperatures combined with leaf wetting).

## *Common Bacterial Diseases*

### Angular Leaf Spot of Cucumber (*Pseudomonas syringae*, *lachrymans* strain)

**Damage and Symptoms:** Infestations on leaves of cucumber plants appear as dark, angular spots bordered by leaf veins. During most conditions, tearlike droplets of bacteria may ooze from the infested tissues. As the leaf dries and shrinks, irregular tears in the leaf are formed. Similar symptoms occur on squash and watermelon, although infested areas are typically surrounded by a small yellow halo on these plants. Fruit may also be infested, resulting in circular spotting. This is usually confined to the surface of the fruit but can allow soft rotting bacteria to enter.

**Disease Cycle and Development:** Angular leaf spot is caused by the bacteria *Pseudomonas syringae*, *lachrymans* strain. They overwinter within the garden on old, infected plant debris. Movement into plants is through wounds or natural openings (stomates). Since the bacteria require water to move and persist, wet conditions in early morning, when stomates are open, can be particularly important in infections. After the disease has

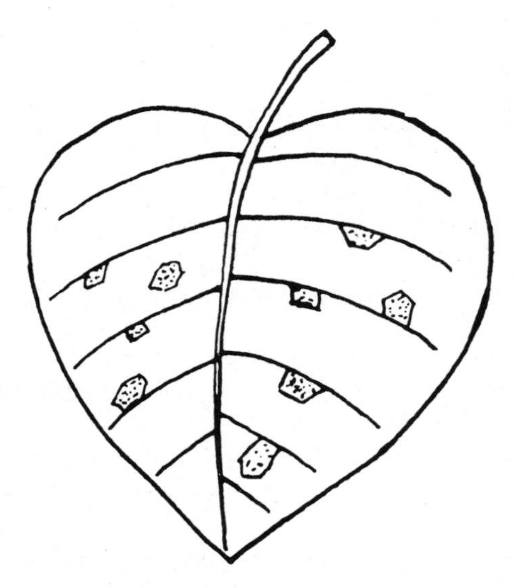

*Bacterial leaf spots. Spots are often angular because they are limited by the small leaf veins. The color is usually uniform. The tissue may first appear oil- or water-soaked when fresh but on drying becomes translucent and papery.*

started to develop, it can spread easily on tools, clothing and hands during gardening operations.

**Cultural Controls:** Since the bacteria survive for only a limited period, crop rotation can be very useful. Tillage or removal of overwintering plant debris can further decrease the amount of surviving bacteria. Leaf wetting is very important in spreading disease. Furrow or drip irrigation can retard the disease greatly, in contrast to sprinkler watering. It is particularly important to water in a way that allows the leaves to dry during the early part of the day. Nitrogen fertilizers should be used moderately, since excessive nitrogen can increase the severity of the disease. Some cucumber varieties have been developed with resistance or tolerance to the disease, including Stokes Early Hybrid, Bounty, Carolina, Score and Liberty Hybrid.

**Chemical Controls:** Copper-based bactericides can protect plants, and several are registered for use on cucumbers, melons and squash. However, frequent use of these materials can damage cucurbits, particularly in high temperatures.

**Miscellaneous Notes:** One of the related bacteria, *Pseudomonas syringae*, *phaseolicola* strain, infect beans, causing a disease known as halo blight. Controls are similar to those for angular leaf blight.

## Crown Gall (*Agrobacterium tumefaciens*)

**Damage and Symptoms:** Infections with crown gall cause warty, tumorlike growths to develop on stems and roots. The growths (galls) are generally rounded with a rough surface. At first the growth may be soft and light-colored, but it becomes woody and darkens with age. Crown gall can infect a wide range of plants, including rose, apple, raspberry, cherry, plum, peach and walnut. Several garden flowers, such as chrysanthemum, daisy and aster, may be infected. Most plants show some stunting or slowed growth from infection with crown gall. However, response to infection is variable, with some plants showing little damage while others may stop producing completely.

**Disease Cycle and Development:** Crown gall results from infection of plants by the bacteria *Agrobacterium tumefaciens*, which enter through wounds in the bark caused by pruning, grafts, cultivation injury or natural causes such as frost heaving. Once inside the plant, the bacteria transform some of the cells into tumor-producing cells. These transformed cells continue to grow without the bacteria. The bacteria develop between the tumor cells at or near the surface of the gall. As the gall breaks down, or is cut during gardening work, the bacteria are released and may spread to new plants.

**Cultural Controls:** New plants should be carefully examined before purchase and planting so that infected plants are not introduced into a yard. Diseased plants should be promptly removed. If possible, the soil surrounding the plant should

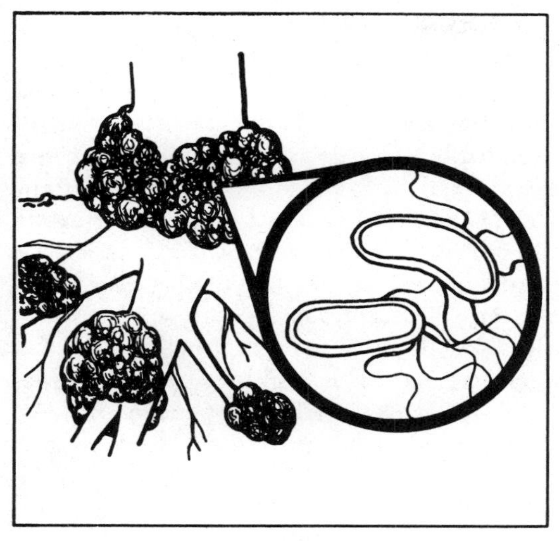

*Crown gall*

also be removed or sterilized, since the bacteria can remain in the soil for up to two years.

**Biological Controls:** Another type of bacteria, *Agrobacterium radiobacter* (strain K84), can inhibit the crown gall bacteria and prevent infection. These beneficial bacteria are sold under the trade name Galltrol. However, it is not readily available and is sold only by some specialty suppliers of the nursery industry for use during plant propagation.

**Chemical Controls:** Paints of the product Gallex can control existing galls by drying them out. Gallex is a kerosene-based product available from some specialty suppliers of the nursery industry.

**Miscellaneous Notes:** Because of the tumor growth, crown gall has been used extensively as one of the first cancer research models.

## Fire Blight (*Erwinia amylovora*)

**Damage and Symptoms:** Areas of inner bark are killed by the disease, producing a canker. Branches, twigs, leaves and fruit often die back rapidly beyond the cankers, appearing dried and blackened as if scorched by fire. In severe infections, entire trees are killed. Many plants of the rose family are susceptible, including crabapple, apple, pear, cotoneaster, hawthorn and mountain ash. Among the most susceptible fruit varieties are Bosc and Bartlett pears and Rome, Lodi and Jonathan apples.

**Disease Cycle and Development:** Fire blight is caused by infection with the bacteria *Erwinia amylovora.* Infected areas of bark produce bacteria-rich ooze during favorable periods of growth. The bacteria are most often splashed by rain and enter the plant through wounds or natural openings such as blossoms. Spread of the disease by insects, including pollinators such as honeybees, is also common. Problems with fire blight can increase greatly if cool temperatures lengthen blossoming time. The bacteria then moves into the inner bark and multiply. They kill the infected areas, producing a canker injury, which weakens or girdles the branch or twig. The plant dies back to the area where girdling cankers have formed. The bacteria continue to grow and spread during favorable conditions, moving to progressively larger areas of the tree. They grow most rapidly in temperatures of 70°F to 81°F. Very little growth occurs when temperatures drop below 65°F or exceed 90°F. Rainfall or high humidity also favor bacteria growth and allow increased infection.

*Fireblight on apple. Note the symptom of wilting leaves.*

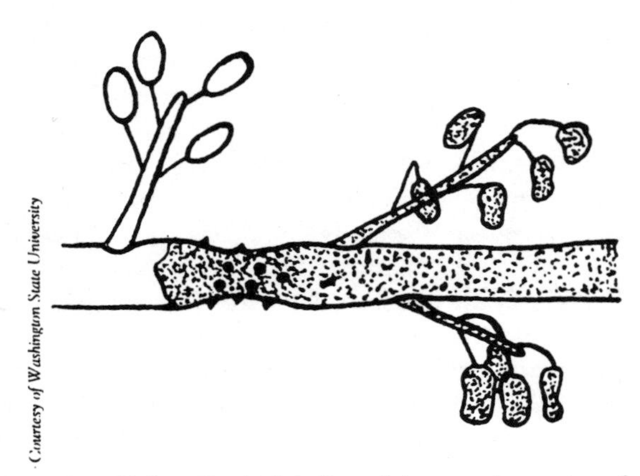


*Shoot blights. Gradual decline of shoots and retention of dead leaves may indicate a parasitic disease. The margin between affected and healthy tissue is often irregular and sunken. There may be small, pinlike projections of bumps over the surface of dead bark. These are spore-producing structures of disease organisms.*

**Cultural Controls:** Some varieties are partially resistant to infection, including apple varieties Red Delicious, Winesap and Haralson; pear varieties Magness, Moonglow and Starking Delicious; and crabapple varieties Vanguard and Red Vein Russian. Trees should not be excessively fertilized with nitrogen, since succulent growth is susceptible to infection. Watering and fertilizing just enough to sustain growth can slow fire blight infection.

**Mechanical and Physical Controls:** Pruning infected branches can prevent spread of fire blight. However, several precautions should be taken when pruning, since the disease is highly infectious. Prune during dry periods in winter, when the trees are dormant. Cuts should be cleanly made, at least 12 inches below visible infections in twigs; 6 inches in branches. If darkened wood is observed at the cutting site, cut farther back. When multiple, new blossom infections occur, pruning may not be possible without destroying the tree. In these cases, it is advisable to hold off on some pruning until the following season to determine if the infection is spreading. Many blossom infections die out in winter. Throughout all pruning, tools should be continuously sterilized to avoid transferring bacteria to new wounds. Sterilization can be done with a disinfectant (either bleach at 1:4 dilution with water or 70 percent alcohol) after each cut. Lysol spray is an alternative disinfectant. Allow at least 30 seconds for adequate sterilization to occur. Avoid any other tree wounding, since this can provide sites for new infections. It is possible, but difficult, to prune individual canker wounds on larger limbs. Bark of infected areas can be scraped out during periods of warm, dry weather. All the outer bark and cambium should be removed at the darkened canker and at least 1 inch beyond the infected area.

**Chemical Controls:** Most infections occur during blossoming or following periods of tree wounding, such as hailstorms. Applications of the antibiotic streptomycin or copper sprays during these high-risk periods can prevent new infections. Rotation of streptomycin and copper sprays can retard development of fire blight bacterial strains that are resistant to these pesticides.

## Bacterial Wilt of Cucurbits (*Erwinia tracheiphila*)

**Damage and Symptoms:** Bacterial wilt causes wilting and death of leaves and stems of various squash family plants. Cucumbers and muskmelons are particularly susceptible, followed by pumpkins and squash. Initial wilting symptoms occur on single leaves and are followed by progressive wilting and death of shoots and, eventually, entire plants. Symptom development is much more rapid on younger plants. Bacterial wilt is confined to the region east of the Rockies. It is generally little problem in areas receiving less than 18 inches of rain annually. There are several causes for wilting of squash family plants. Bacterial wilt can usually be distinguished from other causes by cutting a wilting leaf; often a white ooze is visible at the cut.

Wait a few minutes and then touch the cut end; fine strands may appear when the finger is slowly removed.

**Disease Cycle and Development:** Bacterial wilt is caused by infection with the bacteria *Erwinia tracheiphila*. The bacteria and their spread are intimately related to the striped cucumber beetle (see page 95). These bacteria only overwinter, and can even multiply, in the digestive tract of the beetles. In spring, the beetles return to feed on seedlings, making gouging wounds in the plant. The bacteria are periodically excreted and can enter plants through wounds that the beetles produce. They then enter the water-conducting tissues of the plant (xylem). The initial symptom is a darkening of the leaf area around the infected feeding wound. As the bacteria grow, they plug these vessels, producing the wilt. The overwintered beetles produce the first (primary) cycle of infection. Later in the summer, beetles further spread the disease (secondary infection) by feeding on newly infected plants and moving to healthy plants.

**Cultural Controls:** Wilting plants should be removed immediately to prevent later spread of the disease. This may be further improved by purposefully overseeding and then thinning excess plants. Young plants can be protected by row covers to exclude the overwintered beetles from the planting.

**Chemical Controls:** Insecticides applied against the first-generation beetles as soon as they are observed can provide very good control of the disease.

## Bacterial Soft Rot (*Erwinia caratovora, Erwinia chrysanthemi*)

**Damage and Symptoms:** Bacterial soft rot can produce many different symptoms on garden vegetables. It is one of the most common causes of decay of root crops (carrots, onions, potatoes) both during the growing season and after harvest. The bacteria may also infect leafy vegetables, such as lettuce. Usual symptoms include a slimy gray or brown rot, with the outer layer of the plant remaining intact. Often other organisms may invade the wound and produce a foul-smelling decay. Bacterial soft rot is also involved in decay of potato seed pieces, combined with the effects of certain *Fusarium* fungi.

**Disease Cycle and Development:** Bacterial soft rot is produced by infection with a strain of the bacteria *Erwinia caratovora* or by a related species, *Erwinia chrysanthemi*. The bacteria are widespread in soils and survive on crop debris. They also multiply on roots of crop plants and weeds. The bacteria must enter through wounds in the plant. Sunscald, wind or hail damage, wounding during cultivation and insect injuries are common points where the bacteria enter plants. Bacteria spread primarily by splashing or moving water but can also be spread by certain insects, such as sap beetles and root maggots. The root maggots (seedcorn maggot, onion maggot, cabbage maggot) are particularly important, since they physically wound the plant as well as carry the bacteria on their body.

**Cultural Controls:** Because soft rotting bacteria enter through wounds, avoid bruising or injuring plants during garden cultivation and harvest. Use special care when pulling plants with foliage (e.g., "greentop" carrots) or with thin skin (e.g., "new" potatoes), since these are particularly susceptible to infection. Since the bacteria that cause bacterial soft rot are very common in soils, crop rotation is of limited benefit. Among the better plants for use in rotations to reduce this problem are beans, peas, spinach, beets and sweet corn. Populations of bacteria that survive on discarded plant materials can be reduced by promoting rapid decomposition through fall tillage or composting. To reduce problems with seed piece decay of potatoes, make sure that the cut surfaces of seed potatoes have healed (suberized) before planting. Planting in warm (above 50°F), moist soil will further speed wound healing and reduce decay problems. To reduce infections in stored vegetables after harvest, keep temperatures just above freezing (32°F) and relative humidity below 90 percent.

**Chemical Controls:** No effective bactericides are available to control bacterial soft rot in gardens. Insecticide seed treatments can kill root maggot flies that spread the disease to some plants, such as seed potatoes and germinating beans.

**Related Strains:** A strain of *Erwinia caratovora* commonly infects potatoes, producing a disease known as blackleg. Symptoms of this disease include a black discoloration and wilting of the shoots. The best way to avoid problems with blackleg in a garden is to buy certified seed potatoes.

*Aster yellows symptoms on cosmos flower*

## Aster Yellows

**Damage and Symptoms:** Aster yellows infects a wide range of vegetables and flowers. Most common hosts are in the aster family (Compositae), including lettuce, carrots, aster, cosmos, marigold and zinnia. However, species in many other plant families can be infected, such as beets and onions. Aster yellows is much more prevalent east of the Rockies. The most common symptom of infection is a twisting and distortion of new growth. For example, head lettuce leaves are highly distorted and fail to form a head. In carrots, top growth becomes very bushy and shortened. Yellowing or reddening of foliage are also related to infection. On carrots, hairy roots are a common symptom. Symptoms on flowers also vary. Commonly, the petals distort to a leaflike form (phyllody). Symptoms vary depending on the stage of the plant at the time it becomes infected. Aster yellows infection can kill young plants, while older plants may show only mild effects and little yield loss.

**Disease Cycle and Development:** Aster yellows is transmitted only by the feeding activities of certain species of leafhoppers that have acquired the disease from infected plants. Within the region, the aster (sixspotted) leafhopper (see page 150) is the primary vector of the disease. Neither the aster leafhopper nor the aster yellows organism overwinters in the northern United States and Canada. The leafhoppers are restricted to areas around the Gulf of Mexico during winter. In late spring and early summer, the leafhoppers fly and disperse throughout most of the United States and southern Canada. Aster leafhoppers acquire the aster yellows disease organism (a mycoplasma-like organism) while feeding on an infected plant. A period of approximately two weeks must pass before the insect is capable of transmitting the disease to new plants. Symptoms of infection appear about three weeks after the disease organism is transmitted. Peak infection typically occurs in late July and early August.

## X-Disease of Stone Fruits (Other Names: Western X-diseases, Peach X-diseases, Yellow Leaf Rot)

**Damage and Symptoms:** X-disease can infect all stone fruits but is particularly damaging to sweet cherries, peaches and nectarines. The most severe response is rapid wilting and death of the tree, usually following high temperatures before harvest. Sometimes a zipperlike, irregular darkening area is noticeable at the graft union. Irregular fruit ripening, gradual decline and dieback are symptoms on other varieties. Red spotting on leaves and a yellowing similar to iron chlorosis are additional leaf symptoms. Fruit often drop prematurely and peaches may show swelling at the flower end. Increased susceptibility to winter kill is also associated with X-disease infection. At present, incidence of X-disease in the region is limited to areas west of the Rockies.

**Disease Cycle and Development:** X-disease is caused by infection with a bacteria, similar to a mycoplasma. The disease is spread by several species of leafhoppers, including the mountain leafhopper *Collandonus montanus*. Leafhoppers acquire the X-disease mycoplasma when feeding on an infected plant. It moves into the blood of the leafhopper and multiplies in the insect vector. After a six- to eight-week latent period, the leafhopper can transmit the disease organism to healthy plants for the remainder of its life.

**Mechanical Controls:** No cure for X-disease is known. Since infected trees can be sources of additional infection, trees should be promptly destroyed.

**Chemical Controls:** The disease is spread by leafhoppers, so effective control of the leafhopper vectors theoretically could control transmission to new trees. However, the species of leafhoppers that are important in X-disease spread are poorly known, and effective insecticide treatments have not been identified. Regardless, in the absence of strict sanitation practice that removes infected trees, any chemical controls are likely to be only minimally effective at best. Sprays of iron chelate fertilizers can help extend the life of producing trees infected with X-disease.

# *Common Environmental Disorders*

## Tipburn

**Damage and Symptoms:** Edges of the internal leaves brown on cabbage, Brussels sprouts and lettuce. Symptoms often are not visible from the outside. Soft-rotting bacteria may invade the damaged tissues, causing the heads to decay.

**Disease Development:** A deficiency of calcium in the growing plant can induce tipburn symptoms. This often results from nutrient imbalances that occur during periods of high temperature when soils have excess nitrogen. Fluctuating water and humidity can also aggravate the problem.

**Cultural Controls:** Moderate rates of nitrogen fertilizers and watering practices that avoid periodic water stress can reduce the incidence of tipburn. Susceptibility to tipburn is also related to varietal differences.

## Blossom-end Rot

**Damage and Symptoms:** Blossom-end rot begins as a small watery bruise on the blossom end of tomatoes and peppers. The spot enlarges, darkens and becomes depressed as the fruit grows. Fruits damaged by blossom-end rot often ripen earlier. Blossom-end rot is differentiated from sunscald and fungal infections on tomato by its restriction to the blossom end of the fruit. Flat tomato varieties tend to be most susceptible.

**Disease Development:** Blossom-end rot results from a deficiency of calcium in the developing fruit. This causes actively growing cells to leak water and nutrients and die. Calcium deficiencies can be induced by deficient soil calcium levels, imbalances of magnesium or nitrogen, excess soil moisture, compacted soils or other fluctuating conditions during plant growth. Excess levels of salts in the root zone may contribute to blossom-end rot. Planting in cold soils, poor drainage and high temperatures can also increase problems.

**Cultural Controls:** Blossom-end rot can be reduced if garden culture allows for moderate, steady growing conditions. This includes providing adequate water and avoiding root injury by hoeing. Few soils in the region are deficient in calcium. However, calcium may not be readily available to plants if soil imbalances (nutrients, pH, salts, water) occur. Sprays of calcium chloride or calcium nitrate have reduced symptoms of blossom-end rot. Some tomato varieties are less susceptible to blossom-end rot.

## Edema

**Damage and Symptoms:** Edema results in wart-like outgrowths on the undersides of leaves. Many succulent vegetables can be affected, such as potatoes, tomatoes, muskmelons, cauliflower and beans, but it is most common on cabbage. As the outer layer of cells bursts, the damaged area becomes brown and corky.

**Disease Development:** Edema is caused by a water imbalance in the plant created when water is taken up by the roots faster than it is lost through transpiration. This typically occurs during periods when cool nights follow warm, humid days and soil moisture is high. The engorged cells burst from the water pressure.

**Cultural Controls:** During periods of weather that favor edema, watering should be reduced and air circulation around the plants should be increased. There is also a range in resistance to edema, so susceptible varieties should be replaced in future plantings where edema is a common problem.

**Miscellaneous Notes:** On some cabbage varieties, feeding by onion thrips (see page 136) can cause a reaction similar to edema.

## Sunscald

**Damage and Symptoms:** Sunscald often appears as large, pale-colored, wrinkled or sunken areas. They are most noticeable on fruit but can also occur on trunks, stems and leaves. Damaged areas often become papery as they dry out. On raspberry plants, sunscald (solar injury) appears as white areas on the fruit. Sunscald injuries are limited to areas of the plant receiving the most intense sun exposure.

*Sunscald*

**Disease Development:** Sunscald is caused by exposure to direct sunlight, particularly during conditions of high temperature. Previously shaded plant parts that suddenly become exposed are particularly susceptible to injury. In addition to direct injury caused by high temperature, several insects, fungi and bacteria can enter the damaged area.

**Cultural Controls:** Plants that produce sufficient foliage to shade fruit and other susceptible tissues can limit sunscald injuries. Also, varieties with fruit that grows downward and underneath foliage are less likely to be injured by sunscald. When harvesting and working around plants, avoid breaking branches or otherwise exposing fruits. Control of diseases that kill leaves can prevent sunscald injuries. Shading cloths can be used to protect some fruits and vegetables from sunscald. The trunks of susceptible fruit trees can be protected by use of reflectant paint applied to south and west sides. White latex paint diluted 1:4 with water is commonly used for trunk protection. Older trees that have thicker bark and bark protected by shading show less damage.

## Cold Injury

**Damage and Symptoms:** Cold temperatures can produce many different types of plant disorders, including:

- Cessation of plant growth and seed germination;
- Death of leaves, plants, blossoms, shoots and developing fruit;
- Prevention of pollen production;
- Prevention of pollinator activity;
- Catfacing of fruit (e.g., strawberries, tomatoes);
- Blistering of leaves and fruit;
- Shrinking and pitting of fruit;
- Production of "off" tastes;
- Skin russeting (sheet pitting) of fruits and vegetables; and
- Cracking of sun-exposed bark in winter.

Exposure to cold temperatures can also affect the growth response of transplants, causing them to flower prematurely (bolt). Many garden vegetables (onions, cabbage, broccoli, etc.) are biennial plants, which require two years to complete development by producing flowers and seeds. Normally these crops are harvested the first season, before flowering occurs. However, cold temperatures after transplanting can produce the signal to the plant that the first season is over, initiating flowering.

In addition, cold damage is a common indirect source of wounds or damaged tissues, which allow invasion of disease organisms. The reduced vigor of plants by prolonged exposure to cold can also favor damage by insects and diseases that are well adapted to cold conditions, such as seedcorn maggots and damping-off fungi.

**Disease Development:** Cool temperatures slow the growth rate of all plants. This effect is greatest on warm-season plants, such as muskmelons and peppers, which stop growing and ripening when temperatures drop below 60°F. More cold-hardy plants, such as onions, spinach and lettuce, may not stop growing until temperatures have dropped below 35°F.

Even temperatures above freezing can cause problems with some cold-sensitive plants. Pitting can occur in beans, melons and cucumbers when temperatures reach 45°F; a more generalized russeting of skin (sometimes called sheet pitting) occurs on peppers and eggplant. Temperatures at which many fruits and vegetables show bursting of cells and cell death are usually slightly below freezing, around 30°F. Fruit blossoms are somewhat more tolerant, requiring a hard freeze of approximately 28°F. These effects can be further dependent on hardening, or previous exposure to cool temperatures, so the timing of cold exposure is often more important than temperature in determining plant injury.

**Controls:** Several different types of controls can be used to avoid damage from the cold, including the following.

- Plant cold-resistant varieties that can tolerate freezing injury.
- Plant cold-dodging varieties of perennial plants that emerge late and avoid damage by spring frosts.
- Plant bulbs deeply to avoid early emergence in spring.
- Plant easily damaged perennials in shaded areas, such as north-facing slopes, to avoid premature emergence in spring.
- Avoid planting susceptible plants in low, sheltered areas where cool air settles. Orchards are often best located on slopes.
- Mulch perennial plants after ground has frozen to retard emergence in spring.

## SUSCEPTIBILITY OF FRUITS AND VEGETABLES TO COLD INJURIES

**Most susceptible**

| | | | |
|---|---|---|---|
| Apricots | Asparagus | Basil | Beans |
| Cosmos | Cucumbers | Eggplant | Grapes (wine) |
| Morning glory | Okra | Nectarine | Peaches |
| Peppers | Petunia | Potatoes | Summer squash |
| Sweet cherries | Sweet potatoes | Tomatoes | Zinnia |

**Moderately susceptible**

| | | | |
|---|---|---|---|
| Ageratum | Broccoli | Carrots | Cauliflower |
| Celery | Cosmos | Geranium | Grapes (native) |
| Lettuce | Marigold | Onions | Pears |
| Peas | Plums | Radishes | Raspberries |
| Rosemary | Strawberries | Tart cherries | (summer-bearing) |
| Winter squash | | | |

**Least susceptible**

| | | | |
|---|---|---|---|
| Apples | Beets | Brussels sprouts | Cabbage |
| Currants | Garlic | Gooseberries | Kale |
| Kohlrabi | Oregano | Parsley | Raspberries |
| Rutabagas | Sage | Spinach | (fall-bearing) |
| Turnips | | | |

- Harden seedlings in spring before transplanting by gradually exposing them to outdoor conditions.
- Protect seedlings with row covers, hotcaps or other protectant screens that retain heat.
- Cover susceptible vegetables and flowers during early freezing periods in fall.
- Do not apply nitrogen fertilizer to perennial plants late in the season so that plants can harden off for winter protection.
- During late summer, reduce watering of peaches, grapes and other sensitive woody perennials so that they can harden off.
- Provide water to perennial plants in fall to reduce damage by winter drying. This should be done after sensitive species have hardened.
- Do not handle frozen plants until they have thawed.

## 2,4-D Injury

**Damage and Symptoms:** 2,4-D, and the other growth regulator herbicides, produce characteristic symptoms, such as twisting of the plant stems and leaf petioles. Developing leaves do not expand normally and show crinkled or highly elongated growth. Leaves may curl upward (cup) from effects of some herbicides. Leaf veins are prominent and light-colored. If herbicide exposure is severe, plants may die. Most garden vegetables are susceptible to 2,4-D injury. There is a range in susceptibility, however, with beans, grapes, tomatoes, cucumbers and squash being among the most sensitive. Some of the effects of herbicide injury

*2,4-D injury to tomatoes*

also turn into the gas form (volatilize) during very warm weather and move to garden plants. Herbicide-treated grass clippings used as garden mulch can be a source of herbicide residues. Accidental garden injury is also possible by using lawn fertilizers that contain herbicides.

**Controls:** Use of 2,4-D around gardens should be avoided whenever possible to avoid potential problems. Sprays should never be applied in hot or windy conditions. If herbicides are employed, a coarse setting spray should be used to avoid producing fine droplets, which drift much more easily in wind than larger droplets. When herbicides are applied, treatments should not be made during very high temperatures (above 90°F). High temperatures can allow some of these chemicals to volatilize and the resulting gas can drift and harm other plants in the area. Grass clippings suspected of containing herbicide residues should not be applied directly to the garden. If used, they should first be composted thoroughly to allow decomposition of the herbicide.

are similar to those of other types of garden disorders, including infection with viral diseases, aster yellows or potato psyllids. They are distinguished from these other injuries by looking at the pattern of symptoms. Herbicide injury is generally spread among many different garden plants. Other diseases are typically limited to a single type of plant.

**Disease Development:** 2,4-D and the other growth regulator herbicides (e.g., MCPP, dicamba, mecoprop) are widely used in lawns and other areas for weed control. Most are selective herbicides that have little effect on grasses. However, many garden and flower crops are broadleaf plants very susceptible to injury. Herbicide injury typically results from drift of the herbicide onto desirable plants. This can occur during applications of sprays in windy conditions. These herbicides may

## Nitrogen Deficiency

**Damage and Symptoms:** Nitrogen deficiencies appear as a general yellowing (chlorosis) of the entire plant, with symptoms particularly evident on older leaves. If severe, older leaves die and drop from the plant. Deficient plants are stunted and yield poorly. An excess of nitrogen can occur with overfertilization. High levels of nitrogen stimulate top growth at the expense of balancing root growth. Such plants show a delay in flowering and are more susceptible to many diseases.

**Disease Development:** Nitrogen is a basic element of proteins and amino acids, being used throughout the plant. It is also used in the chlorophyll molecule, which captures energy from the sun for use by the plant. Nitrogen is constantly being recycled within the plant for various growth needs and is concentrated in the newer growth. Nitrogen is usually the most limiting plant nutrient, and deficiencies can be common. Problems are aggravated by repeatedly removing plant matter and not fertilizing. Temporary deficiencies can occur in cold soils, which slow the activity of soil microbes that release nitrogen. The addition of carbon-rich plant matter, such as straw or sawdust, will also cause deficiencies as the microbes remove nitrogen for decomposition. Since soil nitrogen is very water-soluble and moves readily in the soil, it is easily leached if excess water is applied. Recently there has been increased attention given to groundwater pollution by the nitrate form of nitrogen.

**Controls:** Nitrogen is continuously being lost from soils because of plant removal and natural activities of soil bacteria. The addition of nitrogen is needed to maintain productivity of a garden. Many nitrogen-containing chemical fertilizers are readily available. Nitrogen is also available in most organic materials, being particularly abundant in most dried animal manures. These organic sources are more slowly released, since they require the activity of microbes to release nitrogen to the plant. Some plants "fix" nitrogen from the air through special nitrogen-fixing bacteria on the plant roots. This ability is common to all the legumes (beans, peas, etc.), and legumes can be used to increase the amount of nitrogen in the soil. However, since these plants move almost all of the nitrogen to their seeds, there is much more nitrogen benefit if legumes are tilled into the soil without harvesting.

## Iron Deficiency

**Damage and Symptoms:** Iron deficiencies appear as a general yellowing of leaves, accompanied by dark green veins (interveinal chlorosis). Because iron does not move in plants, symptoms are most pronounced on the new leaves. Iron-deficient plants are stunted and grow poorly. Where deficiencies are moderate, plants may recover later in the season. However, seedlings are particularly sensitive to iron deficiency and plant yields may later be decreased following short periods of seedling deficiency. Iron deficiencies may also vary during the season, being most severe in cool, damp springs. Peach, corn, bean, potato, petunia and dahlia are among the garden plants that are especially sensitive to iron deficiencies. Because deficiencies are also affected by the chemical environment around the roots, different rootstocks or plant varieties can also affect iron deficiency.

**Disease Development:** Problems with iron deficiencies are most common in soils that have high pH or excess calcium carbonate which can produce chemical reactions that inhibit the availability of iron to the plant roots. Since iron is absorbed in the area immediately behind the growing root tip, cultural conditions that slow root growth also contribute to iron deficiency. Compaction, excess water or other soil conditions that cause oxygen starvation of the root zone can increase iron deficiency problems.

**Controls:** Because iron deficiencies are greatly aggravated by high pH or high levels of calcium carbonate, correcting these soil conditions is the best long-term solution. Addition of sulfur or amending with manure or sewage sludge may help improve iron availability. Inorganic sources of iron, such as iron sulfate, applied to the soil will not correct iron deficiencies in soils that are excessively alkaline or high in calcium carbonate. Only the iron chelate EDDHA (Sequestrene 138 Fe) can remain in a usable form in alkaline soils (pH above 7.0). Where soil conditions cannot be adequately altered, less susceptible plants should be considered. Improvement of soil structure can also help alleviate iron deficiency by providing better conditions for root growth. Sprays of iron chelates or iron sulphate on leaves can restore green color to plants and temporarily correct iron deficiencies. However, if soil sources remain deficient, reapplication is required.

## Zinc Deficiency

**Damage and Symptoms:** Visible symptoms of zinc deficiency vary among different plants. Zinc-deficient plants tend to be stunted. Upper leaves of some plants, such as beans, may be yellow, while lower leaves show bronzing. A soil test is needed to confirm zinc deficiencies.

**Disease Development:** Zinc deficiency is most common in soils that are alkaline and low in organic matter. This is particularly likely where topsoil has been removed. Excessive levels of phosphorus may also increase zinc deficiency. It is most severe in cold, wet spring growing conditions.

**Controls:** Zinc deficiencies can easily be corrected by several types of zinc-containing fertilizers. Animal manure is a good source of zinc; zinc chelates or zinc sulphate are also good zinc fertilizers.

CHAPTER 8

# Weed Control in Flower and Vegetable Gardens

Sometimes a gardener's green thumb gets a little out of hand as plants we don't consciously plant start to grow. These unwanted plants, or weeds, can do more than merely upset our aesthetic sense. Perhaps most important is that they compete with vegetables or flowers for light, nutrients and water. Some weeds, such as quackgrass, also produce chemicals from their roots, which inhibit the growth of nearby plants.

You may work hard to grow Kentucky bluegrass in your lawn. But if it creeps into your garden, it becomes a weed. Even things we intentionally plant in the garden, such as horseradish, Jerusalem artichokes and many flowers, can become weeds if they spread to areas where they aren't wanted. On the other hand, plants that are widely recognized as weeds can be beneficial at some sites. For example, quackgrass growing on a steep bank prevents erosion; thistles are important foods for butterflies and birds.

## Characteristics of Weeds

Weeds can be classified in several ways, but most commonly they are described by the length of their life cycle. Using this method, weeds can be *annuals* (one-year life cycle), *biennials* (two-year life cycle) or *perennials* (multi-year life cycle).

Annual weeds are plants that germinate from seed, grow, flower and produce new seed in less than a twelve-month period. However, among the annual weeds there are further varieties of life cycles. *Spring* or *summer annuals* are plants that germinate in spring or early summer, flower and produce seed during summer and die in summer or fall. Many of the most common yard and garden weeds are spring or summer annuals, such as lambsquarters, pigweed, crabgrass and purslane. *Winter annuals* are less common in gardens. These are plants that germinate in fall or early winter and survive the winter as low-growing plants that resume growth and flower in the spring. Common winter annual weeds are mustards, such as flixweed, blue mustard and sheperd's purse.

Biennial weeds produce a compact cluster (rosette) of leaves the first year. Biennials survive the winter as a rosette, resuming growth the following spring, and then bolt, flower and die during this second season. Biennial weeds (e.g., mullein, musk thistle) are not common in gardens, but are usually found along roadsides and in pastures. However, many of our garden plants (carrots, cabbage, rutabagas, Brussels sprouts, etc.) are biennial plants that we harvest the first season.

Perennial weeds live for several years and are among the most difficult weeds to control. Bindweed, quackgrass, dandelion and Canada thistle are examples of common perennial weeds. Many perennial weeds produce extensive root systems that allow the plants to resist destruction by practices that kill only aboveground parts of the plants. Perennial weeds reproduce by seeds, and most infestations originate in this manner. However, once established, many perennial weeds can spread by underground structures such as runners (spreading roots, stolons or rhizomes), bulbs and tubers.

Weeds can be particularly damaging to garden crops early in the season. When plants are small, they are easily displaced by crowding, and a tender-spirited seedling may never fully recover from the trauma of being pushed aside by a bullying weed.

The effect of weeds as competitors is greatest with plants that root and grow in the same area of the soil. Several garden weeds are close cousins of plants we now choose to grow in the garden and compete strongly with them (e.g., pigweed and spinach). Garden weeds also affect other pest management practices. Attempts to maintain a garden rotation plan can be upset by weeds that also support the fungus we are trying so hard to squeeze out. Thick weed cover is a favored egg-laying area for cutworm moths in late summer.

Other gardening efforts determine the importance of weeds in a garden. Since weeds compete for water, light and nutrients, vigorously growing garden plants may outgrow the competition with little effect. Fertility and water conditions in the garden that provide enough for all further reduce the effects of competition.

Weeds in the garden often have a more offbeat nature than weeds in typical agricultural settings. Kentucky bluegrass creeps beyond the boundaries we set for the lawn and over the edge of flower plantings. Seeding trees, such as Siberian elm, can be a major garden headache as the seeds blow into the garden and sprout. A major source of garden weeds is often those we planted earlier that have since "gone wild" and naturalized.

## *Ways Weeds Spread*

More than any other type of garden pest, common garden weeds are the result of accidental, and sometimes purposeful, introduction by humans. Almost all garden weeds (as well as our cultivated garden plants) are not native to North America. Over 700 different species of plants now considered to be weeds were introduced into North America from Europe and Eurasia. A considerably smaller number of plants that have become garden weeds originated from Asia or tropical America, or happen to be native to the region.

Gardeners further distribute weeds by moving soil or other materials infested with weed seeds or viable plant parts. Frequently seeds move into a yard via straw, hay or animal manures brought into the garden. These common and useful garden materials are often infested with weed seeds, so it is important for the gardener to know their source. Hay or straw mulches grown in areas with serious weed problems, or manure from animals fed weed-infested feed, should not be directly introduced into the garden until composted at high temperatures.

Windblown weeds, such as dandelions and sow thistle, are more difficult to permanently control. If possible, try to prevent seeds from forming on plants in and around the garden by mowing or removing the weeds. Periodically checking the garden to kill germinating weeds may also be necessary to control occasional migrants.

One of the positive things about weed control is that a gardener really can make satisfyingly obvious progress against this type of pest. If weed control practices kill off existing weeds and preventive efforts restrict unwanted migrants, problems with weeds gradually decline.

## *Weed Controls*

### Mulches

Mulches are an easy solution to many garden weed problems. Properly placed, mulches exclude light and smother seedlings. Although they are considerably less effective against established perennial weeds such as bindweed, mulches can even help by making these weeds easier to pull and further delaying their regrowth.

Mulches have other effects on the garden environment. Moisture conditions around the plants are greatly changed by mulches that reduce evaporation from the soil. (Some mulches may

ORIGIN OF COMMON GARDEN WEEDS OF THE WEST

| Weed | Area of origin |
|---|---|
| Crabgrass | Eurasia |
| Green foxtail | Eurasia |
| Fall panicum | North America |
| Annual bluegrass | Eurasia |
| Barnyardgrass | Europe |
| Quackgrass | Europe |
| Redroot pigweed | Tropical America |
| Lambsquarters | North America/Eurasia |
| Wild buckwheat | Europe |
| Curly dock | Europe |
| Purslane | Western Asia |
| Wild mustard | Europe |
| Common ragweed | North America |
| Sunflower | North America |
| Prickly lettuce | North America |
| Annual sow thistle | North America |
| Dandelion | Eurasia |
| Common mallow | Europe |
| Russian thistle | Central Asia |
| Kochia | Europe |
| Field bindweed | Eurasia |
| Canada thistle | Eurasia |

also prevent water from moving into the soil.) This can help avoid fluctuating soil moisture and allow more even growth of the plants.

On the other hand, increased soil moisture can sometimes lead to garden problems. Moisture-loving garden pests, such as slugs, thrive in a well-mulched garden. Mulches also can restrict air circulation in a planting, contributing to some disease problems.

Changes in garden temperatures can be caused by mulching. Most organic mulches, such as grass clippings, straw or sawdust, are light-colored and reflect heat. This may have a cooling effect on the soil. Dark-colored plastic mulches have a reverse effect—heating the soil. As a result, these are sometimes used to accelerate the growth of warm-season plants such as peppers and melons. (The effects of specific mulches follow.)

*Grass clippings*

Grass clippings are one of the most readily available sources of mulch. Used in the compost or garden, they eliminate a major source of landfill trash and recycle the nutrients. Grass clippings can produce a dense mulch that will smother seedlings of most weeds and conserve soil moisture. They decompose more quickly than most organic mulches, primarily because of their small size. However, grass clippings have a tendency to mat, which prevents water percolation. Regular stirring of the clippings can limit matting. Occasionally there are problems with herbicide contamination of grass clippings. Several herbicides used in lawn care, such as 2,4-D and dicamba, can be carried on grass clippings for several days or even weeks after an application and subsequently injure susceptible garden plants. Where herbicide contamination is suspected, compost the grass clippings first to decompose the herbicide or try to find another source.

***Hay or straw***

Hay or straw is a readily available and often cheap source of mulching material, particularly when spoiled bales are used. However, these mulches are coarse, so it can be difficult to achieve a uniform smothering effect when thinly applied. (On the other hand, this same quality makes these mulches excellent for protecting garden plants from winter injury.) The main drawback of hay and straw is that they may inadvertently be a source of weed seeds in a garden. The thick mulches created by these materials also may decrease air circulation around the plants and increase some disease problems. This can be avoided, in part, by keeping the mulch away from the base of garden plants. As hay and straw decomposes, soil nitrogen may be temporarily drawn from the soil. This problem is most likely with more nitrogen-poor materials such as straw. Alfalfa hay is fairly high in nitrogen and can largely decompose without additional nitrogen sources.

### *Leaves*

Leaves are usually a readily available source in fall. They are weed-free and can provide an effective mulch to smother annual weeds. However, use of leaves is often limited by problems with matting, which prevents water movement into the garden. Because of their high carbon content, decomposing leaves may temporarily remove nitrogen from the garden. Because of these problems, it is usually best to compost leaves before using them in the garden.

### *Sawdust*

One may have access to sawdust in areas near sawmills or other woodworking operations. This may be a very beneficial mulching material, but it can also cause garden problems. Sawdust mulches applied directly to the garden require application of additional nitrogen to prevent soil deficiencies as the sawdust decays. Fresh sawdust can cause plant injury by acids produced during decomposition. Where large amounts of sawdust are used, they should be well aged and largely decomposed before being put in the garden.

### *Compost*

Well-decomposed compost can be a very effective mulch in a garden. However, if improperly prepared so that high-temperature decomposition does not occur, compost may contain viable weed seeds. This problem can be avoided by using only composting materials that are free of weed seed.

### *Black plastic*

Dark plastic mulches are among the most effective garden mulches—and they are weed-free. A thick plastic mulch that is impermeable to light can even challenge the livelihood of a persistent perennial weed such as bindweed. Black plastic mulch warms the soil, which can help the growth of peppers, melons and other warm-season plants that may need some extra help to make it through the gardening season. The plastic barrier further prevents direct contact of fruit with the soil, reducing problems with fruit-rotting organisms. However, because black plastic mulch is so effective at blocking water out of the soil, it can be too effective at preventing water movement *into* the soil. Where sprinklers are used, plastic mulches may need to be punctured to allow watering, although this reduces their effectiveness by permitting light to penetrate. Drip systems placed under the plastic can avoid this problem. Although well-maintained plastic mulches can often be used for several years, they eventually must be discarded. This end use of the plastic is the main reason many gardeners opt for a more biodegradable mulching material. Recently, thick, dark-colored paper mulches have been marketed that can be tilled into the garden at the end of the year to decompose.

### *Landscape mulches*

In recent years, several fabric weed barriers have appeared on the market for use around

*Pulling weeds*

landscape plants. These are constructed to prevent light penetration while allowing water to percolate. Typically, they are covered with an organic mulch, such as wood chips or bark. Landscape fabrics are rather expensive but very durable. When properly placed, they can suppress weeds for several years so long as the fabric remains intact. To use them for flower gardens, small planting holes must be cut. They are not appropriate for most berry or vegetable gardens where plants spread or are placed in new areas each year.

***Miscellaneous mulches***

For the creative gardener, mulching materials may be found in many forms. For example, old newspapers or discarded carpet laid between rows can give the garden that "lived-in" look.

## Hoeing and Pulling

The most common means of battling garden weed problems is "one-on-one" handpulling or hoeing. Although this activity allows the gardener to become intimately familiar with the garden environment, it is generally not appreciated.

The primary aims of hand hoeing and pulling are fairly brutal: (1) kill the weeds and (2) kill all their progeny (seeds, rhizomes, etc). Ideally, the weeds will ultimately be banished from the garden—and this is not entirely unrealistic. Unlike some of the more migratory insect and disease garden pests, weeds can steadily decline as problems following a sustained effort to control them. Your good efforts in weed control *will* be rewarded by reduced weed problems in the future.

Plants vary in susceptibility to hoeing. Environmental conditions can help or hinder their recovery. To begin with, try to control weeds when they are young. Seedlings have poorly developed root systems and few energy reserves that would allow regrowth. For example, the annual weed purslane cannot regrow from the roots until the plants are 3 to 5 inches tall. Even seedlings of the persistent perennial field bindweed need ten to fifteen weeks of growth (about the five-leaf stage) before its roots have developed sufficiently to bounce back from a hoeing attack. A single cultivation usually means the end of these young upstarts.

With established perennial weeds (e.g., quackgrass, field bindweed, Canada thistle), weed control is substantially more difficult. Since food reserves exist in the roots of the plants, the aim is to

*Hoeing prevents weed spread.*

prevent food manufacture by the plant. This is done by constantly (at approximately two- to three-week intervals) forcing the weed to regrow, speeding use of its food stores.

Hoeing, handpulling or tillage breaks off terminal buds of these plants, an exercise similar to pinching back plants to make them bushier. As a result, numerous side shoots form. This regrowth causes the plant to temporarily use more energy producing the new leaves than it recovers from photosynthesis.

Typically, this regrowth process requires two to three weeks, after which time the plants once again begin to replenish root reserves. Therefore, repeated, frequent attacks on the plants are most effective. Since regrowth is usually more rapid early in the season, attention to perennial weeds is particularly important at this time.

## Principles of Effective Hoeing

Plants should also be hoed or pulled during periods that hasten their demise. Early morning hoeing is often best, since it allows the heat of the day to finish off the weeds. This is important for killing rhizomes and roots brought to the surface during hoeing or pulling. Delaying irrigation for a day or so after hoeing can further prevent rerooting by a particularly hardy survivor.

Depth of hoeing is often not very important for weed control. Most annual weeds can be killed by cutting them at the soil line. Deep-rooted perennial weeds, such as Canada thistle and field bindweed, can recover rapidly from cuts that occur several inches below ground. Deep cultivation is more useful for control of some shallow-rooted perennials (e.g., quackgrass) to bring the rhizomes near the surface. This is then followed by shallow hoeing of sprouts as they emerge.

However, deep hoeing or tillage should usually be avoided for weed control purposes. Deep burial of surface layers during spading or rototilling can merely prolong weed problems by delaying germination. Many weed seeds can survive for years in a dormant state underground, germinating when favorable conditions of warmth and light occur as they are returned to the surface. A shallow cut of the hoe that barely disturbs the soil surface is usually the best technique.

Each gardener has his or her own favorite gadgets for hoeing weeds. Almost any hoe can do the job—so long as the gardener maintains the interest to work it. (My personal favorite is a "Glide n' Groom," which slides just under the soil surface and cuts the plants.) Regardless of type, hoes should be kept sharp to ease cutting and reduce the effort.

But even hoes can do serious damage to the garden if wielded inexpertly. The blade can cut both crops and weeds with equal ease, selectivity being in the hands of the gardener. (A parallel situation, known as cultivator blight, is a phenomenon of mechanized agriculture.) More subtly, overzealous hoeing may prune roots of plants, retarding growth and reducing their resistance to plant pests.

A sustained, season-long weed control effort is particularly important in light of the need to prevent reproduction. A temporary lapse, particularly late in the season, will allow the garden to be rapidly repopulated in following years by young "garden urchins." And don't be misled by the size of the plants. Many weeds can produce hundreds of seeds when only a few weeks old and a few inches in size if conditions are favorable to flowering (e.g., short and declining day length). Remove plants that have begun to flower and set seed.

Since weeds can also come from our cultivated plants, remove spent flowers before they reproduce. This often has the additional benefit of stimulating renewed flowering.

## Herbicides

Using herbicides in and around vegetable and flower gardens poses special difficulties, since they must be handled with extreme care to avoid damage to desirable plants. Because of these potential problems or because they wish to avoid pesticides in

## *Soil Sterilants—Accidents Waiting to Happen*

Some herbicides are used to destroy all vegetation and prevent regrowth for an extended period. These soil sterilants are most commonly used under pavement or in pavement cracks and other areas where no plant growth is wanted. Advertising claims typically promote their ability to "control vegetation for up to one year." Pramitol (Triox), ureabor and bromacil are among the widely available soil sterilants.

Unfortunately, the soil and growing conditions of the high plains and Rocky Mountain region often allow these materials to persist for a long time. Degradation of soil sterilants is slowed by the high soil pH and dry weather common to the West, allowing them to persist for many years. This very long persistence means greatly increased hazards of accidental plant injury.

Furthermore, many soil sterilant herbicides can move horizontally long distances through the soil, carried by water. This allows them to be carried into the root system of desirable trees and shrubs. These root systems often extend well beyond the drip line of the plants and may be concentrated under pavement or waste areas where soil sterilants are used.

The effects of these soil sterilants are nonselective and long-lasting. The end result of their use can too often be devastating to garden and landscaping plants. Because of their high potential for misuse, some states have moved to restrict their use to licensed professionals. However, so long as these products remain on the shelf, users must be especially aware of the special hazards whenever applying soil sterilants around desirable plants.

general, many gardeners opt not to use herbicides and limit weed control to mechanical practices such as hoeing. Also, there are few herbicides that are registered for use in gardens compared to those available for lawns or agricultural crops. However, they should be discussed, since in some instances herbicides can be used effectively in the garden to reduce weed problems and gardening effort.

Effective use of herbicides requires that the gardener learn how they act to kill or suppress unwanted plants. Herbicides work in many different ways. (See the following table.) Several types, the *preemergent herbicides*, work by killing newly germinated seedlings. On the other hand, *postemergent herbicides* are applied to plants after they have emerged, and some do not affect later germination of seeds.

These effects determine the *selectivity* of the herbicide. Most herbicides can harm many desirable plants as well as weeds. How these products are applied determines how effectively they can be used. For example, most preemergent herbicides are fairly selective in that only seedlings are affected. Use of preemergence herbicides around transplants or established plants allows them to work selectively on the germinating weeds. However, if the vegetable garden is to be seeded, the herbicides may also kill the vegetable plants.

Most postemergent herbicides are selective in what types of plants are affected. The newly introduced selective grass killers, such as fluazifop-butyl (Grass-B-Gon), are only active against grasses and can sometimes be used to remove grass from around certain flowers or perennial plants. Other herbicides, such as 2,4-D (not a garden herbicide), tend to have the opposite effect and are mostly used to kill broadleaf plants. However, even these selective herbicides can cause unwanted injury if they are used in too high a concentration or drift onto desirable plants.

| COMMON GARDEN HERBICIDES | | |
|---|---|---|
| **Herbicide** | **Pathway into the plant** | **Activity in the plant** |
| DCPA (Dacthal) | Absorbed only through the roots and shoots of seedlings. | Unknown |
| trifluralin (Treflan) | Absorbed only through the roots and growing shoots of seedlings before soil emergence. | Inhibits cell division in growing cells. |
| fluazifop-butyl (Grass-B-Gon) | Absorbed through leaves of actively growing grasses. | Interferes with the production of lipids needed for cell membranes to function. |
| 2,4-D | Primarily absorbed through leaves and green stems of established plants but has some residual effect in soil. | Acts as a plant hormone and affects many plant actions; most obvious is an overstimulation of growing cells, producing distorted growth and starving roots. Broadleaf plants are much more susceptible than grasses. |
| glyphosate (Kleenup, Roundup) | Absorbed through leaves or green stems of established plants. | Disrupts the production of certain amino acids used by plants. |
| herbicidal soaps (Sharpshooter) | Contact action on cells of leaves and stems. | Causes cells to leak ions used in normal function and die rapidly. |

One commonly used nonselective herbicide is glyphosate (Roundup, Kleenup), which can kill most plants to which it is applied. Selective use of this material requires that it be very carefully placed, in spot treatments, to prevent injury to garden plants. Shielding desirable plants and painting the herbicide onto the weeds are methods used to avoid contaminating and damaging desirable plants. Glyphosate should not be used in or around edible plants. Characteristics of specific herbicides are also discussed in Appendix II.

## *Common Weeds in Flower and Vegetable Gardens*

### Quackgrass (*Agropyron repens*)

**Weed Description and Growth Pattern:** Quackgrass is a perennial grass that commonly reproduces by seeds and creeping rhizomes. It grows extensively during the cool weather of early spring and fall. Flowers are produced throughout

Agropyron repens. *Quackgrass*.

most of the garden growing season. Seeds can remain dormant for many years. Quackgrass roots produce toxic chemicals that can inhibit the growth of surrounding vegetation. The pointed rhizomes can penetrate most soft plant tissues and damage root crops.

**Mechanical Controls:** Quackgrass germinating from pieces of rhizomes takes about sixteen weeks to regrow roots and produce new rhizomes, so young plants are the easiest to control. Established plants should be tilled or raked to expose the shallow roots to kill rhizomes and root fragments. Repeated cultivation at short intervals is particularly useful, since it can fragment the roots and rhizomes, causing the plant to use up stored energy to produce leaves.

**Chemical Controls:** In flower gardens, quackgrass can be killed by the selective herbicide fluazifop-butyl (Grass-B-Gon). The nonselective herbicide glyphosate (Roundup, Kleenup) can systemically move in the plant and kill roots, although desirable plants are also easily damaged by this herbicide.

## GreenFoxtail/Yellow Foxtail (*Setaria viridis/ Setaria glauca*)

**Weed Description and Growth Pattern:** Green and yellow foxtails are summer-annual grasses that reproduce by seeds. They are marked by the dense, bushy spike the plant produces. Flowering begins in June, and seeds may be produced continuously through the summer and early fall. Foxtail seeds are less persistent than those of other weeds. Most seeds germinate the season after they are produced.

A) Setaria faberi. *Giant foxtail;* B) S. viridis. *Green foxtail;* C) S. glavca. *Yellow foxtail.*

**Mechanical Controls:** Young plants are very susceptible to hoeing. Older plants can easily reroot, so the root system must be thoroughly disrupted. Mulching can effectively control foxtails in a garden. Since most seeds germinate the following season, a sustained weed-control effort can eliminate most foxtail problems in one to two years.

**Chemical Controls:** The preemergence herbicides DCPA (Dacthal) and trifluralin (Treflan) can control plants if applied in spring before foxtail seeds germinate. In flower gardens, the selective postemergence grass killer fluazifop-butyl (Grass-B-Gon) can kill younger seedlings.

Setaria viridis. *Green foxtail.*

## Crabgrass (species of *Digitaria*)

**Weed Description and Growth Pattern:** Crabgrass is a summer-annual, grassy weed most commonly associated with lawns. It thrives in hot, dry weather and may invade flower and vegetable gardens. Although its ground-hugging growth does not often compete for light, crabgrass can spread rapidly, removing water and nutrients used by other plants. Most seeds germinate during the early part of the growing season. Areas of warmer soils may begin to produce seedlings by mid-spring. However, some seeds may continue to germinate through August, particularly when watered after a dry spell. Flowers and seeds are produced on spikelets from July through October. With cold weather, crabgrass turns purplish, dying after hard frosts. Two species of crabgrass are found in the West. Smooth crabgrass (*Digitaria ischaemum*) tends to occur in more northern areas; large crabgrass (*Digitaria sanguinalis*) is a more southern spe-

Digitaria sanguinalis. *Large crabgrass.*

Digitaria ischaemum. *Smooth crabgrass.*

cies. However, both occur throughout much of the region, and controls are similar.

**Mechanical Controls:** Germinating seeds and small seedlings are easily controlled with a mulch. Hoeing or cultivating can kill small seedlings. This method is particularly effective on low-growing weeds, such as crabgrass. However, larger plants can often reroot unless the root system is exposed to drying conditions.

**Chemical Controls:** Preemergence herbicides that prevent seed germination, such as DCPA (Dacthal) and trifluralin (Treflan), can control crabgrass. But garden use is generally impractical because of the timing of crabgrass germination. Very early-season treatments necessary to control early crabgrass do not persist in controlling later-germinating seeds. The selective grass killer fluazifop-butyl (Grass-B-Gon) can be used to spot treat smaller crabgrass plants around some flowers and ornamentals. Nonselective herbicides, such as herbicidal soaps (Sharpshooter) and glyphosate (Kleenup, Roundup), can kill crabgrass but must be used carefully to prevent injury to desirable plants. Cultivation is usually far simpler and safer for crabgrass control in established gardens.

*Seedling lambsquarters*

## Lambsquarters (*Chenopodium album*)

**Weed Description and Growth Pattern:** Lambsquarters is a spring-annual broadleaf plant that reproduces by seeds. Leaves are generally light green and covered with a white powder; stems are often marked with red and green streaks. Under favorable conditions, lambsquarters may grow several feet in height. Lambsquarters seed germinates early in the season. Plants may grow rapidly and easily outgrow most garden seedlings. The plant produces flowers in ball-like clusters from June through October. The small black seeds are formed continuously, beginning in early summer.

**Mechanical Controls:** Lambsquarters is susceptible to hoeing, although very large plants may require handpulling. Thick mulches can prevent seedling establishment.

**Biological Controls:** Lambsquarters is fed on by many of the same insects that attack spinach and beets, including the spinach leafminer, leafhoppers and cutworms.

**Chemical Controls:** Lambsquarters is susceptible to preemergence herbicides such as DCPA (Dacthal) and trifluralin (Treflan). However, these need to be applied very early in the season to kill early-germinating seeds and are usually impractical or undesirable for gardening.

**Harvest Controls:** When picked young, lambsquarters is an edible green similar in taste to spinach.

Chenopodium album. *Common lambsquarters.*

## Redroot Pigweed (*Amaranthus retroflexus*)

**Weed Description and Growth Pattern:** Redroot pigweed is a spring-annual broadleaf plant marked by a distinctly red root. It produces oval to lance-

Amaranthus retroflexus L. *Redroot pigweed.*

shaped, light green leaves and may reach several feet in height under favorable conditions. Redroot pigweed is a vigorously growing weed that is highly competitive with other plants. Flowers are produced in bristly, irregular clusters at the top of the plant. Seeds are produced continually after flowering starts.

**Mechanical Controls:** Redroot pigweed is very susceptible to hoeing during early stages of growth. However, older plants are more fibrous and can regrow from roots. Since redroot pigweed can begin flowering when only a few inches tall, it should be controlled early.

*Pigweed*

**Chemical Controls:** Redroot pigweed is susceptible to the preemergence herbicides DCPA (Dacthal) and trifluralin (Treflan). Emerging plants can be spot treated with herbicidal soaps (e.g., sharpshooter).

**Miscellaneous Notes:** Redroot pigweed is a close relative of the cultivated amaranth.

## Spurge (species of *Euphorbia*)

**Weed Description and Growth Pattern:** Prostrate spurge (*Euphorbia supina*) and spotted spurge (*Euphorbia maculata*) are relative newcomers to much of the West. They are annual broadleaf weeds with a low, spreading growth habit. Like other members of the plant family Euphorbiaceae, they produce a milky sap. They are related to leafy spurge (*Euphorbia esula*), which is an extremely aggressive perennial weed of pastures that has spread rapidly throughout the region. The plants are highly tolerant of drought and poor soil, which have allowed them to thrive in cracks of driveways and sidewalks, as well as in gardens and lawns. Seeds begin to germinate in late spring as soil temperatures rise above 55°F, and germination

Euphorbia supina. *Prostrate spurge.*

may continue over several months. The plants grow rapidly in warm weather and produce small flowers, usually at the base of leaves. Seeds are shed from early July until the plants are killed by frost.

**Mechanical Controls:** The seeds are small and seedlings are easily smothered by mulches and hoeing. Older plants, which are easily hand-pulled or hoed, should be removed before seeds are produced.

**Chemical Controls:** Spurges are not very susceptible to most herbicides. The preemergence herbicide DCPA (Dacthal) is one of the most effective treatments but may need to be reapplied because

*Spurge*

of the extended period over which spurge seeds germinate. Spurges are also quite tolerant of contact herbicides such as glyphosate (Roundup, Kleenup), herbicidal soaps (Sharpshooter) and 2,4-D.

## Purslane (*Portulaca oleracea*)

**Weed Description and Growth Pattern:** Purslane is an annual broadleaf weed that reproduces by seeds. It has succulent leaves and a low-growing, spreading growth habit and is related to the garden flower known as portulaca or moss rose. The seeds require fairly warm soil temperatures to germinate. Flowers appear in mid- to late summer. Because of its low-growing habit, purslane does not compete for light with garden plants as much as other common weeds. However, it uses a lot of water during growth and can form a mat that will inhibit small plants.

**Mechanical Controls:** Hoeing is most effective on smaller plants, since purslane can regrow from the roots after it reaches 3 to 5 inches. Because of the succulent growth, pulled plants may reroot if the soil remains moist. After flowers have begun to form, purslane should be removed from the garden, since seeds can mature on drying plants. Because of its low-growing habit, purslane is easy to control with mulch.

**Chemical Controls:** DCPA (Dacthal) and trifluralin (Treflan) can control germinating purslane. However, because seed germination is rather late, applications of these herbicides early in the season to control other weeds may no longer be effective against later-germinating purslane.

**Harvest Controls:** Purslane is an edible plant, and cultivated varieties are grown in many areas of the world. It was originally introduced into North America as a vegetable crop. The younger leaves have a tart, acidic taste.

*Purslane*

## Dandelion (*Taraxacum officianale*)

**Weed Description and Growth Pattern:** Since its introduction into North America, dandelion has spread widely throughout much of the continent. Presently, it is one of the most conspicuous weeds in lawns because of its bright yellow flowers. Dandelion can also be found in pastures, mountain meadows, roadsides and gardens, since the wind-borne seeds grow rapidly after landing on moist, bare soil. Dandelion is a perennial, producing a long taproot that may penetrate 1 to 2 feet in depth. The leaves are flattened and often grow close to the soil. Because of this low-growing habit, dandelions generally do not compete greatly with garden plants for sunlight and are not aggressive garden weeds. However, they remove nutrients and water from garden soils and may crowd out some garden plants. Most flowers are produced in early spring and fall.

**Mechanical Controls:** Although seedling plants are easily pulled, established plants can resist cutting or pulling. Unless the taproot is completely removed, which is sometimes possible in very loose soils, the plants may regrow, requiring repeated hoeing until the root reserves are depleted. Since the plant can also regrow from root fragments, rototilling may spread dandelions in a garden. Because of its low-growing habit, dandelions are more easily controlled with mulches than are more aggressive perennials, such as field bindweed.

**Chemical Controls:** Dandelions are most susceptible to the herbicides 2,4-D and glyphosate during early spring (before flowering) and fall. However, these herbicides can cause injury to most garden plants, so chemical control of dandelions is often undesirable.

**Harvest Controls:** Dandelions were deliberately introduced into North America and are cultivated as food. All parts of the plant are edible, and it is a good source for a number of vitamins (A and C) and minerals (iron, copper and potassium).

## Field Bindweed/Wild Morning Glory (*Convolvulus arvensis*)

**Weed Description and Growth Pattern:** Field bindweed, also known as morning glory or creeping jenny, is one of the most persistent and difficult weeds to control. Once established, field bindweed produces roots that may penetrate 15 to 30 feet or more, and it tenaciously regrows following hoeing. Field bindweed produces a vine that may climb or spread across the garden. From June through September, the plant produces showy white or pink flowers, miniatures of the cultivated morning glory. The flowers only last a short while (typically one day), and field bindweed produces highly persistent seeds throughout the growing season. Field bindweed can also reproduce and spread by underground roots.

**Mechanical Controls:** Seedling plants (less than the five-leaf stage) can be killed by hoeing a single time. After this point, the plants can regrow from the root system. Regrowth from established plants starts about two weeks after cutting, before the leaves begin to send food back to the roots. By cutting at 14- to 18-day intervals, the plants can gradually be starved. Individual plants in a garden can be "canned." Inverting a large can over plants that emerge following hoeing can smother them. Soil solarization can also kill plants during periods of high temperature and light intensity.

**Biological Controls:** No biological controls appear to greatly affect field bindweed. Among the more conspicuous insects associated with the plant is the brightly colored golden tortoise beetle, which chews holes in the leaves. An eriophyid mite is currently under development by the USDA as a potential future control.

competitive plants, such as squash, improve control.

**Miscellaneous Notes:** Field bindweed is often confused with annual buckwheat, a weed with lancelike leaves and a similar vine habit. Annual buckwheat does not produce the showy flowers of bindweed and it is not perennial.

## Canada Thistle (*Cirsium arvense*)

**Weed Description and Growth Pattern:** Along with field bindweed, Canada thistle is one of the more challenging garden weeds to control. Once established in the garden, this deep-rooted perennial spreads rapidly by creeping roots and rhizomes. Eradication is not only a long-term project, but the spiny leaves continue to remind the ungloved hand of their presence. Initial infestations can occur through windblown seed, although fertile seeds are rarely shed in the thistledown. However, the plants can easily regrow from roots, which may remain dormant for several years if buried deeply. Plants tend to die back after flowering, regrowing in late summer and early fall.

Convolvulus arvensis. *Field bindweed.*

**Chemical Controls:** Field bindweed is quite resistant to control with herbicides. Although several herbicides can kill the top growth, the extensive root system often remains, allowing the plant to regrow. Treatments of glyphosate (Roundup, Kleenup) are most effective in fall before the bindweed freezes off. Combining the herbicide treatment with regular hoeing and rotating with highly

**Mechanical Controls:** Root reserves of plants can be exhausted by repeated hoeing. Hoeing should be done at least every two weeks to prevent plants from recovering energy reserves. Because the plants can grow from small root pieces, deep cultivation, including rototilling, can spread them.

Cirsium arvense. *Canada thistle*.

**Biological Controls:** Several insects feed on Canada thistle, most prominently the thistle caterpillar, alter ego of the painted lady butterfly. Aphids, lacebugs and planthoppers occasionally suck sap from plants. Rust disease can individually destroy susceptible plants. Unfortunately, these natural controls appear to have little permanent effect on Canada thistle. The USDA, in conjunction with some state agencies, is currently releasing new species of insects for biological control of Canada thistle.

**Chemical Controls:** Canada thistle is most susceptible to herbicides when plants are small or have been stressed by repeated pulling and hoeing. It is also more easily controlled if the root system is periodically disrupted by tilling the soil, inhibiting the production of lateral roots. Plants in the early bud stage are most susceptible to the herbicide 2,4-D. After bloom is the recommended time for application of glyphosate (Roundup, Kleenup). However, since these herbicides will also kill or damage many garden plants and cannot be employed around food crops, they must be used with extreme care. Selectively brushing or painting the herbicide is one of the easiest means of selective treatment.

*Canada thistle*

Malva neglecta. *Common mallow.*

## Common Mallow/Cheeseweed (*Malva neglecta*)

**Weed Description and Growth Pattern:** Common mallow (cheeseweed) is a broadleafed plant with a spreading habit of growth. It often invades lawns and gardens and, although not highly aggressive, can be difficult to eliminate from a planting. Common mallow is highly variable in growth and may occur as a leafy annual, biennial or even perennial weed, depending on growing conditions. The plant produces pale-colored, five-petal flowers. Time of flowering depends on the age of the plant but can last throughout the growing season in most areas.

Common mallow later forms a sectioned ("cheesewheel") seed pod that scatters seeds as they dry. Seeds most often germinate in late spring, but germination can occur through much of the summer.

**Mechanical Controls:** Seedlings and young plants can be controlled easily with hoeing. However, where common mallow has grown for several months, it produces a deep taproot that makes it difficult to kill. These taproots must be pulled or cut well below the underground crown area to prevent rapid regrowth. Mulches can prevent seedling emergence. They are also useful to prevent regrowth following pulling and hoeing.

**Chemical Controls:** No effective chemical controls are available for use in vegetable plantings. Spot treatment with glyphosate (Roundup, Kleenup) can provide control in flowers and around fruit trees but can damage most other plants that it contacts.

*Common mallow*

CHAPTER 9

# Managing Garden Problems with Mammals and Birds

Peter Rabbit is only one of the uninvited vertebrate pests that may visit the garden patch. Deer and elk can demolish a vegetable garden in short order, favoring the tender shoots or most succulent parts of the vegetables. Coyotes may move in late in the summer to feast on melons, for which they show a decided preference. Ripening fruit is devoured by raccoons, robins and blackbirds. Sometimes even neighborhood pets wreak havoc digging around the plants.

There are special problems in controlling vertebrate pests. They are highly mobile and intelligent and can be difficult to exclude from a garden. Lethal controls (shooting, poisoning) and even trapping are often strictly limited by game laws and concerns of possible injury to pets, desirable wildlife or children.

## Mammals

### Voles (species of *Microtus*)

**Damage:** Voles (also called meadow mice) injure fruit and ornamental trees and shrubs by gnawing on the bark. Extensive injury produces girdling wounds that can kill smaller plants. The small size (about 1/8 inch wide) of the chewing wounds differs from those of other gnawing animals, such as jackrabbits. Vole damage to tree bark is most frequent during the winter. Voles will eat almost anything green, including grasses, tubers, bulbs and garden plants. They also commonly damage lawns by clipping grass while making surface runways.

**Life History and Habits:** Voles are rodents, related to the white-footed and deer mice that sometimes invade homes. There are several species of voles in the region, including the pine vole, prairie vole, long-tailed vole, montane vole and meadow vole. The pine vole has been particularly damaging to orchards. Voles actively feed and reproduce throughout the year. Peak breeding occurs during spring and summer. About three to six young are produced in each litter. They develop rapidly and become full-grown in about 45 days. Their life span is fairly short, averaging less than one year. Population numbers fluctuate greatly, with peak injury to trees and shrubs associated with periodic outbreaks. Voles produce and maintain runways and shallow tunnels that cross the soil surface. These burrow systems may support several voles. The home range area used by voles varies during the year.

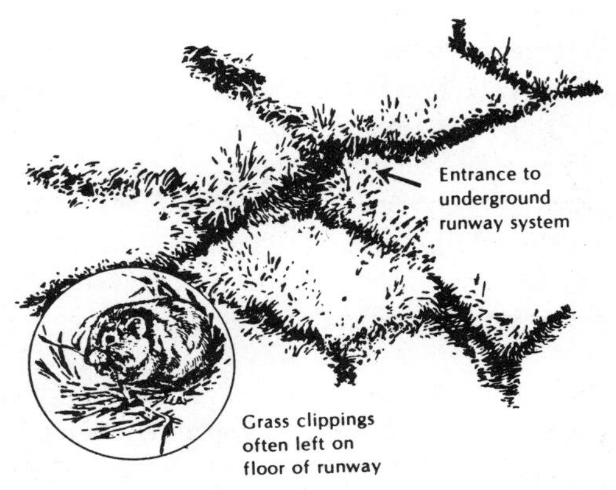


*Surface runway system of the prairie vole*

*Pine vole*, Microtus pinetorum *(left)*, *and prairie vole*, M. ochrogaster *(right)*

**Biological Controls:** Natural predators of voles include large snakes, owls and weasels, which normally will not be found in yards. Cats commonly feed on voles.

**Cultural Controls:** To occur in large numbers, voles require cover and ready access to food. Mowing lawns and weedy areas, and periodic spading or tilling of gardens, can reduce favorable cover used by voles. Mulches, which also provide cover for voles, should be cleared away at least 3 feet from the base of susceptible trees and shrubs to reduce bark feeding. If vole injury is serious to overwintered perennials, such as strawberries, mulching should be delayed until after the ground has frozen.

**Mechanical Controls:** Individual trees and shrubs can be protected from vole injury by tree guards or wire mesh screening, 1/4 inch or less in diameter. Because voles can tunnel, the barriers should be buried 6 inches. Placing small, sharp pebbles in planting holes for bulbs can deter voles. During the winter, compacting the snow around trees and shrubs can act as a barrier to vole tunneling. Mouse traps, placed with the trigger along the runways, can be used to kill voles. Baiting is not necessary if traps are properly placed in the runway.

**Chemical Controls:** Hot pepper sauce and thiram-based repellents have been registered for control of meadow voles. Their effectiveness is considered questionable.

## Eastern Fox Squirrel (*Sciurus niger*)

**Damage:** The eastern fox squirrel commonly runs around yards and gardens east of the Rockies. Active throughout the year, it is usually an interesting and valued member of backyard wildlife. But raids of gardens, trees and bird feeders sometimes cause people to have second thoughts about squirrels. Fox squirrels prefer to feed on nuts and fruits, which can include the plums and cherries that were planted for another purpose. More important, they chew tree bark during winters when food is scarce, causing dieback of branches. Squirrels occasionally feed on sweet corn in the vegetable garden.

**Life History and Habits:** Fox squirrels reproduce rapidly. There are two breeding seasons (December to January and June) during which the male squirrels noisily chase females and each other around the yard. Young are born about 45 days later. Nesting may occur in leaf nests lined with bark and other plant materials. However, fox squirrels prefer to nest in a cavity, such as a tree hollow or even an attic. Litters average about three young squirrels, which are weaned in two and one-half to three months. Fox squirrels can be highly migratory and will travel for miles if food is scarce. They do not hibernate and are active all year.

**Legal Status:** Fox squirrels are considered to be game animals under some state wildlife laws. Check

*Tree squirrel*

local regulations if squirrels are to be trapped or killed.

**Biological Controls:** Although squirrels may live up to ten years in captivity, about half of all squirrels typically die from natural causes each year. Insect and mite parasites are very common. Predators such as hawks, owls and house cats can also take many squirrels, although this is not considered to be very important in regulating populations. House cats can discourage nesting of squirrels in attics and other areas of a home.

**Cultural Controls:** If high numbers of squirrels occur during winter and threaten to cause bark-chewing injuries, providing food can limit this damage.

**Mechanical Controls:** Squirrels can be excluded from individual trees by metal collars. These should be 2 feet wide and placed high on the trunk—preferably at least 6 feet above ground. However, fox squirrels are excellent jumpers and all other access must be eliminated (e.g., fences, nearby trees, roofs). Squirrels nesting in homes can be kept out by 1/2-inch mesh hardware cloth. They are also easily trapped in box or cage live traps. Nuts, seeds, apples and peanut butter make good attractants. However, trapping generally has little effect on populations, since new squirrels migrate readily and reproduction is rapid.

**Repellents:** Moth repellents containing naphthalene or paradichlorobenzene can repel squirrels from nesting in attics. However, the value of these fumigant-type materials outdoors is questionable. Thiram-based animal repellents may prevent gnawing damage to fruit trees when applied in the dormant season. However, registrations and product availability are limited.

**Miscellaneous Notes:** Several types of ground-dwelling squirrels, such as the Richardson ground squirrel (*Spermophilus richardsoni*) and thirteen-lined ground squirrel (*S. tridecemlineatus*), are common in the region. These feed on crop seedings and damage pastureland in less developed areas, but rarely occur in high numbers around yards and gardens. Unlike fox squirrels, ground squirrels hibernate during winter. Ground squirrels are most often controlled by trapping and use of poison baits. They are not generally protected by law, although local wildlife agencies should be consulted before using these methods.

## Raccoon (*Procyon lotor*)

**Damage:** Raccoons feed on a wide variety of plant and animal foods, including fruits and vegetables. Most commonly damaged is sweet corn. Plants are knocked over and ears partially eaten just as they begin to ripen. Raccoon damage to sweet corn is characterized by husks that have been pulled back. Raccoons will also eat melons, digging a small hole in the rind and scooping out the contents. Some tree fruits, such as cherries, and berries are eaten by raccoons. Raccoons will kill young birds, including poultry. They also feed on insects and commonly dig lawns in search of white grubs, cutworms or other large insects.

**Life History and Habits:** There are raccoons throughout the West, but they are most common near wooded cover and water. They are active at night and may roam 1/2 mile or more from dens. Adult males, which remain solitary, have the

*Electric fencing can be very effective at excluding raccoons from sweet corn or other crops. Two wires are recommended, but one wire 6 inches above the ground may be sufficient. Electric fence charges are available at farm supply dealers. The fence can be activated at dusk and turned off after daybreak.*

largest ranges. Raccoons usually give birth in April or May, occasionally later. Litter size is typically three to five. The young often stay with the mother for the first year, separating the following spring. In winter, raccoons do not hibernate, but remain inactive in the dens during unfavorable weather.

**Legal Status:** Raccoons are classified as protected furbearers in most states. Individual raccoons that are causing damage can be killed or trapped only with permission from state wildlife agencies.

**Mechanical Controls:** Raccoons can be excluded by fencing. Electric fences are very effective for protecting sweet corn and other garden vegetables. Fence wires should be 6 inches high, with a second lower wire increasing effectiveness. Wire mesh can also exclude raccoons. However, raccoons are excellent climbers. Mesh screening should have an overhang or be left unattached (floppy wire fence) to deter climbing.

Frightening devices sometimes provide temporary control of raccoons, but rarely have given long-term control. These include radios, dogs, wind chimes and other noise-making devices. Similarly, nighttime lighting may deter raccoons, but they will begin to revisit the site after they become accustomed to the change.

**Repellents:** Repellents have given little control of raccoons in open-air settings, such as gardens.

**Traps:** Raccoons are best trapped using cage or box-type live traps. Traps used for capturing raccoons need to be large (at least 10 by 12 by 32 inches) and of strong construction. Baits of fish-flavored cat food, chicken or fish are attractive.

## Cottontail Rabbits (species of *Sylvilagus*)

**Damage:** Rabbits feed on and can destroy many different garden plants. Despite their reputation for carrots (promoted largely by Bugs Bunny), other vegetables, such as beans and peas, appear to be even more preferred. Several bulb crops, notably tulips, are commonly damaged. During winter, cottontails will gnaw fruit tree bark, raspberries and many ornamental plants. (Evergreens are particular favorites.) Succulent shoots of young trees are also browsed by rabbits, clipped at snow height. Rabbit damage is characterized by a sharp, clean 45° angle feeding cut. When there are lots of rabbits, other signs, including round droppings and footprints, can be used to separate rabbit injury from that of other gnawing mammals, such as voles and squirrels. Several species of cottontail rabbits occur in the region. The eastern cottontail (*Sylvilagus floridanus*) is generally most abundant east of the Rockies, although several other species occur in the West. Cottontail rabbits prefer to nest in areas with brush or other cover, and landscaping of developed neighborhoods is ideal for this activity. Cottontails are rarely found in dense forests or open rangeland. In open areas of the plains, jackrabbits (species of *Lepus*) may predominate. These rabbit relatives (actually classified as hares) are particularly damaging when drought or overgrazing force them into yards and gardens.

*Eastern cottontail rabbit*

**Life History and Habits:** Despite the well-known exploits of Peter Cottontail (actually a European hare), cottontail rabbits do not dig underground nests. During warmer months, they form shallow hollows in dense vegetation for cover; for winter protection, natural cavities or burrows are used. Rabbits breed like—rabbits. Cottontails typically produce two to three litters per year and each litter contains about three to five young. The young rabbits can leave the nest about three weeks after birth and are sexually mature within a few months. Populations can increase rapidly when food is abundant. Cottontails do not hibernate and are active throughout the winter. They can walk on snow and feed on plants at the level the snow cover allows.

**Legal Status:** In most of the region, cottontail rabbits are classified as a game animal. As such, they fall under regulations of state wildlife agencies, which generally restrict hunting and trapping to specific seasons, requiring a license. However, exemptions may be granted by these state agencies for rabbits that are damaging property.

**Biological Controls:** Rabbits suffer from predation by many animals in the wild, including owls,

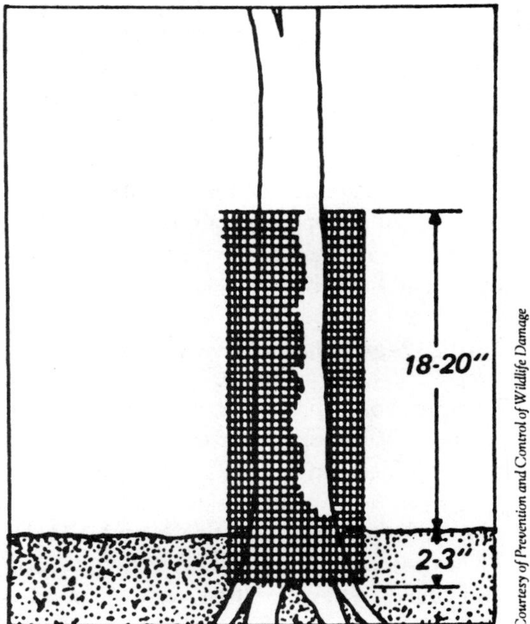


*A cylinder of hardware cloth or other wire mesh can protect trees from rabbit damage.*

*Courtesy of Prevention and Control of Wildlife Damage*

foxes, snakes and hawks. Rabbits also succumb to various diseases. Cottontail rabbits rarely live for more than fifteen months under natural conditions. In areas of dense human habitation, house cats are important predators, feeding on young rabbits in nests. Dogs will also deter rabbits from roaming in yards, although they rarely kill them.

**Mechanical Controls:** Rabbits are easily excluded from gardens by fencing. A 2-foot-high chicken wire fencing, buried shallowly in the soil, can prevent most rabbits from entering gardens during the summer. Plastic tree wraps and wire mesh are effective for preventing winter damage to trees and shrubs if they are drawn 1 to 2 feet above snow line. Holes in fencing should not be larger than 1/2 inch to exclude small rabbits.

**Repellents:** Various types of odor repellents marketed as animal repellents, such as naphthalene moth balls, are not effective for rabbit control outdoors. Taste repellents, usually involving the fungicide repellent thiram, are effective for preventing rabbit feeding. However, thiram is a toxic material to other mammals, such as humans, and cannot be applied to plants that are to be eaten.

**Traps:** Rabbits can be trapped easily in box live traps baited with apples, carrots, ears of corn or cabbage. Traps should be placed close to areas of cover used by rabbits.

## Mule Deer (*Odocoileus humanus*)

**Damage:** Throughout much of the region, no pest is more destructive and difficult to manage than the mule deer. In spite of the residual "Bambi syndrome," most gardeners quickly lose patience after losing a vegetable or flower garden to deer. (The white-tailed deer, *Odocoileus virginianus*, is also found in much of the region and causes similar plant injury.) Deer feed on a very wide variety of plants, depending on the season and the availability of alternative foods. Many vegetable crops may be browsed by deer feeding; the chewing of sweet corn ear tips and nipping of the center from heads of broccoli and cauliflower are examples. Deer also chew on twigs of fruit and ornamental trees, sometimes causing severe damage to younger trees. Feeding damage by deer is characterized by a ragged wound. Deer (and elk) lack upper incisor teeth, which produce the clean cuts of other gnawing animals, such as rabbits.

*Mule deer*

**Life History and Habits:** Deer prefer areas of mixed vegetation and have been described as creatures of the forest edge. Ideal habitat for deer includes areas near shrubby plants that provide year-round food of leaves, twigs and buds and where denser growth is available for cover. This mix is sometimes produced by landscaping activities, which can attract deer into neighborhoods. Where deer have lost their fear of humans, they may prefer the plantings found in yards and become serious pests. Most deer feed during dusk and dawn. They may wander a half mile or more from resting areas to search for food. In the more northern areas, deer often gather in areas of dense cover for winter protection. Deer breed during late fall and bear young about 202 days later; peak numbers of births occur during May and June. Twin births are most common. The young fawns develop rapidly and some does may be able to reproduce when only six months old. Reproduction of deer is very dependent on the amount of available food. Deer are long-lived, with some individuals living almost twenty years. Hunting is the most important factor in deer life expectancy in most areas.

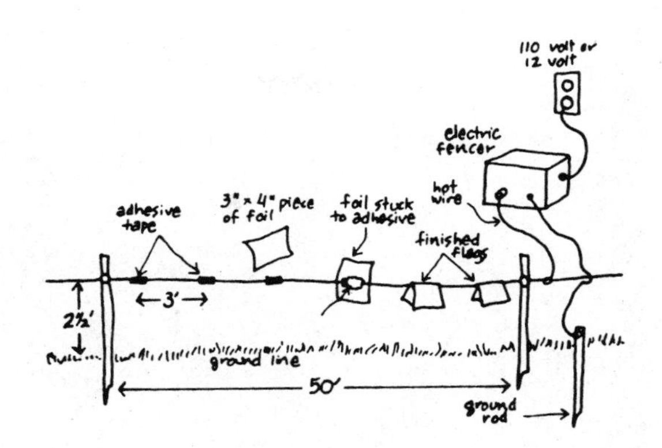


*Minnesota electric deer fence*

**Legal Status:** Deer are protected animals and are only to be killed during specified hunting periods by licensed hunters. Deer hunting within many communities is further restricted by local ordinances. In some cases, special permits may be issued by state wildlife agencies to farmers with specific deer problems.

**Mechanical Controls (Exclusion):** In areas where deer damage is severe, fencing provides the only satisfactory control. However, deer are excellent jumpers that can easily get over a typical 6-foot yard fence. They can also climb through or under fence openings less than a foot wide. Construction of a deer-proof fence is a substantial project. Among the various designs are the following.

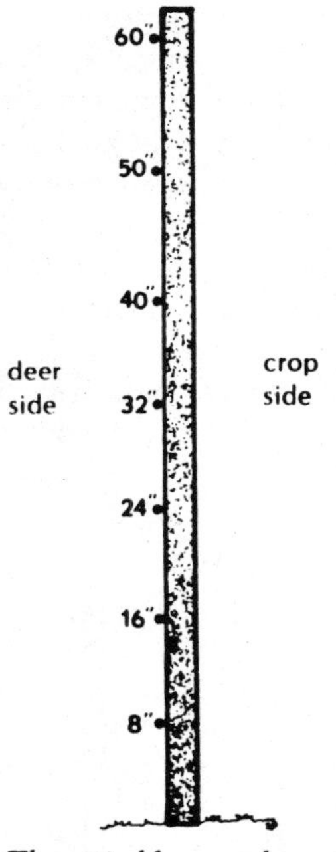


*The vertical fence can be constructed with five or seven wires. The spacing of the wires is critical to the effectiveness of the fence.*

*Basic deer fence.* A properly constructed fence built at least 8 feet tall will exclude deer. A thick wire mesh is best for this purpose and highly durable. Lighter mesh materials (chicken wire, plastic netting) can also be used, but these degrade rapidly. Unless well marked, the light mesh may not be seen by the deer, which may then accidentally push through. The fence should reach to the soil line to prevent deer from crawling under it. Solid wood or brick fences, through which the deer cannot see, may be somewhat shorter (6 to 8 feet) and still deter deer from jumping into the yard.

*Slant fence.* An outward-slanting fence can be more simply and cheaply constructed than an 8-foot vertical fence. For example, a 7-foot fence angled at 45° can usually deter deer. Deer will tend to approach and travel under the fence, then be unable to jump through it. However, the area under the fence should be kept mowed to encourage the deer to move under it.

*Electric fence.* A variety of effective electric fence designs have been proposed for ex-

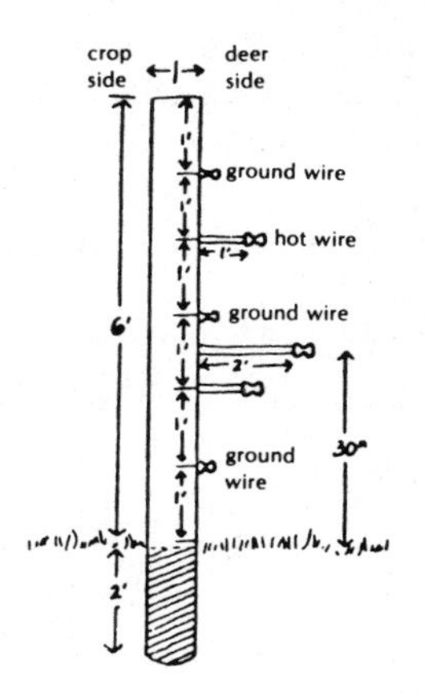


*An electric deer fence of traditional design*

*Deer fence*

cluding deer. Although not generally appropriate for most yards, these are much cheaper and easier to construct than fences built to physically exclude the deer. Deer will generally try to go under or through a fence, even if they could easily jump over it. Most electric fence designs encourage this habit so that the deer will touch the fence. After this "shocking experience," deer will often learn to avoid the fence and stay several feet away. Since they usually jump from a point very close to the fence, their avoidance of the fence vicinity reduces their inclination to jump relatively low fences. A typical electric fence design would be about 5 feet in height with wires placed at 8- to 12-inch spacings. Usually wire spacing is shorter at the lower end of the fence. Alternatively, two shorter electrified fences, spaced 3 feet apart, can be used to exclude deer.

*Repellents:* Many different repellents have been attempted for deterring deer. Some of these are contact repellents, which act by taste. Contact deer repellents include hot pepper sauce (commercially available as well as home brews) and thiram (a fungicide and animal repellent). An important limitation of these contact repellents is their inability to protect the favored new growth that emerges after treatment. To make some of the less residual sprays, such as hot pepper, last longer during rainy periods, it is suggested to mix them with additives that reduce evaporation (e.g., Vapor-Guard, Wilt-Pruf). The commercially available contact repellents cannot be applied directly to food crops.

Area repellents protect plants by odor, and dozens of materials have been suggested. For example, human hair or blood meal are often suggested as repellents, although the general consensus of the deer-plagued is that these are marginally effective at best. Bags of deodorant soap or the manure and urine of large cats (provided by your

Courtesy of Prevention and Control of Wildlife Damage

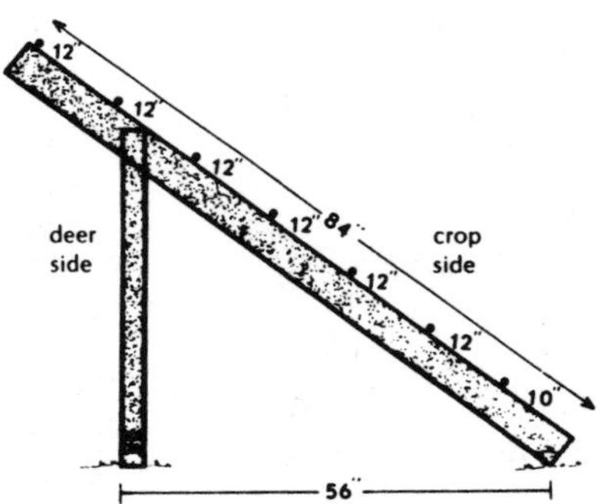


*An 8-foot standard hog wire fence, using 12-foot posts, can exclude deer (left). A slanted deer fence can be constructed as an electric or nonelectric fence (right).*

obliging neighborhood cougar) or coyotes are more widely recognized as having fairly good repellent activity. Some commercial repellents appear to be the most promising. These include fermented egg solids (Deer-Away, MGK-BGR, Big Game Repellent) or ammonium soaps produced from certain fatty acids (Hinder). The latter is one of the few repellents that can be applied directly to plants that are to be eaten. A highly effective homemade repellent is a mixture of eggs blended with water (a 1:4 egg/water ratio is one mixture), sprayed on plants. Dogs kept around garden areas will repel deer. However, unrestricted dogs are often not allowed in many residential areas, since they also cause severe damage to livestock and other wildlife.

# *Birds*

**Damage:** Several types of birds feed on ripening fruits. Cherries and berries are particularly favored. The fruits may be entirely eaten or merely pecked at, allowing growth of rotting organisms. Birds also feed on seeds. Although not often a problem in home gardens, sunflowers and other garden plants may be eaten.

*Bird netting*

**Life History and Habits:** Among the most damaging birds to fruit crops are robins, starlings and the various blackbirds. All of these commonly nest in and around yards and gardens. Although more protein-rich foods (worms, insects) are usually eaten while raising young in late spring, many birds will switch to ripening fruit as it becomes available. Late in the season, flocking birds may become particularly abundant and damaging to ripening crops.

**Legal Status:** With the exception of the starling, pigeon and English sparrow (the latter two are not garden pests), all birds commonly found around yards are protected by the Federal Migratory Bird Treaty Act and should not be trapped or harmed. Where serious damage is threatened, local wildlife agencies may allow limited trapping.

**Cultural Controls:** Injury to fruit crops may be reduced if other fruiting plants occur in the area when the garden fruits are maturing. Among the ornamental plants that can provide alternative foods for fruit-feeding birds are elderberry, mulberry, hawthorn and mountain ash.

**Mechanical Controls:** Several bird-repellent devices are sold through garden catalogs, including "scare-eyes," owl replicas and vibrating wires. Typically these provide only short-term protection because the birds readily become adapted to their presence. The most effective control is netting to exclude birds from fruit trees and berries. The netting must completely enclose the plant, since the birds will easily find openings. It should also extend a few inches beyond the plant to prevent birds perched on the netting from reaching fruit. Mesh bird netting, as well as cheesecloth or shade cloth, can be used.

APPENDIX I

# Sources of Biological Control Organisms

A wide variety of beneficial organisms are offered for sale by several suppliers. Numbers supplied under "Sources" refer to listings under "Suppliers of Biological Control Organisms" below. (Descriptions of these "bugs for hire" are also included in more detail in Chapter 2.)

## Biological Control Organisms

**Lady beetles**—Lady beetles (ladybugs) are one of the most common and beneficial kinds of insects found in the yard and garden. Naturally occurring lady beetles can provide adequate control of a wide range of insect pests, including aphids, scales, small caterpillars and beetle larvae. Purchased lady beetles are collected from aggregation sites where they occur in tremendous numbers during some periods of the year. They are marginally reliable since they typically disperse from the release site.

Sources: 2, 3, 4, 5, 9, 17, 18, 20, 23, 24, 25, 27, 29, 30, 31, 32, 33, 34, 35, 36, 38, 41.

**Green lacewings**—Various species of green lacewings are commonly found in the West, feeding on a wide variety of insect pests. Two species (*Chrysoperla carnea*, *Chrysoperla rufilabris*) are also available from commercial suppliers. Green lacewings are sold as eggs, usually in lots of several thousand, the eggs being shipped in a packing material such as rice hulls. Distribution of these eggs in a planting can provide some control of pest insects as the developing lacewing larvae feed and develop. Because of the wide host range of green lacewings and because the larvae will not disperse out of the planting, green lacewings are probably the best general purpose predator. After lacewings reach the winged adult stage, they can be expected to disperse from an area unless high numbers of aphids and similar plant pests remain. The larvae are cannibalistic and will eat each other if left together too long in the original container at room temperature.

Sources: 3, 4, 5, 7, 9, 10, 15, 23, 24, 27, 29, 30, 31, 33, 34, 35, 38, 41.

**Praying mantids**—Praying mantids are general predators of insects, and the presence of a few of these interesting insects can liven up the garden. However, they provide little benefit for control of pest insects in the garden, since their feeding preferences are too general and they are rarely very abundant. Praying mantids are one of the most widely available insect predators. They are sold as egg masses, which hatch in late spring. In much of the West, purchased species of praying mantids do not survive winter conditions, but they can become permanently established in some of the milder areas.

Sources: 3, 4, 5, 9, 17, 19, 24, 25, 30, 31, 34, 36, 41.

**Aphid predator midge**—A predator of aphids, the cecidomyid midge *Aphidoletes aphidimyza*, has shown promise in greenhouse trials in recent years. The greatest effectiveness has been with potato aphids, but the common green peach aphid is also a host for the midge. Although more efficient at higher temperatures, the aphid predator midge can be effective even if average temperatures drop into the high 60s. However, during winter, supplemental lighting must be provided to maintain a minimum of 12 hours of daylight or the predators become dormant. Use of aphid predator midges has been quite limited and must still be considered experimental. Availability is also a little erratic but is likely to improve with increasing demand. *Aphidoletes aphidimyza* is native to the region and occasionally occur naturally in gardens and greenhouses.

Sources: 2, 3, 8, 21, 24, 35, 38, 40.

**Predatory mites**—At least four species of mites that feed on plant-feeding spider mites are available commercially. Also, some strains of these predatory mites are resistant to certain insecticides. Most of these predatory mites do not occur naturally in the region, although native species do exist.

Sources: 2, 3, 5, 8, 14, 21, 24, 25, 26, 27, 28, 30, 31, 32, 33, 34, 36, 38.

**Spined soldier bug**—Spined soldier bugs are a native species of stink bug that is predatory on many types of caterpillars and leaf beetle larvae. Experimental work with the species is limited, although naturally occurring populations have often been reported as useful biological control agents.

Sources: 3, 29.

**Trichogramma wasps**—Several species of trichogramma wasps exist, all of which attack and kill various kinds of insect eggs. Insect larvae already hatched are not susceptible to trichogramma attack. Most of the eggs parasitized by trichogramma are from insects in the Lepidoptera order, which includes cutworms, codling moths, cabbageworms and armyworms. Although some trichogramma wasps are naturally present in the West, they are usually found in low numbers. Commercially available trichogramma wasps are often used as a form of a biological insecticide where they are expected to eliminate most of the developing eggs of pests shortly after release. High levels of control are not often achieved in practice, but the wasps may effectively supplement exisiting controls. Multiple releases of trichogramma wasps are recommended, since persistence of the parasites may be short-term. Several different species of trichogramma wasps are produced (e.g., *T. minutum*, *T. planteri*, *T. pretiosum*), and they have different habits. The more sophisticated suppliers will provide advice on which species is most appropriate for the intended crop and pest.

Sources: 3, 5, 9, 10, 11, 15, 17, 22, 23, 24, 27, 29, 30, 31, 33, 34, 36, 38, 39, 41.

**Whitefly parasite**—A small wasp, *Encarsia formosa*, attacks and develops within immature whitefly nymphs. Introduction of this parasitic wasp has proven useful for whitefly management in greenhouses where average temperatures remain at least 72°F. In cool greenhouses, this parasite will not be effective for whitefly control. The whitefly parasite is less attracted to yellow sticky boards than is the whitefly and is compatible with whitefly trapping for control.

Sources: 2, 3, 6, 8, 21, 24, 25, 27, 28, 29, 31, 32, 33, 35, 36, 38, 39, 41.

**Mexican bean beetle parasite**—*Pediobius foveolatus* is a small, parasitic wasp that develops within immature stages of the Mexican bean beetle. Several studies have demonstrated Mexican bean beetle suppression following artificial releases of the parasite at the appropriate time. Releases must be made annually, since the parasite does not successfully survive winters in the region.

Sources: 3, 33.

**Bacillus thuringiensis**—*Bacillus thuringiensis* (Bt) is a bacterial disease organism that has been formulated into a number of microbial insecticides. Most Bt products are highly effective for control of leaf-feeding Lepidoptera (webworms, cabbageworms, leafrollers, tussock moths, cutworms, etc.). Trade names include Dipel, Thuricide, Steward, MVP and Biological Worm Spray. More recently, a strain of the bacteria (*Bacillus thuringiensis* var. *israelensis)* has been developed for control of mosquito larvae in water. Claims of effectiveness against fungus gnat larvae are on some labels. This product is sold under various names, including Gnatrol, Mosquito Attack, Vectobac, Bactimos and Teknar. Experimental strains (var. *tenebrionis/ san diego*) with activity against elm leaf beetles and Colorado potato beetles have recently become available through limited outlets.

Sources (Lepidoptera larvae): 1, 3, 18, 23, 27, 30, 33, 35, plus most garden shops.

Sources (mosquito, fungus gnat larvae): 1, 3, 18, 23, 24, 27, 30, 31, 32, 36, plus some garden shops.

Sources (Colorado potato beetle, elm leaf beetle): 3, 30, 33.

**Parasitic (predatory) nematodes**—Insect-parasitic nematodes have recently become commercially available. Because they are sensitive to heat and drying, they have been successfully used primarily for soil insects, such as cutworms, white grubs and certain borers.

Sources (species of *Steinernema*): 3, 12, 13, 16, 17, 24, 25, 29, 30, 36.

Sources (species of *Heterorhabditis* ): 12, 15, 29, 36.

# *Suppliers of Biological Control Organisms*

The following list includes several suppliers of biological control organisms. Many of these businesses are producers of these organisms; others are merely distributors. Prices range considerably. When purchasing large numbers of these organisms, it is often desirable to make prior arrangements with a producer to get the most timely service at the most favorable price.

1. All Pest Control
   6030 Grenville Lane
   Lansing, MI 48910
   (517) 646-0038

2. Applied Bionomics
   8801 East Saanich Road
   Sidney, British Columbia
   Canada V8L 1H1
   (604) 656-2123

3. Arbico
   P.O. Box 4247
   Tuscon, AZ 85738
   (800) 305-2847

4. Bob Bauer
   26A Ford Road
   Howell, NJ 07731

5. Beneficial Insectary
   245 Oak Run Road
   Oak Run, CA 96069
   (916) 472-3715

6. Beneficial Insects, Inc.
   P.O. Box 40634
   Memphis, TN 38174
   (901) 276-6879

7. Beneficial Insects, Ltd.
   P.O. Box 154
   Batna, CA 95304

8. Better Yield Insects
   P.O. Box 3451
   Tecumseh Station
   Windsor, Ontario
   Canada N8N 3C4
   (519) 727-6108

9. Bio-Control Company
   Box 337
   Berry Creek, CA 95916
   (916) 589-5227

10. Bio-Fac
    Box 87
    Mathis, TX 78368
    (512) 547-3259

11. Biogenesis, Inc.
    Box 36
    Mathis, TX 78368

12. BioLogic
    418 Briar Lane
    Chambersburg, PA 17201
    (717) 263-2789

13. Biosys
    1057 East Meadow Circle
    Palo Alto, CA 94303
    (415) 856-9500

14. Biotactics, Inc.
    7765 Lakeside Drive
    Riverside, CA 92509
    (714) 685-7681

15. Bo-Biotrol
    54 South Bear Creek Drive
    P.O. Box 2495
    Merced, CA 95340
    (209) 722-4985

16. B.R. Supply Company
    P.O. Box 845
    Exeter, CA 93221

17. Burpee Seed Company
    300 Park Avenue
    Warminister, PA 19874

18. Farmers Seed and Nursery
    2207 East Oakland Avenue
    Bloomington, IL 61701

19. Foothill Agricultural Research
    510 West Chase Drive
    Corona, CA 91720
    (714) 371-0120

20. Fountain's Sierra Bug Company
    P.O. Box 114
    Rough and Ready, CA 95975

21. Gerharts
    P.O. Box 146
    North Ridgeville, Ohio 44039
    (216) 327-8056

22. Gothard Seed and Nursery
    P.O. Box 370
    Canutillo, TX 79835

23. Gurney Seed and Nursery
    Yankton, SD 57079

24. Harmony Farm Supply
    P.O. Box 451
    Grafton, CA 95444
    (707) 823-9125

25. Hydro-Gardens
    P.O. Box 9707
    Colorado Springs, CO 80932
    (719) 495-2266

26. Indiana Insectaries
    8302 Tequista Circle
    Indianapolis, IN 46236
    (317) 823-0432

27. Integrated Fertility Management
    333-B Ohme Garden Road
    Wenatchee, WA 98801
    (509) 662-3179

28. IPM Labs, Inc.
    Main Street
    Locke, NY 13092
    (315) 497-3129

29. M & R Durango, Inc.
    P.O. Box 886
    Bayfield, CO 81122
    (800) 526-4075

30. Natural Gardening Research Center
    P.O. Box 149
    Highway 46
    Sunman, IN 47041

31. Natural Pest Controls
    8864 Little Creek Drive
    Orangevale, CA 95662
    (916) 726-0855

32. Nature's Control
    P.O. Box 35
    2518 Stewart Avenue
    Medford, OR 97501
    (503) 899-8318

33. Necessary Trading Company
    Box 305
    Main Street
    New Castle, VA 24127

34. Organic Control, Inc.
    5132 Venice Boulevard
    Los Angeles, CA 90019

35. Organic Pest Management
    Box 55267
    Seattle, WA 98155
    (206) 367-0707

36. Peaceful Valley Farm Supply
    11173 Peaceful Valley Road
    Nevada City, CA 95959
    (916) 265-3276

37. Phero-Tech, Inc.
    1140 Clark Drive
    Vancouver, British Columbia
    Canada V5L 3K3
    (604) 255-7381

38. Rincon Vitova Insectaries, Inc.
    P.O. Box 95
    Oakview, CA 93022
    (805) 643-5407

39. Spaulding Laboratories
    760 Printz Road
    Arroyo Grande, CA 93420
    (805) 489-5946

40. Troy Hygrow Systems
    4096 Highway ES
    East Troy, WI 53120
    (414) 642-5928

41. Unique Insect Control
    P.O. Box 15376
    Sacramento, CA 95851

APPENDIX II

# Characteristics of Common Garden Pesticides

A large number of different pesticides are in use in agriculture. A considerably smaller number of these are registered for use around yards and gardens. These yard and garden pesticides have a wide range of uses, chemistries and biological effects. A discussion of all the different pesticides used in yards and gardens is beyond the scope of this publication, but a brief summary is included.

## Garden Pesticides

This section is organized using the common names of *active ingredients*, arranged in alphabetical order. Each active ingredient is discussed as listed below.

**Common Name:** The accepted common name of the active ingredient of the pesticide. (For some older pesticides, this is the same as the trade name.)

**Trade Names:** The names of products sold containing the active ingredient.

**Scientific Name:** The complete chemical name, included on the label in small print as an active ingredient.

**Pesticide Class:** The chemical class of the active ingredient, if any.

**Toxicity:** Primarily the acute toxicity of the compound, based on laboratory animal studies testing effects of exposure following ingestion or skin exposure. (Toxicity and hazard are also reflected on label instructions. Highly toxic or hazardous chemicals have the signal words *Danger-Poison*; moderately toxic chemicals have the signal word *Warning*; lower toxicity chemicals use the signal word *Caution*.)

**Uses:** Uses of the product in formulations sold for yard and garden use. (Agricultural or commercial uses of the same active ingredient may be different.)

**Notes:** General comments as to special properties or cautions associated with the pesticide.

## Acephate (Orthene)

**Trade Names:** Orthene, Isotox (various formulas), Orthenex

**Scientific Name:** O,S-Dimethyl acetyl phosphoramidothioate

**Pesticide Class:** Organophosphate insecticide

**Toxicity:** Moderate to low, although some breakdown products are highly toxic.

**Uses:** Control of aphids, caterpillars, thrips and other insects on ornamentals. Control of sod webworms in turf. Grasshopper control in noncrop areas. (*Note: Homeowner formulations are not labeled for use on food crops*. Certain commercial formulations are labeled for use on head lettuce, beans and some cole crops.)

**Notes:** Acephate has systemic activity when applied to leaves, making it highly effective against concealed insects such as leafminers and leaf-curling aphids. When applied to the soil, it has some systemic activity but is not currently registered for this use. The environmental persistence of acephate and its breakdown products is longer than with most insecticides, with preharvest intervals of 7 to 21 days for vegetables. *Note: Acephate is not labeled for any garden food crop (fruit, vegetable, herb) uses*. Lack of registration on food crops has caused frequent problems with accidental drift onto gardens. Acephate may injure poplar, cottonwood, willow, American elm, redbud and sugar maple trees. A primary breakdown product of acephate, methamidophos, is highly toxic to mammals.

## Bacillus Thuringiensis (Bt)

**Trade/Scientific Names:** *kurstaki* strain: Dipel, Caterpillar Attack, Sod Webworm Attack, Vegetable Garden Attack, Thuricide, MVP, Steward; *israelensis* strain: Mosquito Attack, Vectobac, Gnatrol, Bactimos, Teknar; *san diego* strain: M-One, M-Trak; *tenebrionis* strain: Foil, Trident

**Pesticide Class:** Microbial insecticide

**Toxicity:** Considered essentially nontoxic and can be used on most food crops up to harvest.

**Uses:** Control of leaf-feeding caterpillars on vegetables and fruit (*kurstaki* strain). Control of Colorado potato beetles and elm leaf beetles (*san diego* and *tenebrionis* strains). *Israelensis* strain used for control of mosquitoes and blackflies in water and fungus gnats in greenhouses.

**Notes:** A naturally occurring bacterial disease, *Bacillus thuringiensis* is produced by several manufacturers for use as a microbial insecticide. It must be eaten to be effective (stomach poison) and is highly selective in its activity. (Selectivity differs according to various strains.) Affected insects stop feeding within hours after feeding but may not die for 2 to 3 days. Persistence on foliage less than one week if exposed to sunlight. Its selective activity allows it to conserve the natural enemies of insects and integrate with existing biological controls. Bt is more perishable than most pesticides but will remain effective for at least two years if kept dry and out of direct sunlight.

*Tiger swallowtail visiting zinnia*

## Benomyl

**Trade Names:** Benlate, Tersan 1991

**Scientific Name:** Methyl 1-(butylcarbamoyl)-2-benzimidazole carbamate

**Pesticide Class:** Systemic fungicide

**Toxicity:** Low mammalian toxicity

**Uses:** Lawn and garden formulations include some vegetable and fruit crop uses. It is more commonly used in yards to control powdery mildew on flowers and woolly ornamentals or patch diseases on turf.

**Notes:** Benomyl is registered for a wide variety of agricultural uses on fruit, vegetables and nut crops. It also has registered uses on ornamental plants and turf grass. However, formulations sold for garden use are not commonly available. Benomyl has some systemic activity that can allow control of existing disease (eradicant). It is highly toxic to earthworms. There have been some problems with benomyl formulations being contaminated by other pesticides. As a result, the availability of benomyl may be limited.

## Bordeaux Mixture

**Trade Names:** Bordeaux mixture, Bor-Dox, Bordeaux powder

**Scientific Name:** A mixture of hydrated lime and copper sulfate

**Pesticide Class:** Inorganic fungicide

**Toxicity:** Low mammalian toxicity but may be irritating to skin

**Uses:** Primarily used to control fungal diseases on fruits, particularly in combination with dormant oil sprays. Registered crops include tomatoes, strawberries, apples, peaches and grapes.

**Notes:** Bordeaux mixture is one of the oldest known fungicides, discovered over 100 years ago. In cool, damp weather it may cause plant injury. It has a long persistence but acts only as a preventative fungicide, so it must be applied before infection. It is also repellent and insecticidal to some insect species. Young leaves of apple and peach trees may be injured by sprays.

## Captan

**Trade Names:** Captan, Orthocide

**Scientific Name:** N-trichloromethylthio-4-cyclohexene-1,2-dicarboximide

**Pesticide Class:** Heterocyclic nitrogen fungicide

**Toxicity:** Acute toxicity is very low, but dusts may be irritating. However, recent studies indicate captan may be a carcinogen, which has raised concerns and led to restriction on use.

**Uses:** Commonly available alone or in garden fruit/vegetable mixtures combined with insecticides. Captan is also a common seed treatment used to prevent seedling diseases.

**Notes:** Because of safety concerns, there have been recent restrictions on label use instructions. Gloves and other protective equipment, as indicated in the label instructions, should always be used when handling captan. This includes planting captan-treated seed. Captan is still used as an active ingredient in many general vegetable and fruit sprays as a nonsystemic, protectant fungicide, but some newer labels restrict handling treated plants for 4 days after application. Captan is incompatible for use with oil sprays or strong alkalies, such as lime sulfur.

## Carbaryl (Sevin)

**Trade Names:** Sevin, Sevimol, Savit, Tree and Ornamental Sprays, many garden dusts, baits

**Scientific Name:** 1-Naphthyl N-methyl carbamate

**Pesticide Class:** Carbamate insecticide

**Toxicity:** Low toxicity to mammals and birds. Highly toxic to honeybees and earthworms.

**Uses:** Carbaryl has registrations on most vegetables and fruit crops as well as flowers, turf and shade trees. Carbaryl is generally effective against most chewing insects, such as caterpillars, sawflies, beetles, grasshoppers and earwigs. It has some activity against slugs and snails.

**Notes:** Carbaryl is mostly active as a stomach poison. It is generally ineffective against most sucking insects and may aggravate problems with certain aphids and spider mites. Not systemic. The environmental persistence is fairly short, with typical preharvest intervals of 1 to 3 days (14 days on leaf lettuce). Carbaryl can be very hazardous to honeybees and should not be applied to blooming flowers or weeds. Carbaryl can cause excessive blossom shed of apple trees and is used to thin them. It can injure certain vines (Virginia creeper, clematis, etc.).

## Chlorothalonil

**Trade Names:** Daconil 2787, Bravo

**Scientific Name:** Tetrachloroisophalonitrile

**Pesticide Class:** Unclassified fungicide

**Toxicity:** Very low toxicity to mammals but can cause eye irritation

**Uses:** Broadly labeled for use on vegetables, fruits, turf and ornamentals, chlorothalonil controls a wide variety of leaf-spotting fungi, some rusts and Botrytis.

**Notes:** Chlorothalonil is the most widely used fungicide in commercial vegetable production and is commonly available in yard and garden products. Its use is likely to increase with recent restrictions on the EBDC fungicides, such as maneb. Chlorothalonil is not systemic in its action and is most effective when applied as a protectant before disease develops. Most food crops have preharvest intervals of 0 to 14 days.

## Coppers, Fixed

**Trade Names:** Microcop, Tri-basic copper, Kocide 101, included in Fertilome Triple Action and other garden mixtures

**Scientific Name:** Metallic copper, copper oleate

**Pesticide Class:** Inorganic fungicide, bactericide

**Toxicity:** Low mammalian toxicity and exempted from food crop tolerance requirements.

**Uses:** Controls a wide variety of fungi and bacteria on fruits and vegetables. Has some bactericidal activity.

**Notes:** Copper-based pesticides can cause plant injury, such as burning of leaf edges, and should not be used on cabbage family plants. Should not be mixed with many other pesticides or plant injury may result. Federal agencies have exempted copper compounds from tolerance requirements, so they can be used up to harvest. The compounds are often accepted for use in "certified organic" production.

## DCPA(Dacthal )

**Trade Names:** Garden Weed Preventer, Dacthal

**Scientific Name:** Dimethyl tetrachlorterephthalate

**Pesticide Class:** Chlorinated benzoic acid herbicide

**Toxicity:** DCPA is considered to have very low acute toxicity to mammals and is nonirritating. However, it is an older pesticide and several "data gaps" are noted in a recent registration review by the Environmental Protection Agency. Some contaminants (notably hexachlorobenzene) are produced as by-products during DCPA manufacture and are under safety review.

**Uses:** DCPA is used to control germinating weeds in flower gardens and around some fruit and vegetable crops.

**Notes:** DCPA acts by inhibiting cell division of root tips. It has little effect on established plants. It has occurred as a groundwater contaminant in several surveys. It should not be used in areas (e.g., high water table, sandy soils) where groundwater contamination is likely. Future uses may be restricted because of environmental concerns.

## Diazinon

**Trade Names:** Diazinon, Spectracide, Knox-out PT-265

**Scientific Name:** O,O-Diethyl O-(2-isopropyl-4-methyl-7-pyrimidinyl) phosphorothioate

**Pesticide Class:** Organophosphate insecticide

**Toxicity:** Moderate toxicity to mammals; extremely toxic to birds

**Uses:** Diazinon is most widely used for control of soil insects on turf and many vegetables. It has broad spectrum activity as a foliar spray against exposed sucking and chewing insects on fruits, ornamentals and some vegetables.

**Notes:** Diazinon is a contact and stomach poison without systemic activity in plants. Persistence in the environment moderately long with typical preharvest intervals of 7 days for vegetables (longer on root crops) and 10 days for fruit. Toxicity of diazinon to birds is approximately 100 times greater than to mammals, and its uses have recently been restricted or canceled for many sites where bird injury is likely. Because of costs associated with reregistering diazinon, the primary manufacturer has decided to cancel many minor uses of this insecticide.

## Dicofol

**Trade Names:** Kelthane, Red Spider and Mite spray

**Scientific Name:** 4,4-Dichlor-alpha-trichloromethylbenzhydrol/1, 1-bis(chlorophenyl)-2,2,2-trichloroethanol

**Pesticide Class:** Chlorinated hydrocarbon insecticide

**Toxicity:** Moderate to low

**Uses:** Spider mite control on ornamental plants, some vegetables and fruits.

**Notes:** Dicofol is a selective miticide that has little effect on most insects. It has moderate persistence on plants, with preharvest intervals on registered food crops ranging from 2 to 30 days. Contamination of dicofol with residues of the chemically similar insecticide DDT has been a recognized problem in recent years, which has limited supplies. During the mid-1980s, the Environmental Protection Agency sharply restricted use of dicofol in the United States until manufacturing techniques eliminated these residues. Current manufacture appears to have met these restrictions, but garden formulations of dicofol remain generally unavailable. Nicotine, hexakis (Vendex) and acephate (Orthene) are the ingredients that have been most commonly used to replace dicofol.

## Disulfoton

**Trade Names:** DiSyston, Systemic Insecticide Granules, Systemic Insecticide Bug Dart

**Scientific Name:** O,O-diethyl S-((2-(ethylthio) ethyl)) phosphorodithioate

**Pesticide Class:** Organophosphate insecticide

**Toxicity:** Disulfoton is very highly toxic to mammals and the most toxic insecticide available to homeowners.

**Uses:** Disulfoton in gardens is sold as a soil-applied granule or plant stake, often formulated with fertilizer. Almost all products containing disulfoton limit use to ornamental plants, including roses. However, some disulfoton granules (1 to 2 percent) continue to be sold for control of some vegetable insects as well as for ornamental plants.

**Notes:** Disulfoton is a systemic insecticide that is picked up by plant roots and translocated to leaves and other actively growing plant tissues. It is generally effective against most leaf-feeding insects and mites but less effective against borers and scale insects. Homeowner formulations are in the form of low-percentage granules or stakes that reduce hazard. *Note: Despite safeguards, this product should be used with extra caution because of the high toxicity of the active ingredient.* Disulfoton is highly water-soluble, so use with house plants should be avoided if watering allows leaching out of pots.

## Egg Solids, Putrescent Whole

**Trade Names;** Deer-Away, Big Game Repellent, MGK BGR

**Pesticide Class:** Unclassified, produced from putrefying egg solids

**Toxicity:** Very low

**Uses:** Animal repellent for ornamental trees and shrubs, nonbearing fruit trees and bearing fruit trees when fruit is not present.

**Notes:** This group of animal repellents is sold as a two-part concentrate designed to be mixed and sprayed. To humans it has a slight odor of fermented eggs but is a strong odor repellent to deer and other animals. It cannot be used on fruit trees when flowers or harvestable fruit are on the tree. Homemade mixtures of eggs diluted in water (1:4 egg/water dilution) have also shown repellency to mule deer.

## Endosulfan

**Trade Names:** Thiodan, Endocide, various garden vegetable sprays

**Scientific Name:** 6,7,8,9,10,10-Hexachloro-1,5,5a,6,9,9a-hexahydro-6, 9-methano,2,4,3-benzodioxathiepin-3-benzodioxathiepin-3-oxide

**Pesticide Class:** Chlorinated hydrocarbon insecticide

**Toxicity:** Moderate to high to mammals

**Uses:** Agricultural uses of endosulfan are widely labeled for control of chewing and sucking insects on most vegetables (except root crops). Use on ornamental plants is also allowed with some formulations.

**Notes:** Endosulfan has broad spectrum activity and has been particularly effective against certain aphids and thrips. Persistence is not usually long, with typical preharvest intervals of approximately 3 days. However, residues concentrate in underground parts of many plants, so root crops are excluded from labeling. Endosulfan is not generally available to homeowners.

## Fluazifop-Butyl

**Trade Names:** Grass-B-Gon, Over-The-Top-Grass Killer, Fusilade

**Scientific Name:** RS butyl 2-[4-(5-trifluoromethyl-2-pyridol oxy) phenoxy] propinoate

**Pesticide Class:** Postemergence herbicide

**Toxicity:** Low toxicity to mammals but irritating

**Uses:** Control of grass around ornamental plants. Some agricultural formulations, but no yard and garden formulations allow use for grass control around strawberries and raspberries.

**Notes:** Fluazifop-butyl is a selective herbicide affecting grasses that acts systemically. Because of its selective action, it can be used over the top of broadleafed plants. It is most effective when applied against young, actively growing plants. Typical symptoms following treatment include slowing of growth, yellowing and death of plants after one to four weeks. Plants should not be cut or tilled after treatment. Fluazifop-butyl is water-soluble and should not be applied in light soils where contamination of groundwater is likely.

## Glyphosate

**Trade Names:** Roundup, Kleenup, Knock-out

**Scientific Name:** N-(phosphonomethyl) glycine

**Pesticide Class:** Nonclassified herbicide

**Toxicity:** Relatively nontoxic to mammals, birds and fish

**Uses:** Glyphosate is a nonselective, broad-spectrum systemic herbicide. It is used in spot treatment or directed treatment in noncrop areas.

**Notes:** The systemic activity of glyphosate allows control of several perennial broadleaf and grassy weeds. It does not have soil activity (e.g., root uptake) and rapidly degrades in soils. Drift onto desirable plants is a common problem and can injure woody plants if immature bark or foliage is treated. Glyphosate is not volatile, but droplets can easily drift by wind action during application.

## Lime Sulfur (Calcium polysulfides)

**Trade Names:** Polysul, Dormant spray, Dormant disease control

**Scientific Name:** Calcium polysulfides

**Pesticide Class:** Inorganic fungicide/miticide

**Toxicity:** Highly corrosive to eyes and harmful if swallowed or absorbed through skin. If mixed with acids, can cause release of toxic hydrogen sulfide gas.

**Uses:** Lime sulfur is primarily used as a dormant application to control fungal diseases on fruit trees, raspberries and roses. Some insects and mites that overwinter on the plants are also controlled by these applications. Lime sulfur is sometimes used as foliar or summer applications, although this use is limited because of potential to cause plant injury.

**Notes:** One of the oldest fungicides, lime sulfur is produced by boiling sulfur and lime in water. Formerly used widely, current availability is quite limited. It gives some control of mites and scale insects that may overwinter on plants. Its corrosive activity can damage painted surfaces. To reduce chance of injury to any crop, do not apply if temperatures are expected to exceed 80°F during the following 24 hours. Lime sulfur has a strong odor, similar to that of rotten eggs. Its use is often allowed in "certified organic" production.

## Malathion

**Trade Name:** Malathion, found in many combination pesticides

**Scientific Name:** O, O-Dimethyl dithiophosphate of diethyl mercaptosuccinate/O, O-Dimethyl phosphorodithioate of diethyl mercaptosuccinate

**Pesticide Class:** Organophosphate insecticide

**Toxicity:** Low to moderate to mammals; higher to birds

**Uses:** Broadly labeled for insect control on vegetables, fruits and ornamentals. Primary target insects are aphids and mites, but has broad spectrum activity against most chewing insects. Still frequently used to kill adult mosquitoes.

**Notes:** Malathion has more labeled uses than any insecticide other than carbaryl (Sevin). Persistence is short, with typical preharvest intervals of 1 to 3 days. Despite label claims, effectiveness against most spider mites is marginal, since resistant strains are common. Because of existing gaps in safety data, registrations of malathion are currently under review by the Environmental Protection Agency. Registered uses of malathion will be more limited in the near future. Malathion has a strong, unpleasant odor. It is not systemic. Liquid formulations may injure aspen, viburnum and some varieties of juniper. Some formulations (notably ultra-low-volume formulations sprayed by aircraft) can injure paint finishes.

## Maneb

**Trade Name:** Maneb, found in a large number of general-purpose garden protectant sprays

**Scientific Name:** Manganese ethylene bisdithiocarbamate

**Pesticide Class:** Dithiocarbamate fungicide (EBDC fungicide)

**Toxicity:** Low mammalian toxicity. However, there are serious health concerns surrounding breakdown products of maneb (see Notes below).

**Uses:** Control of blight, anthracnose and leaf-spotting fungi on vegetables, fruits, ornamentals and turf.

**Notes:** Availability to homeowners is limited and manufacturers have cut production. Most vegetables have a preharvest interval requirement of 4 to 10 days. Maneb has protectant fungicide activity only, so it cannot control existing infections. It is not systemic in the plant. Maneb, and other ethylene bisdithiocarbamate (EBDC) fungicides, have recently come under serious public review. This is due to potential health concerns about breakdown products that occur during natural decomposition and, particularly, after cooking. Some of these (e.g., ethylene thiourea) are considered to be possible carcinogens based on laboratory studies.

## Metaldehyde

**Trade Names:** Bug-geta Snail and Slug Bait, Slug Bait, Deadline, Corry's Snail and Slug Death

**Pesticide Class:** Unclassified, molluscicide

**Toxicity:** Low toxicity to humans and other mammals. Accidental poisoning of pets usually results in a temporary stupefaction, resembling drunkenness.

**Uses:** Slug control around food crops and vegetables.

**Notes:** Metaldehyde is the most commonly used slug bait. It kills slugs by causing irritation and dehydration. Baits become attractive after wetting, but effectiveness is reduced if wet conditions persist after application. They break down rapidly in sunlight and should be applied late in the day. Metaldehyde baits are generally not fed upon by birds or pets but have caused poisoning of dogs.

## Methoxychlor

**Trade Name:** Methoxychlor, almost always found in garden vegetable or fruit spray mixtures combined with malathion and a fungicide.

**Scientific Name:** 2,2-bis (p-methoxphenol) 1,1,1-trichloroethane

**Pesticide Class:** Chlorinated hydrocarbon insecticide

**Toxicity:** Low

**Uses:** Methoxychlor is very rarely sold alone for yard and garden use, but many fruit spray mixtures contain methoxychlor-malathion combinations. Most current use is for control of elm bark beetles (which transmit Dutch elm disease) and livestock pests.

**Notes:** Although a chlorinated hydrocarbon, methoxychlor is short-lived and not as fat-soluble in animal tissues as other DDT-related chemicals. Activity is primarily against beetles, although other chewing insects are often controlled. Preharvest intervals on certain fruits and vegetables are typically 7 to 21 days.

## Neem Extract

**Trade Names:** Margosan-O, Neemisis

**Scientific Name:** Extract of the neem tree; primary active ingredient azadirachtin

**Pesticide Class:** Botanical insecticide (derived from seed pods of the tree *Azadirachta indica*)

**Toxicity:** Neem extracts appear to have very low toxicity to mammals and actually have several pharmaceutical uses (e.g., ingredient in certain toothpaste products).

**Uses:** Insecticides containing extracts of neem are just recently being marketed for control of insect pests in North America. Original registrations were for control of leafminers and whiteflies on greenhouse crops. However, neem extracts appear to be particularly effective for control of various chewing insects, such as caterpillars and leaf beetles. At present, all neem-based insecticides are only registered for use on ornamental plants. There are no food crop uses.

**Notes:** Neem can affect insects both as a feeding deterrent and as a growth regulator of their development. Typically, neem extracts have fairly slow acting effects, since sensitive insects are killed as they attempt to molt or at some other change in stage. Treated leaf surfaces are often avoided.

## Nicotine Sulfate

**Trade Names:** Black Leaf 40, Nicotine Sulfate, Tender Leaf Plant Insect Spray, included as miticide in some general-purpose pesticide mixtures

**Scientific Name:** 3-(1-Methyl-2-pyrrolidyl) pyridine

**Pesticide Class:** Botanical insecticide

**Toxicity:** Nicotine is highly toxic to warm-blooded animals and is readily absorbed through the skin. It is the most toxic garden pesticide used in a liquid form. (Disulfoton toxicity is higher but is only formulated as granules or plant stakes.)

**Uses:** Control of spider mites, aphids and other insects on ornamentals, some vegetables and fruits. Current labels require a 7-day preharvest interval between application and harvest.

**Notes:** Nicotine is readily absorbed through the body covering of insects as well as through the spiracles. It is a nerve poison that excites the nerves in low concentrations, resulting in tremors, and causes paralysis and death in higher concentrations. Nicotine is more effective at higher temperatures, but breaks down more rapidly. Roses and some other ornamental plants are sensitive to nicotine sulfate injury. When using nicotine sulfate formulations, the addition of soaps, hydrated lime or other alkaline agents will increase activity of the spray mixture by helping to release free nicotine. Nicotine is incompatible in mixtures with several inorganic pesticides. Nicotine is rapidly broken down upon exposure to light, and toxic residues rarely remain after a few days following application. Future registrations will be limited during reregistration of this older pesticide.

## Pyrethrins, Pyrethrum

**Trade Names:** A variety of flying insect aerosols, some vegetable crop sprays, Natures Insecticidal Spray, Pyrenone, etc.; sometimes combined with other insecticides

**Scientific Name:** Pyrethrins

**Pesticide Class:** Botanical insecticide (derived from flowers of *Chrysanthemum cinaeriofolium*)

**Toxicity:** Pyrethrins have very low toxicity to most mammals, making them among the safest insecticides in use. Asthmatic-like reactions have been reported from individuals of broad allergic backgrounds. Some breeds of cats are sensitive to pyrethrins.

**Uses:** The fast degradation and safety of pyrethrins allow them to be used for more agricultural purposes than any other insecticide. Pyrethrins can be used legally in production of almost all food crops as well as on ornamentals. They are among the only insecticides labeled for use as a postharvest treatment on fruits and vegetables. Pyrethrins are also labeled for use in food handling establishments and other indoor areas. Pyrethrins, or related compounds (see tetramethrin, resmethrin, sumithrin), are extremely fast-acting, "knockdown" insecticides for flying insects and irritant flushing agents for cockroaches. Some formulations are available for vegetable insect control. Indoor fogging formulations are also available.

**Notes:** Pyrethrins are naturally derived insecticides extracted from flowers of the pyrethrum daisy, which thrives as a cultivated plant in many areas of the region. Pyrethrins have some unusual activities as insecticides. Most insects are highly susceptible to pyrethrins at quite low concentrations. Perhaps most striking is the rapid knockdown effect, which causes most flying insects to drop almost immediately upon exposure. Pyrethrins are also highly irritating to insects, allowing their use as a flushing agent. Because of high cost, pyrethrins are almost always formulated with the synergist piperonyl butoxide, which greatly increases insecticidal action. Pyrethrins rapidly break down in light, so persistence is less than 1 day in most circumstances. Pyrethrum (powdered formulations of the flower) or extracted pyrethrins can be used in "certified organic" production in many areas, if synergists are not included.

## Ro-Pel

**Trade Name:** Ro-Pel

**Scientific Name:** A mixture of benzyldiethyl [(2,6 xylyl carbomoyl) methyl] ammonium saccarhide and thymol

**Pesticide Class:** Unclassified animal repellent

**Toxicity:** Low

**Uses:** Animal feeding repellent for use on ornamental plants.

**Notes:** Ro-Pel is an extremely bitter material. Use is not permitted on edible plants, indoor plants or seeds of edible plants.

## Rotenone

**Trade Names:** Rotenone, sometimes found in mixtures with other garden pesticides such as pyrethrins (e.g., Red Arrow)

**Scientific Name:** Rotenone and other cube resins (rotenoids)

**Pesticide Class:** Botanical insecticide

**Toxicity:** Rotenone is moderately toxic to most mammals but exhibits a wide range of animal sensitivity. Some animals, notably swine, are very susceptible. Rotenone is an extremely potent fish poison and is widely used to poison "trash" fish during restocking projects.

**Uses:** Most use of rotenone is for control of various leaf-feeding caterpillars and beetles, such as cabbageworms and Colorado potato beetles, on vegetables. Some insects with sucking mouthparts, such as aphids, are also susceptible to rotenone. Thrips are also commonly listed as insects that can be controlled with rotenone. Rotenone has been used to control external parasites of cattle. Rotenone is fairly slow-acting, often requiring several days to kill most susceptible insects. However, insects stop feeding shortly after exposure.

**Notes:** Commercially produced rotenone is derived from either species of *Derris*, found in Malaysia and the East Indies, or species of *Lonchocarpus* (cube), from South America. Presently, almost all rotenone sold in the United States is from the latter area, primarily Peru. Rotenone is used most commonly as a dust prepared by grinding the plant roots or by extracting the rotenone with a solvent and coating dust particles. Rotenone degrades fairly rapidly upon exposure to light and air, so it is best applied during cloudy weather. Residues of rotenone on foliage typically lose their effectiveness within a week. Rotenone should not be mixed with alkaline materials or breakdown is greatly accelerated.

## Sabadilla

**Trade Names:** Sabadilla Dust, Red Devil Sabadilla Dust

**Scientific Name:** Ground seeds of a South and Central American plant, sabadilla (*Schoenocaulon officinale*), which contain veratrine alkaloids

**Pesticide Class:** Botanical insecticide

**Toxicity:** Generally considered to have low toxicity but highly irritating to mucous membranes of the nose and throat, inducing a sneezing reaction. Active ingredient is a powerful suppressor of blood pressure.

**Uses:** Used to control caterpillars, thrips and bugs on certain vegetables, including squash, cucumbers, melons, beans, turnips, mustard, collards, cabbage, peanuts and potatoes.

**Notes:** Since the 1940s and early 1950s, little research on control of insects has involved sabadilla dust. It is both a contact and a stomach poison and has shown promise against several of the true bugs (Hemiptera), such as squash bugs, chinch bugs, harlequin bugs and stink bugs. Effectiveness against leaf-feeding caterpillars, Mexican bean beetles and thrips has also been demonstrated. The toxins in sabadilla are rapidly destroyed by light, although extended storage in dry, dark conditions reportedly does not reduce insecticidal activity. Uses on food crops are poorly specified but appear to require that a 24-hour interval elapse between treatment and harvest on registered crops.

## Soaps, Ammonium

**Trade Name:** Hinder

**Scientific Name:** Ammonium soaps of higher fatty acids

**Pesticide Class:** Fatty acid salt (soap)

**Toxicity:** Very low

**Uses:** Deer and rabbit repellent on fruit trees, ornamental plants, vegetables and field crops.

**Notes:** Hinder is designed to be sprayed or painted on plants. It is the only animal repellent registered to be applied directly to food crops.

## Soaps, Herbicidal

**Trade Name:** Sharpshooter

**Scientific Name:** Potassium salts of saturated fatty acids

**Pesticide Class:** Fatty acid salt (soap)

**Toxicity:** Very low

**Uses:** A nonselective, contact herbicide with allowable uses around food crops as well as ornamental plantings.

**Notes:** Herbicidal soaps are nonselective and will injure foliage of desirable plants that are treated or receive drift. Activity of herbicidal soaps is most rapid during warm, dry weather. They can rapidly (within hours) burn foliage of many treated plants but do not move systemically to kill plant roots. Several grasses, including Kentucky bluegrass and quackgrass, do not appear to be very susceptible. Soaps leave a visible, but temporary, white residue on concrete or asphalt when used to kill weeds in cracks.

## Soaps, Insecticidal

**Trade Names:** Safers Insecticidal Soap, Insecticidal Soap

**Scientific Name:** Various potassium salts of fatty acids

**Pesticide Class:** Fatty acid salt (soap)

**Toxicity:** Essentially nontoxic to humans. Can damage some plants.

**Uses:** Control of small, soft-bodied insects and mites on trees, shrubs, flowers, vegetables and fruits. Also controls moss and algae in greenhouses.

**Notes:** High degree of safety to applicators has recently made use of insecticidal soaps more popular. Acts strictly as a contact insecticide with no residual activity so thorough application is essential. Reacts with minerals in hard water and will be more effective when used with soft water. Can be used on food crops. Insecticidal soaps work best if applied when environmental conditions favor slow drying. Several plants are injured by soap sprays, so always test on a small area before treatment.

Nasturtiums, sweet peas, gardenias, certain zinnias, euphorbias, horse chestnut, coleus, certain maples and mountain ash are among the plants that may be injured. Several liquid dishwashing detergents are almost equally effective for insect control. Insecticidal soaps are usually acceptable for use in "certified organic" production.

## Streptomycin

**Trade Names:** Streptomycin, Streptomycin sulfate

**Pesticide Class:** Antibiotic bactericide

**Toxicity:** Low toxicity to mammals but may cause allergic skin reaction. Toxic to fish.

**Uses:** Control of fire blight in apples and pears; crown gall control of roses.

**Notes:** Use directions do not allow application when fruit is present. Many bacteria have developed resistance to streptomycin where use is frequent. Rotation with alternative bactericides, such as copper-based products, is often recommended.

## Sulfur

**Trade Names:** Sulfur, Flowable Sulfur, Wettable Sulfur

**Scientific Name:** Sulfur

**Pesticide Class:** Inorganic fungicide and miticide

**Toxicity:** Essentially nontoxic but highly irritating to skin and eyes.

**Uses:** Sulfur is used primarily to control powdery mildew, rust, brown rot and other fungal diseases on fruits, vegetables, roses and ornamentals. It is also effective and labeled for use against spider and rust (eriophyid) mites on several crops. Sulfur dusts are highly effective for control of potato (tomato) psyllids.

**Notes:** Sulfur is the oldest pesticide, having a recorded use of over 2,000 years. It can cause plant injury, particularly if applied during very high temperatures. (Labels vary, suggesting treatments be avoided during temperatures in the range of 85°F to 95°F or above.). Sulfur reacts with oils to cause plant injury on many fruit crops. Labels recommend that sulfur not be applied for two to four weeks after oils have been applied. Several plants are sulfur-shy and can be injured by sulfur. These include cucurbits (melons, cucumbers, etc.), nut trees, viburnum, spinach and apricots. Since residual sulfur can affect processed fruits and vegetables (e.g., canned tomatoes, wine grapes), use should be discontinued for two or more weeks before processing. Due to its irritation properties, applications of sulfur currently require a 24-hour waiting period (reentry interval) before gardeners can return to treated areas of the yard and garden.

## Thiram

**Trade Name:** Thiram

**Scientific Name:** Tetramethyl thiuram disulfide

**Pesticide Class:** Organo-sulfur fungicide; animal feeding repellent

**Toxicity:** Thiram has moderately low toxicity to mammals but is irritating to skin. Thiram is chemically similar to the drug Antabuse, used in treatment of chronic alcoholism. Ingestion of alcohol after exposure to thiram will similarly produce sickness. Thiram is an older fungicide, and several safety studies are considered deficient based on current registration standards.

**Uses:** Fungicide seed protectant. Animal taste repellent for use on trees, shrubs and dormant fruit crops.

**Notes:** On fruit trees, thiram may be applied only during the dormant season. Thiram is generally used for a deer repellent, applied as a dilute spray. If rainfall is frequent, addition of a latex sticker or the specialty nursery protectant Wilt-Pruf will increase the retention of thiram on twigs. However, addition of wetting agents may produce severe plant injury.

## Trifluralin

**Trade Name:** Treflan

**Scientific Name:** a,a,a-Trifluoro-2,6-dinitro-N,N-dipropyl-p-toluidine

**Pesticide Class:** Dinitroaniline herbicide

**Toxicity:** Trifluralin has very low acute toxicity to mammals. However, some studies have raised concerns about possible carcinogenicity and mutagenicity, apparently because of impurities produced during manufacture. Trifluralin is highly toxic to earthworms and fish. It is considered nontoxic to most other animals.

**Uses:** A preemergence herbicide used to control newly germinated garden weeds and widely used in agriculture. Trifluralin can be used for weed control in and around seeded carrots, okra, beans and southern peas. Other permitted uses include transplanted (but not seeded) broccoli, tomatoes, Brussels sprouts, cabbage, cauliflower, peppers and potatoes. Trifluralin also may be used for weed control around many seeded and most transplanted garden flowers.

**Notes:** Trifluralin inhibits germination of seeds. Existing weeds are not controlled and must first be removed. Trifluralin is not water-soluble and does not move far in soil. Effective use requires that the herbicide be mixed with the upper soil during application. It can persist in soil for several months before decomposing.

## Triforine

**Trade Names:** Funginex, Orthenex (combined with acephate)

**Scientific Name:** {N,N'-1,4-piperazinediylbis [2,2,2,-trichloro-ethylidene]-bis-(formamide)}

**Pesticide Class:** Unclassified systemic fungicide

**Toxicity:** Active ingredient has low toxicity to mammals, but Funginex label indicates moderate to high hazard (*Warning*, *Danger* signal statements).

**Uses:** Control of black spot, powdery mildew and rust on roses and many other flowering plants.

**Notes:** Triforine moves systemically a short distance within the plant. It has eradicant and protectant activity.

## Zinc Phosphide

**Trade Names:** Mole and Gopher Bait

**Pesticide Class:** Inorganic rodenticide

**Toxicity:** Highly toxic to mammals, although most formulations are sold as baits with a low percentage of the active ingredient.

**Uses:** Used as a pelleted bait to control moles and other ground-burrowing rodents to protect bulbs and other plants.

**Notes:** Not for use in vegetable gardens or other areas where food plants may be contaminated. Dead animals should be collected and buried after treatment. Zinc phosphide can react with acids to form highly toxic phosphine gas. Many formulations are "Restricted Use Pesticides," which require licensing by users.

APPENDIX III

# Sources of Insect Traps and Other Pest Management Products

The following companies are involved in the manufacture or distribution of insect traps or other pest management supplies. Prices vary, and minimum orders are required by some distributors.

AgriSense
4230 West Swift Avenue
Suite 106
Fresno, CA 93722
(209) 276-7080
Manufacturer and distributor of several insect traps

All Pest Control
6030 Grenville Lane
Lansing, MI 48910
(517) 646-0038
Distributor of several lines of pheromone traps, visual traps and other pest management supplies

*Painted lady butterfly*

Consep Membranes
213 SW Columbia
P.O. Box 6059
Bend, OR 97708
(503) 388-3705
Producer of BioLure line of pheromone traps and CheckMate line of insect mating disruptants

Great Lakes IPM
10220 Church Road, NE
Vestaburg, MI 48891
(517) 268-5693
Distributor of several lines of pheromone traps, visual traps and other pest management supplies

Harmony Farm Supply
P.O. Box 451
Grafton, CA 95444
(707) 823-9125
Primarily a supplier of fertilizers, biological controls and other garden supplies but also distributes pheromone traps

Integrated Fertility Management
333-B Ohme Garden Road
Wenatchee, WA 98801
(509) 662-3179
Primarily a supplier of fertilizers, biological controls and other garden supplies but also distributes pheromone traps

Natural Gardening Research Center
P.O. Box 149
Highway 46
Sunman, IN 47041

Primarily a supplier of fertilizers, biological controls and other garden supplies but also distributes pheromone traps

Necessary Trading Company
Box 305
Main Street
New Castle, VA 24127

Primarily a supplier of fertilizers, biological controls and other garden supplies but also distributes pheromone traps

Olson Products, Inc.
P.O. Box 1043
Medina, OH 44258
(216) 723-3210

Producer of sticky tape for control of whiteflies and thrips in greenhouses

Peaceful Valley Farm Supply
11173 Peaceful Valley Road
Nevada City, CA 95959
(916) 265-3276

Primarily a supplier of fertilizers, biological controls and other garden supplies but also distributes pheromone traps

Pest Management Supply
P.O. Box 938
Amherst, MA 01004
(413) 253-3747

Distributor of several lines of pheromone traps, visual traps and other pest management supplies

Scentry, Inc
P.O. Box 426
Buckeye, AZ 85326
(602) 233-1772

Manufacturer of the Scentry line of pheromone traps

Trece, Inc.
P.O. Box 5267
Salinas, CA 93915
(408) 758-0205

Manufacturer of the Pherocon line of pheromone traps

APPENDIX IV

# Attracting Insectivorous Birds to the Yard and Garden

Many people wish to encourage birds to nest and feed in their yard and neighborhood because birds are a source of enjoyment. They may also wish to benefit from their insect-eating habits so that pest problems can be reduced. Many common species of birds can help control insect pests. Insectivorous (insect-feeding) birds that can be easily attracted to the yard include the house wren, American robin, blue jay, northern flicker and European starling.

The percentage of insects in the diet of birds varies greatly. Some birds feed almost exclusively on insects; others have vegetarian and seed-feeding habits. However, even many seed feeders feed high-protein insect foods to their young.

The nesting requirements of most birds are quite specific, but a few common species, such as robins, chickadees and house wrens, are fairly adaptable. These can be found in a variety of rural and urban habitats. Other birds may have more discriminating needs that will limit their ability to nest in a backyard or neighborhood. However, proper provision of the environmental needs of a bird species will help to increase its nesting incidence in an area.

## Nesting Needs of Birds

Like all other wildlife, birds have three basic requirements: adequate food, water and cover. This is true for nesting birds as well as for birds that are transients or seasonal residents. Favorable cover for birds provides nesting habitat, a means of escape from danger and protection from the weather. Weather cover is best achieved with clumped plantings of vegetation that buffer wind and rain. Conifers such as juniper, spruce or pine are ideal for these plantings.

Requirements for nesting cover vary among different bird species. For example, a dense thicket of alder or willow might be ideal for a flycatcher, vireo or warbler but will not attract cavity nesters such as chickadees or flickers. If the bird species is a cavity nester, artificial nest boxes or birdhouses are an acceptable alternative to trees with natural holes. Often the size of the birdhouse cavity and entrance hole will determine its suitability for a species of nesting bird. Consideration should also be given to providing hole sizes small enough to exclude aggressive and less desirable species such as starlings. Birdhouses should be located high above the ground (10 to 20 feet for most species) with open access for easy exit. Other birds, such as robins, barn swallows and phoebes, prefer an open nesting shelf.

Certain important insectivorous bird species, such as the barn swallow and cliff swallow, actually prefer manufactured structures for nesting. Other birds that use buildings for nesting are the western flycatcher and Say's phoebe. Minor problems, such as infestations of barn swallow bugs and mites, may occur with birds nesting on buildings, but these are infrequent and easily managed. Occasional problems also occur from woodpeckers damaging trees and manufactured structures.

Suitable escape cover allows birds more readily to escape attacks from other birds and mammals. Usually this is a mixture of shrubs or brush into which the birds can easily flee when threatened. Escape cover also provides a quiet place for resting, preening and molting.

## *Water and Food Requirements*

Another important feature to provide throughout the year is water. Birdbaths and/or small ponds can attract many species and are easily maintained during warm weather. Overwintering birds also need drinking water; immersion heaters can be used to keep containerized water from freezing. This, along with supplemental feeding, is important in encouraging those species that are marginally adapted to regional winters to live here year-round. Consult with a local birding group before purchasing a heater, as certain brands are safer than others.

Many birds supplement insect food with fruit and seeds (and some, such as robins, may damage garden crops in the process!). The actual type of food consumed usually varies according to local abundance of different insects and other foods. The use of alternate foods is particularly important in winter when even primarily insectivorous species of birds feed on vegetarian foods. These food needs can easily be met by planting a mixture of fruiting trees and shrubs and providing seeds in supplemental feeding. By closely observing the feeding preferences of birds in a yard, one can learn information that will improve the ability to attract and retain birds in the area.

## *Some Common Conflicts*

Some practices are incompatible with attracting birds to a yard. Allowing cats to roam freely in the yard discourages birds since cats commonly prey upon them. In addition, careless use of insecticides can reduce bird populations. These treatments not only reduce the insects on which birds feed but some are also directly toxic to birds that may inadvertently contact the residue. (Diazinon, used widely in lawn care, is among the most toxic insecticides to birds.) A tolerance of some insects and weeds may be essential to attract and maintain the birds that use them for food.

Some species of birds may be considered undesirable because they drive off other species of birds. Starlings, house sparrows and blackbirds are among the birds that become overly numerous in some situations and compete with more desirable species of birds. Blackbirds, robins and starlings may damage ripening fruit, such as cherries. Although undesirable species of birds cannot be eliminated from the yard, provision of shelter specific to the more desirable bird species will limit their establishment. (*Note*: The temporary presence of large numbers of starlings or blackbirds feeding on lawns is typically due to their feeding on turf-feeding insects, such as sod webworms and cutworms. This activity is highly beneficial to the management of these turf insect pests, which are common in the region.)

| INSECTIVOROUS BIRDS AND THEIR NESTING REQUIREMENTS | | | |
|---|---|---|---|
| **Bird species** | **Nesting requirements** | **Part of diet comprised of insects** | **Other food resources, remarks on nesting** |
| American Kestrel | Prefers natural cavities, holes excavated by flickers or manufactured nest boxes. Requires 3-inch entrance diameter. | Large insects like grasshoppers are commonly a major part of diet. | Reptiles, small birds and mammals are also taken. |
| Eastern and Western Screech Owls | Use natural cavities in hardwoods or old flicker holes in cottonwoods and pines. Nest primarily in heavily wooded areas and are rare in towns. | More than one-third of diet is insects, including night-flying moths and beetles. | Small birds, mammals and reptiles are a major part of diet. |
| Chimney Swift | Constructs a nest of twigs and saliva in chimney or barn silo. | Almost entire diet is made up of day-flying insects; occasionally will feed upon caterpillars in trees and shrubs. | Avoid cleaning chimney between May and August if nesting occurs. |
| Downy Woodpecker | Excavates cavities in dead trees. Nest boxes should have entrance diameter of 1-1/2 inches. | About three-quarters of diet is insects, including bark beetles, borer, larvae, ants, caterpillars and weevils. Also feeds on berries, acorns and tree sap. | A year-round resident, easily attracted to wooded areas. Suet and fruits are important winter food supplements. |
| Northern Flicker | Nests in cavities of decayed wood between 10 and 30 feet above ground. | Primarily ants. | A common year-round resident. Grains, fruits and seeds occasionally are eaten. |
| Flycatchers, including Eastern and Western Kingbirds | Construct twig and fiber nests in deciduous trees, commonly in elms and cottonwoods. Sometimes old hawk nests are used. | Diet is almost exclusively winged insects caught as they pass perching area. | Berries and seeds rarely eaten in nesting season but fruit eaten in overwintering grounds. |

| Bird species | Nesting requirements | Part of diet comprised of insects | Other food resources, remarks on nesting |
|---|---|---|---|
| Barn Swallow | Builds mud nests under eaves and bridges and on sides of walls. | Entire diet comprises flying insects. | A nearby source of mud and straw is needed for nest construction. |
| Blue Jay, Steller's Jay | Nests in trees, preferably conifers, approximately 20 feet above ground. | About one-quarter of diet is insects. | A year-round resident. May harass other nesting birds. Nuts, fruits and grains are primary foods. |
| Black-capped and Mountain Chicadees | Use cavities in trees with partially decayed cores but firm shells. Often excavate their own holes or may use old woodpecker holes. | Predominantly insects are eaten, including many insect eggs, scale insects, aphids and small caterpillars. | Common year-round residents that are easy to attract. Fruit, suet and seeds are important in winter diet. |
| White- and Red-breasted Nuthatches | Use cavities in living trees or abandoned woodpecker holes. Prefer birdhouses with entrance diameter of 1-1/4 inches. | In summer, feed exclusively on insects found on tree bark and shrubs, including caterpillars and borers. | Year-round residents that feed upon nuts, suet and sunflower seeds in winter. |
| Brown Creeper | Prefers dense, wooded areas and is often associated with chickadees and kinglets. | Feeds primarily upon small insects it finds while searching bark. | Suet and seeds are important supplemental foods. |
| House Wren | Prefers cavities but will nest in just about any hole, tin cans, hoods of cars. Uses small (1-1/4-inch) openings in birdhouses. | Feeds almost exclusively on a wide variety of insects, including grasshoppers, beetles and caterpillars. | Easily attracted to brushy areas. Will destroy eggs of other birds nesting nearby. |

| Bird species | Nesting requirements | Part of diet comprised of insects | Other food resources, remarks on nesting |
|---|---|---|---|
| European Starling | A cavity nester but highly adaptable and will nest almost anywhere. Can be excluded from cavities with entrance diameter less than 2 inches. | Adult food is almost one-half insects; nestling food entirely insects. Valuable in control of caterpillar insect pests on lawns. | Fruits, grains and seeds comprise remainder of adult diet. May drive off or out-compete other birds attempting to nest in area. Starlings are year-round residents in some western regions. |
| Red-eyed and Warbling Vireos | Prefer upper foliage of willow and other shade trees. | Most of diet is small insects. | Fruits and berries are also eaten. |
| Northern and Orchard Orioles | Build large pendant nest of straw, string and yarn high in trees. Often will return to same nest site. | Much of diet is larger insects, such as caterpillars. | Fruit, berries and nectar are additional foods. |
| American Robin | Nests in crotches of trees and shrubs. May use artificial shelves for nesting. | Feeds primarily on earthworms but will feed on various insects such as caterpillars. | Fruits and berries are common supplemental foods. Can be very damaging to cherries. Needs nearby source of mud for nest construction. Some robins are present year-round. |

APPENDIX V

# *Dilution Rates for Small-quantity Sprayers*

Many pest management products sold in small quantities (e.g., insecticidal soaps, horticultural oils) often state use rates as a percentage dilution or dilution ratio with water. This is not often clear to many users. For the nonmetric world, the following table is prepared to assist in computing the approximate amount to add for gallon, quart or pint quantities of spray.

**DILUTION RATIO APPROXIMATIONS**

| Dilution ratio (ingredient to water) | Percent dilution | Approximate amount to add to water volume: gallon | quart | pint |
|---|---|---|---|---|
| 1:99 | 1 | 2-1/2 Tbsp (-) | 2 tsp (+) | 1 tsp (+) |
| 1:49 | 2 | 5 Tbsp (-) | 4 tsp (+) | 2 tsp (+) |
| 1:32 | 3 | 8 Tbsp (+) | 2 Tbsp (+) | 1 Tbsp (+) |
| 1:24 | 4 | 10 Tbsp (-) | 2-1/2 Tbsp (+) | 4 tsp (+) |
| 1:19 | 5 | 13 Tbsp (+) | 3 Tbsp (-) | 5 tsp (-) |

+ Will produce a solution of a slightly higher concentration than indicated.
- Will produce a solution of a slightly lower concentration than indicated.

60 drops = 1 teaspoon (tsp)

3 teaspoons = 1 tablespoon (Tbsp)

2 tablespoons = 1 fl. oz. = 6 tsp

4 tablespoons = 1/4 cup = 2 fl. oz. = 12 tsp

8 tablespoons = 1/2 cup (teacup) = 4 fl. oz. = 24 tsp

16 tablespoons = 1 cup = 8 fl. oz. = 48 tsp

1 cup = 8 fl. oz. = 16 Tbsp = 48 tsp

2 cups = 1 pt. = 16 fl. oz. = 32 Tbsp = 96 tsp

2 pints = 1 qt. = 32 fl. oz. = 64 Tbsp = 192 tsp

4 quarts = 1 gal. = 128 fl. oz. = 256 Tbsp = 768 tsp

# GLOSSARY

**Abiotic** Nonliving. Typically used to describe plant diseases or insect mortality agents that are due to weather or other aspects of the physical environment. Sometimes called *nonparasitic* diseases when describing plant disease.

**Acaricide** See *miticide*.

**Acidic soil** A soil with a preponderance of hydrogen ions in proportion to hydroxyl ions, measured as having a pH below 7.0.

**Alkali** See *sodic soil*.

**Alkaline soil** A soil with a preponderance of hydroxyl ions relative to hydrogen ions, measured as having a pH above 7.0. Also called a *basic* soil.

**Alliums** Onions or related plants in the genus *Allium*.

**Alternative host** (1) A host organism that allows a parasitic predatory insect to complete its development in the absence of its normal host. (2) One of the two plants that certain fungi (e.g., rusts) or insects (e.g., aphids) require to complete their life cycle.

**Amendment** See *soil amendment*.

**Annual** A plant that completes its life cycle within one year.

**Applied biological control** The purposeful manipulation of predators, parasites or diseases to control a pest.

**Arachnid** A class of arthropods marked by four pairs of legs and typically two distinct body regions, the cephalothorax and the abdomen. Spiders, mites and scorpions are examples.

**Arthropod** The jointed-foot animals, marked by jointed appendages, an exterior skeleton (exoskeleton) made of chitin and growth that involves molting. The largest phylum (Arthopoda) of animals, which includes insects, arachnids, crustaceans, centipedes and millipedes.

**Ascospores** See *spores*.

**Bacteria** A type of one-celled organism without chlorophyll or a formed nucleus that reproduces by fission, currently considered members of the kingdom Procaryotae.

**Bactericide** A chemical substance sold for control of bacteria.

**Biennial** A plant that requires two years to complete its normal life cycle.

**Biological control** The natural population regulation of a species by the action of predators, parasites, pathogens or competing organisms.

**Blight** The rapid death of leaves, stems or other plant tissues due to effects of a plant pathogen.

**Bolt** The production by a plant of a seed-bearing stalk.

**Botanical insecticide** An insecticide with an active ingredient derived from a plant source. Pyrethrum, nicotine and rotenone are examples.

**Broad-leaf plant** Any garden plant that is not a grass.

**Broad-spectrum pesticide** A pesticide that has effects on many different species, sometimes on desirable species. Opposite of a *selective pesticide*.

**Bulbing** The period when bulb-forming plants shift energy from top growth into the storage organs, or bulbs.

**Caneberry** A small fruit, such as raspberry and blackberry, that produces canes.

**Catfacing injury** An overgrowth of a fruit around a damaged area. Insects that chew or damage small areas of a fruit often produce these injuries.

**Chelate** An organic molecule that binds and holds an ion. Iron fertilizers are often in chelated form so that the iron ion does not readily react with other chemicals and is thus more available to the plant.

**Chemical control** The control of a pest by the application of a chemical, such as a pesticide.

**Chemical name** The accepted name of a chemical that describes its structure according to a standard system.

**Chlorosis** Development of abnormal leaf yellowing, usually concentrated between leaf veins.

**Cole crops** Crops grown in the cabbage family Cruciferae. Also called *crucifers.*

**Complete metamorphosis** A pattern of development by certain insects that involves *eggs*, immature stage *larvae*, *pupae* and *adults*. Beetles, butterflies and moths are examples of insects with complete metamorphosis.

**Conidiospores** See *spores*.

**Contact pesticide** A pesticide that enters an organism through the external covering, such as the epidermis.

**Crop rotation** The purposeful planting in a sequence that alternates between unrelated plants, usually for the purpose of avoiding pest problems specific to a crop. For example, a garden rotation might involve tomatoes (Solanaceae), followed the next season by peas (Leguminosae) and then cabbage (Cruciferae).

**Cross-pollination** Transfer of pollen to the receptive female flower between two plants that are not genetically identical, such as two different varieties.

**Crucifer** A member of the cabbage family (Cruciferae), including cabbage, broccoli, kale, collards, radishes, turnips and many other common garden plants. Also called *cole crops.*

**Crustacean** A class of primarily aquatic arthropods that includes shrimp, crayfish and lobsters. Pillbugs and sowbugs are the common terrestrial crustaceans found around gardens.

**Cucurbit** A member of the squash family (Cucurbitaceae), including cucumbers, melons and squash.

**Cultural control** The control of a pest using normal cultural practices (e.g., varietal selection, watering, fertilization) in a purposeful manner.

**Damping off** A plant disease of newly emerged seedlings that causes plants to fall over, wilt and die. Caused by infection by one of several fungi.

**Defoliation** Unnatural loss of leaves by a plant in response to stress, injury or other causes.

**Determinate growth** A habit of plant growth by which flowering and fruiting occur in a set pattern and plant growth does not continue indefinitely. This contrasts with plants that have *indeterminate growth*, which continue to grow so long as conditions allow.

**Diatomaceous earth** Natural deposits of diatoms that are mined for use in insect control and filtration. The highly spined body coverings can abrade the external skeleton of insects.

**Dioecious** A type of flowering by which male and female organs are produced on separate flowers. Several members of the squash family produce separate male and female flowers. Some plants, such as asparagus, produce male and female flowers on separate plants.

**Disease (plant)** A disturbance of normal plant function due to environment, pathogens or other agents.

**Emulsifier** A chemical that allows another substance to form an emulsion. Emulsifiers are commonly added to oil-based pesticides to allow them to emulsify and mix with water.

**Epidermis** The outer layer of cells produced by a plant or animal.

**Eradicant** A chemical substance that destroys a plant pathogen at its source. Commonly used to describe the action of certain fungicides.

**Floating row cover** A lightweight fabric draped on plants for use in frost protection, seedbed warming or exclusion of pest insects.

**Fruit set** Development of fruit ovaries following successful fertilization with pollen.

**Fumigant** A chemical pesticide that moves as a gas, entering the breathing openings of susceptible plants or animals.

**Fungicide** A chemical substance used to control undesirable fungi.

**Fungus** A large and diverse group of lower plants that lack chlorophyll and specialized conductive tissues.

**Gall** A plant swelling or growth produced in response to the action of certain insects, eriophyid mites, fungi or other organisms.

**Grub** The immature (larval) stage of many insects with complete metamorphosis. Commonly used to describe immature Coleoptera (beetles).

**Guttation fluids** Fluids exuded from plants, usually along the edges of leaves.

**Hardpan** A compacted subsoil layer that prevents drainage and inhibits root growth.

**Heading** The period in plant growth when energy is shifted to production of a tight head of leaves (e.g., cabbage) or flower buds (e.g., broccoli).

**Heavy soil** A soil that is dense due to composition by fine soil (usually clay) particles packed close together with few soil pores.

**Herbaceous** A higher plant that does not produce woody tissues.

**Herbicide** A pesticide sold and used to control undesirable plants, usually weeds. A weed killer.

**Honeydew** The sticky, sugar-rich excretion produced by aphids, soft scales and several other insects that suck plant sap.

**Hopperburn** A type of plant injury produced by certain leafhoppers, characterized by a progressive dying of leaves from the margin, which appear scorched.

**Horticultural oil** See *spray oil.*

**Host** A plant or animal on which a parasite feeds.

**Hyphae** Threadlike structures produced by fungi that actively grow, secrete enzymes and absorb nutrients.

**Indeterminate growth** A plant growth habit by which the plant continues to grow, elongate and produce flowers so long as conditions permit. Often used in discussion of tomato varieties. Opposite of *determinate growth.*

**Insecticide** A chemical sold and used for control of undesirable insects.

**Integrated Pest Management (IPM)** A philosophy of managing pests that involves use of all available control techniques in a coordinated (integrated) manner. Often implied in integrated pest management is the ideal that management will be done in a way that minimizes deleterious effects on desirable species and retards development of pest resistance.

**Larva** The immature stage of an insect. Maggots, caterpillars and grubs are examples of immature insects. For insects that have simple metamorphosis, the term *nymph* is sometimes used. The immature stage of a nematode.

**Legume** A plant in the pea family (Leguminosae), including peas and beans.

**Maggot** The immature (larval) stage of many flies (Diptera order), which is marked by the absence of legs and a visible head.

**Mechanical control** The purposeful control of a pest using mechanical methods such as hoeing, handpicking or barriers.

**Mesophyll** The layers of cells beneath the epidermis of plants.

**Metamorphosis** A change in form. Usually applied to arthropods, such as insects, that undergo a series of distinctive changes (e.g., egg, larva, pupa) before reaching the adult form.

**Microbial pesticide** A pesticide derived from a microbe or pathogen of a pest. For example, *Bacillus thuringiensis* is a naturally occurring bacterial disease of insects that has been developed into several pesticides.

**Mirids** Common name for insects in the Miridae family, the plant bugs.

**Miticide** A pesticide sold for control of mites. Nearly synonymous with *acaricide*, a pesticide used to control mites and ticks (Acarina order).

**Molluscicide** A chemical substance used to kill undesirable mollusks, such as slugs and snails.

**Molt** Shedding of the exoskeleton by an arthropod.

**Monoecious** A type of flowering by which male and female organs are produced within the same flower.

**Mosaic** A symptom of plant injury involving a mottling of dark and light green patches, usually produced as a result of infection with a virus.

**Mulch** A protective soil covering used to smother weeds, retain moisture, alter temperature or otherwise change the environment.

**Mycoplasma** A group of bacteria, marked by the absence of a cell wall, that cause some animal diseases. Mycoplasma-like organisms (MLOs) cause some plant diseases such as aster yellows. However, they cannot yet be classified as mycoplasma, since they have not met all the criteria for this group.

**Nematode** Animals in the Nematoda phylum, the roundworms. Most nematodes are microscopic. Include many species that parasitize plants or animals.

**Nitrogen fixing** Ability to convert gaseous nitrogen from the atmosphere and convert it to ammonia nitrogen, which plants use. The property of certain rhizobia bacteria associated with legumes.

**Nontarget organisms** Organisms that may be affected by a pesticide directed at a pest organism. For example, wildlife, honeybees and beneficial insects are desirable nontarget organisms that may be affected by application of an insecticide.

**Nymph** Immature stage of an insect with simple metamorphosis. Often the term *larva* is substituted.

**Organic matter** Material containing carbon compounds derived from living organisms, such as plants.

**Organic pesticide** (1) A pesticide with organic chemistry, containing carbon. (2) A pesticide that is legally acceptable for production in organic gardening and farming, usually on the basis of use history, source and/or impact on the environment.

**Parasite** An organism that develops in or on another living organism, known as a *host*, at the expense of the host. Plant pathogens, tachinid flies and parasitic wasps are examples of parasites common in and around gardens. For insect parasites that kill their host, the term *parasitoid* is more correctly used.

**Pathogen** An organism or virus that can induce disease in another organism.

**Perennial** A plant with a normal life cycle of several years.

**Pest** An organism that is undesirable or detrimental to the interests of humans.

**Pesticide** A chemical substance sold for use to control a pest. Herbicides, insecticides and fungicides are examples of pesticides.

**pH** A measure of acidity, based on the amount of free hydrogen molecules present. From the French *pouvoir hydrogene* or hydrogen power.

**Pheromone** A chemical used to communicate by members of the same species. For example, females of many insects produce sex pheromones to attract male mates.

**Physical control** The purposeful control of a pest using techniques that involve physical means, such as light, temperature and color.

**Pollinizer** A plant that can provide pollen that will successfully fertilize the egg cells in the ovaries of another flower. Usually used to designate types of fruit tree varieties that are purposefully planted to help pollinate a self-unfruitful main variety.

**Postemergence herbicide** A herbicide that affects plants after seedlings have emerged.

**Predator** An animal that feeds on other organisms. The term is usually used to describe insects or other animals that move about, hunt and feed on *prey* to develop. However, the term is also used to describe animals that feed on plants.

**Preemergence herbicide** A herbicide that affects newly germinated and newly developing seedling stages of a plant. It must be applied before seedling emergence to provide control.

**Prey** Animals that are captured and eaten by predators.

**Prolegs** The soft, temporary legs that are attached to the abdomen of caterpillars and the larvae of sawflies but are lost when these insects change to the adult form.

**Protectant** A chemical substance used to protect a plant from a pathogen. Usually used to describe the action of certain fungicides that inactivate fungal pathogens before they infect a plant.

**Protozoan** A one-celled animal of the Protista kingdom, the protozoa or first animals.

**Repellent** A substance that causes an avoidance reaction in another organism.

**Resistance (plant)** Ability of a plant to overcome, completely or partially, the effects of diseases, insects or other factors that deleteriously affect growth. Resistance is often a genetically based characteristic that is purposefully bred in plants. Cultural conditions, such as fertilization and watering, may also affect resistance of a plant to a pest.

**Resistance (to pesticides)** Ability of a pest to survive normally lethal exposures to a pesticide. Pesticide resistance is a genetic trait that develops following exposure to a pesticide that has eliminated genetically susceptible individuals, concentrating genes that confer resistance.

**Rhizobia bacteria** Bacteria that have the ability to fix nitrogen from the atmosphere and are associated with the roots of legumes. Formerly, these were all placed in the genus *Rhizobia*, but they are now considered to include at least three genera.

**Rootstock** The root-bearing part of a plant to which a fruit-bearing *scion* is grafted. All tree fruits commonly sold by nurseries are produced by grafting the fruiting variety to a rootstock. Many roses are grafted.

**Russeting** A brown discoloration, often finely netted, that develops on leaves or fruit in response to cold injury, chemical injury or the action of certain plant pests.

**Salt-affected soil** A soil with high levels of soluble salts that can adversely affect the growth of most plants.

**Scientific name** The name given to any living organism according to the Linnean system of classification, which involves a two-part (binomial) name: the *genus* and the *species*. The scientific name is usually underlined or italicized in print. For example, *Apis mellifera* is the scientific name for the honeybee.

**Sclerotia** Compact masses produced by certain fungi, usually enclosed in a dark covering, that can resist unfavorable environmental conditions. Plural form of *sclerotium*.

**Seeded** Grown by planting seeds directly into the garden. In contrast to *transplanting*, which involves starting plants from seed in a greenhouse or other location and then moving the young plants to the garden.

**Selective pesticide** A pesticide with a very limited range of organisms that it will adversely affect.

**Self-fruitful** Plants that can be successfully pollinated by a plant that is genetically identical, such as by itself or by the same variety. Opposite of *self-unfruitful* which describes plants that require pollination by a different variety.

**Shothole injury** A type of plant injury characterized by small holes in the interior of the leaf.

the shothole fungus *Coryneum*.

The injury is characteristic of flea beetles and the shothole fungus *Coryneum*.

**Simple metamorphosis** A pattern of changes during development by certain insects that includes *eggs*, immature *larvae* (or nymphs) and adults. Grasshoppers, leafhoppers and earwigs are examples of insects with simple metamorphosis.

**Skeletonize** A pattern of feeding damage by leaf beetles and certain other insects by which tissues between the veins are eaten, leaving the skeleton of veins.

**Sodic soil** A soil that contains high amounts of sodium, causing it to develop a cloddy structure. Sodic soil interferes with the growth of most crop plants. Sometimes called *alkali* or, when combined with high amounts of soil salts, *saline-sodic soil*.

**Soil amendment** A substance applied and mixed with soil to improve plant growth. Organic matter and fertilizers are soil amendments commonly applied to garden soils.

**Soil fertility** The ability of a soil to supply the nutrients to support plant growth.

**Soil sterilant** A herbicide that when applied to the soil will kill all living plants at the treated site.

**Soil structure** The arrangement of basic soil particles (sand, clay, humus, etc.) into larger secondary soil units or aggregates.

**Soil texture** The composition of a soil as defined by the relative content of sand, silt and clay particles.

**sp.** Shortened singular form for species. Plural form is *spp*.

**Spores** The reproductive stage of fungi, analogous to the seeds of higher plants. This includes the commonly produced asexual *conidiospores* as well as other specialized sexual spores, such as *ascospores* and *basidiospores*.

**Spray oils** Highly refined specialty oils, such as dormant, superior and supreme oils, that are applied to plants to control insects, mites and other pests. Often called *horticultural oils*.

**Stomach poison** A toxicant that must be ingested to have effect, in contrast to poisons that are externally absorbed (contact poisons) or move into the organism as a gas (fumigants).

**Stomates** Openings on the leaf surface through which air is brought into the plant and water vapor is released.

**Stylets** Mouthparts of insects that are designed for penetrating tissues and sucking fluids. Several plant-feeding insects, such as aphids, whiteflies and squash bugs, have stylets, which they use as piercing-sucking mouthparts.

**Syringing** Washing a plant with a jet of water to physically remove a pest stage or other material.

**Systemic pesticide** A pesticide that can be moved within a plant either following leaf absorption, uptake through the roots or injection into the sap stream.

**Tillage** Working of soil during plant production, such as by plowing or rototilling.

**Trade name** The name of a product (e.g., pesticide) under which the manufacturer markets.

**Trap crop** A crop purposefully planted or maintained to divert attack by plant-feeding pests.

**Vegetative growth** Plant growth involving leaves and other nonreproductive parts.

**Virus** A nonliving particle of protein-covered nucleic acid that can induce susceptible living cells to replicate the particles.

**Weed** A plant that is undesirable because its growth is detrimental in some way to the interests of humans. A plant "out of place."

# INDEX

# NOTES

NOTES

# NOTES

NOTES

## NOTES

## GREENHOUSE GARDENER'S COMPANION

**Growing Food & Flowers in Your Greenhouse or Sunspace**

**SHANE SMITH**

The perfect sourcebook that includes everything you need to know about setting up a healthy growing environment within a greenhouse or sunspace, including a complete guide to growing over three hundred flowers, vegetables and herbs.

ISBN 1-55591-106-4, paper $18.95

## EASY GARDENING

**No Stress — No Strain**

**JACK KRAMER**

Jack Kramer, a well-known gardening writer, offers shortcuts and tips for the gardener who wants to minimize the stress and strain of gardening.

ISBN 1-55591-083-1, 144 pages, paper $14.95

## GARDENING

**Plains and Upper Midwest**

**ROGER VICK**

In this complete, practical and informative guide for a lifetime of gardening on the northern plains, Roger Vick draws on his broad experience as a professional horticulturist and home gardening enthusiast to advise the novice of the best means to successful gardening.

ISBN 1-55591-068-8, 383 pages, paper $16.95

## THE GARDEN AND FARM BOOKS OF THOMAS JEFFERSON

**EDITED BY ROBERT C. BARON**

Thomas Jefferson was at heart a gardener. Here are the complete transcriptions of Jefferson's garden and farm journals and selected horticultural correspondence with his contemporaries.

ISBN 1-55591-013-0, 534 pages, b/w photographs, color plates, cloth $20.00

---

**Fulcrum garden titles are available at your local bookstore or call us directly at 1-800-992-2908 to place an order**

Fulcrum Publishing, 350 Indiana Street, Suite 350, Golden, Colorado 80401-5093